CONSOLE WARS

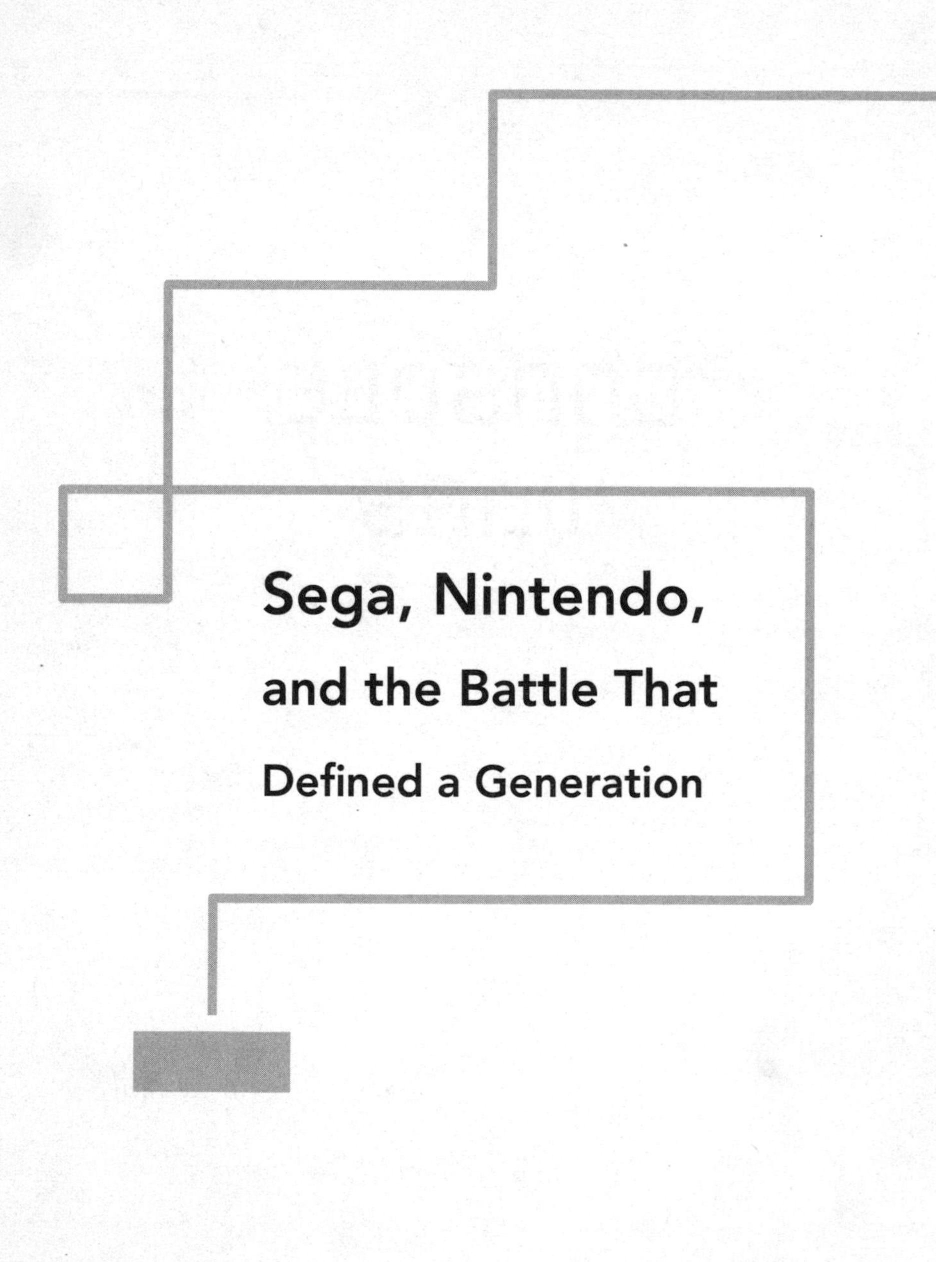

Sega, Nintendo,

and the Battle That

Defined a Generation

CONSOLE WARS

Blake J. Harris

Foreword by Seth Rogen & Evan Goldberg

CONSOLE WARS: SEGA, NINTENDO, AND THE BATTLE THAT DEFINED A GENERATION.
 For information address HarperCollins Publishers, 195 Broadway, New York, NY 10007.

HarperCollins books may be purchased for educational, business, or sales promotional use. For information please e-mail the Special Markets Department at SPsales@harpercollins.com.

Designed by Maria Elias

Library of Congress Cataloging-in-Publication Data

Harris, Blake J.

Console wars : Sega, Nintendo, and the battle that defined a generation / Blake J. Harris.

pages cm

ISBN 978-0-06-227669-8 (hardback)—ISBN 978-0-06-227670-4 (trade paperback) 1. Video games industry—History. 2. Electronic games industry—History. 3. SEGA Entapuraizesu. 4. Nintendo Kabushiki Kaisha 5. Video games—History. I. Title.

HD9993.E452H37 2014

338.7'617948—dc23

2013050668

---

14 15 16 17 18 OV/RRD 10 9 8 7 6 5 4 3 2

For Katie, the girl with the shiny eyes

FOREWORD

BY SETH ROGEN AND
EVAN GOLDBERG

SETH

Hi! Welcome to the foreword for *Console Wars* by legendary author Blake J. Harris!

EVAN

Videogames are great, but books about videogames are even better!

SETH

We grew up as videogames were on the rise, and they played a major role in our upbringing.

EVAN

And that's why we couldn't say no when Blake asked us to write a 2,500-word foreword for this awesome book you will love reading!

SETH

How many words is that?

EVAN

Like, 150.

SETH

Crap. Okay, what next?

Evan thinks intensely and an idea comes to him.

EVAN

Let's talk about what systems we preferred.

SETH

Solid idea, partner-ino!

EVAN

I preferred Nintendo.

SETH

I preferred Sega. I'll never forget the first time I ripped someone's spine out playing Mortal Kombat.

EVAN

Yeah, Sega always seemed to go to a place that Nintendo didn't, and that opened the doors for videogames that weren't just targeted at kids but teenagers and even . . . adults. I don't think games like Grand Theft Auto would even exist without Sega making games that went places Nintendo never would have gone.

SETH

I don't know if that's a good thing or a bad thing. But Mortal Kombat definitely felt like a wonderful step in a new direction at the time. I was awesome at it too. Sub-Zero was my man.

EVAN

Me too. Hey, here's a Sega question: what was up with Sonic and Tails?

SETH

What? It was just your classic platonic speedster hedgehog and two-tail fox relationship.

EVAN

I felt some tension there. Sexual.

SETH

Oh, it was sexual.

Seth and Evan exchange awkward looks in what is clearly a sexually charged moment of their own.

SETH

How many words we at?

EVAN

I'd say 400-ish.

SETH

Use letters. Four-hundred-ish. It takes up way more space.

EVAN

Mega true. I'll say it in letters from now on. Like four-zero-zero.

SETH

Don't hyphenate them, dumbass. It makes them count as one word. [*sigh*] Here's a videogame fact that will take up at least thirty-five words.

EVAN

What is it?

SETH

I used to own a power glove. I got it right when it came out.

EVAN

Sweet petunia bush! Please elaborate using as many words as possible!

SETH

It didn't work that well at all. I remember the bad-ass dude in The Wizard (arguably the most important videogame movie of all time), and mine didn't work worth crap.

EVAN

I was always confused by TurboGrafx-16. As far as I recall, there were only two games for it. Keith Courage and Bonk's Adventure. I only ever played Keith Courage.

SETH

I played Bonk's Adventure. A friend of mine had it, and it truly blew my mind. I also remember renting Sega CD in high school. It had that raunchy horror game with real controversy surrounding it.

EVAN

Yup, that was Night Trap. You had to stop drill-wielding serial killers from impaling sorority girls. That was the first time I remember thinking to myself, "Well they have just gone too far this time." And I was twelve or something . . .

SETH

Then came Sega Saturn, and kind of shat the bed.

EVAN

And then there was Goldeneye.

SETH

I would confidently say the reason I never really had a girlfriend in high school was because of Goldeneye. I specifically remember leaving parties to go play it.

EVAN

Our favorite level was the Facility. We would sit with our buddy Fogell for hours and hours on end and play it.

SETH

I memorized every level. The game was as much about watching your friend's screen as your own.

EVAN

When I went off to college, I met a group of guys from out east that were way better at Goldeneye than we ever were, and it crushed me. They were operating at a whole other level.

SETH

Then you got super into Super Smash Brothers.

EVAN

Yeah. It was on Nintendo 64. My buddies and I would have tournaments that would go for hours: entire evenings. I was the nimble-footed puffball of power, Kirby.

SETH

That game makes no sense. The whole thing is based on a percentage of the likelihood that you're going to fall off a magical island, and it goes up to like 600 percent and that's bad and you're actually trying to keep your score low, which I find confusing and counterintuitive.

EVAN

Well, games are getting continuously confusing. I don't even know what my grandparents would think if they played Grand Theft Auto.

SETH

Remember when Martin Starr and I taught you to drive around LA when you first moved here by playing the game True Crime: Streets of LA, because it had a realistic map of Los Angeles?

EVAN

That was sincerely helpful. It's crazy how they started doing stuff like that.

SETH

I bet soon games will start calling our cell phones and e-mailing us and stuff.

EVAN

Maybe that's how Skynet finally happens and we all end up in a *Terminator/Matrix* nightmare version of the future where mankind is nearly wiped out and machines rule the world.

SETH

Well, I guess it's time to address the elephant in the room—porno. We all know we're going to be getting dirty with our videogames, and if not us, our children, or our children's children.

EVAN

The future can wait. We have to live in the now!

SETH

You're right. These days I mostly like to play games where you shoot people. Call of Duty, GTA5, and such.

EVAN

I'm an iPad tower defense addict. There's something wrong with me. I just love games where things are sent in waves and I get to destroy them with strategy.

SETH

That's a dark want.

EVAN

It's who I am.

SETH

A crazy thing I think about sometimes is that there are teenagers operating videogames connected to deadly drones that fly around the world blowing stuff up. That eighties movie *War Games* is real now. And what to us is a nonthreatening drone will eventually probably turn into, yes, Skynet. All roads lead to Skynet.

EVAN

I think at this point most people would agree that a robot takeover is how things end. I'm personally at peace with that inevitability.

SETH

I still can't get over the whole Nintendo Wii revolution, with these games you have to move around to. When we were growing up, playing videogames made you fat and lazy, not nimble and coordinated.

EVAN

I really got into Wii Fit for a while. It was pretty addictive at first but then it made fun of me too much and mucked up my self-esteem.

SETH

Now kids are getting their self-esteem messed with through videogames way more than when we were kids, thanks to this whole online gaming thing.

EVAN

I try to get into online gaming every now and then, and I constantly find there's young kids out there who are so much better than me I can't even participate.

SETH

Yeah. It's weird to think we are thirty-one and already we can't keep up.

EVAN

Eventually we won't even understand the images on the screen.

SETH

Like how my grandmother would view death metal.

EVAN

I'm not a huge fan of death metal myself.

SETH

"Death Metal" 's a great name for a videogame.

EVAN

I read once that the band Journey had a videogame where you could put your face onto the main character. I want that. It's silly that I can't be the character in the game yet.

SETH

Yeah. And it'd be cool if it used your contact list for names and incorporated your real life a lot.

EVAN

But then, once again, robots would take over the world.

SETH

So what we're realizing here is videogames may not ever get better than they are now, because if they did, robots would take over the world.

EVAN

I think so.

SETH

How many words is that?

EVAN

About 2,300.

SETH

Then let's take this full-circle and connect it to the book our honorable reader is about to read: Nintendo was king of home videogame entertainment systems, then Sega came in and was a contender for the crown. Sega almost toppled Nintendo with their subversive and more adult-oriented games, and these games have led us to a world where GTA and Call of Duty are the top games, and the next step is to have the games incorporate stuff about us and our personal lives, and then sentient technology will inevitably disassociate from mankind and some robot like Skynet will rise up and destroy us all. Hence: the "Console Wars" between Nintendo and

Sega is what began a series of events that will lead to the end of humanity as we know it.

EVAN

Bam! That's what videogames mean to us.

SETH

Damn. I think we nailed this foreword stuff. Our style may have been unconventional, but we ultimately tied it to the downfall of mankind, which is cool.

EVAN

I couldn't agree more. Our next movie should be called *Foreword* and be about this process.

SETH

Or *Foreskin* and be about a circumcision that changed humanity forever.

EVAN

Both good ideas.

SETH

Okay. We should probably get home to our wives now.

EVAN

Yes. We love our wives. Let that be noted.

SETH

See you at work!

EVAN

Ditto!

Seth Rogen and Evan Goldberg are childhood friends, and the writer/director/producers of *This Is the End* and *The Interview*. Together, the duo has also written and produced *Knocked Up*, *Superbad*, and *Pineapple Express*.

AUTHOR'S NOTE

C*onsole Wars* is a narrative account based on information obtained from hundreds of interviews. Re-creating a story of this nature, which draws from the recollections of a multitude of sources, can often lead to inconsistencies; particularly when dealing with industry competitors and especially when dealing with events that took place more than two decades ago. As such, I have re-created the scenes in this book using the information uncovered from my interviews, facts gathered from supporting documents, and my best judgment as to what version most closely fits the documentary record.

In certain situations, details of settings and description have been altered, reconstructed, or imagined. Additionally, most of the dialogue in this book has been re-created based on source recollections of content, premise, and tone. Some of the conversations recounted in this book took place over extended periods of time or in multiple locations, but have been condensed, or reorganized in a slightly different manner, while remaining true to the integrity and spirit of all original discussions.

PROLOGUE

In 1987, Tom Kalinske was at a crossroads. He had spent the better part of his career working at Mattel, where he enjoyed towering success transforming the Barbie line from a niche, has-been series of dolls into a timeless, billion-dollar property. Recognizing his potential, the company groomed him to become their next president. But shortly after taking the reins at only thirty-eight years old, he found himself embroiled in a dangerous game of office politics. With no resolution in sight, Kalinske decided that rather than wage an internal war, he'd prefer to fight an external one. So he ceded control of Mattel to a rival executive, and left the company to become president of a competitor: Universal Matchbox.

Although Matchbox's toy cars had historically gone toe-to-toe with Mattel's Hot Wheels brand, when Kalinske took over, his new company was hemorrhaging money and had recently been placed into receivership. He knew this going in—it was part of what had enticed him—but taking on Goliath eagerly didn't do much to change the incline of the uphill battle ahead. To revive the wounded company from receivership and give Hot Wheels (and, of course, Mattel) a run for their money, Kalinske needed to reorganize Matchbox, and do so in a flash. He spent the next couple of years traveling the globe and implementing ambitious restructuring plans, much of which hinged on moving all production to labor-cheap regions in Asia.

By 1990, his strategies appeared to be working and Matchbox had become relevant again. They were still miles away from bridging the gap with Mattel, but with revenue now over $350 million, the company had managed to turn a profit for the first time in years. Matchbox cars were starting to sell all around the world—except, for some reason, in Spain. So Kalinske went there to find out why not.

After arriving in Barcelona, Kalinske hopped in a cab to meet with the distributor in charge of selling Matchbox cars in the Spanish territories. Instead of embarrassing himself by trying to pronounce the address where he was headed, he simply passed the distributor's business card to the cabdriver. The driver glanced at the card, noticed the Matchbox logo, and nodded.

Kalinske was baffled. His distributor worked out of a tiny office; how could this driver possibly know where to go based on only the logo? Once again, Kalinske tried to present the business card to the driver only to be waved off. "Matchbox, sí," the driver said firmly. Kalinske was confused, but ultimately he decided he wasn't a Spanish cabdriver and it wasn't his job to know such things. He leaned back, took in the panoramic view of a sprawling Barcelona, and tried to remember if you're supposed to tip cabdrivers in Spain or not.

Shortly thereafter, the driver stopped in front of a large yellow building. Kalinske stepped out of the cab and compared the address to the one on the business card. It didn't match. In broken Spanish, Kalinske tried to argue with the driver, but the guy remained insistent that this was the right place. Kalinske finally gave in and told the driver to wait outside as he entered the building. But as soon as he stepped inside, he was greeted by a surprise.

Thousands of surprises, actually. The building turned out to be an industrial factory, churning out tiny die-cast Matchbox cars by the minute. Huh? Kalinske had moved all production to Asia, so why were shiny little cars shooting off a rusty conveyor belt in Spain? Was this why profits had been lagging? Who'd authorized this?

Returning to his broken Spanish, Kalinske asked the first worker he saw if he could use a phone. He whipped out the calling card he'd been using to stay in touch with his wife, and rang up his business partner David Yeh in Hong Kong.

"What's up, Tom?" Yeh asked. "You make it to Barcelona okay?"

Kalinske had no time for small talk. "I thought production had been moved to Asia. All of it."

"It was."

"Then why do we have a factory in Spain?"

"What do you mean? We don't have a factory in Spain."

Kalinske looked around. Cars continued to pop out. "I'm definitely in Spain, and I'm definitely in a factory."

"Oh, shit!" Yeh belted. "Tom, get the hell out of there and call the police."

Kalinske looked up. Dozens of unfriendly eyes were now staring in his direction, and seconds later the gruff group of Spaniards began to approach him. Without even hanging up the phone, Kalinske sprinted out to the street, jumped into the cab, and yelled one of the few Spanish words he knew: "*Vamanos!*"

Kalinske went straight to the police, and shortly after, the illegal factory—set up by the very distributor he was in town to meet—was raided. This unexpected Spanish adventure full of twists, turns, and toy cars led Kalinske to the following thoughts:

1. Hey, at least the cars *were* selling in Spain, even if we didn't make a dime.
2. Good thing for that driver or the mystery might never have been solved.
3. It's probably time for me to leave Matchbox and go do something else.

Though the incident served as a catalyst for his leaving, this was only a small factor in his departure. Ultimately, Kalinske left Matchbox because it would never be Hot Wheels. It would never be number one, and it could never be big enough to prove to Mattel that they were wrong. He needed that chance . . . but where?

Lost, frustrated, and unsure what to do next, Kalinske receded from the world in the safety of his home. He was perfecting the art of becoming a hermit until his wife, Karen, prescribed a remedy for his mood. "We need to get you out of the house," she explained. "So guess what? Congratulations! You're taking your family on a vacation."

Karen, the perpetual voice of reason, was right again. And so with this advice, in July 1990, Kalinske took his family to Hawaii, where Karen enjoyed the beach, their three girls enjoyed building (and then stomping on) sandcastles, and Tom tried his best to stop thinking about a million things at once.

It was exactly what Tom Kalinske needed, but midway through the trip, an unexpected guest showed up . . .

PART ONE

GENESIS

1.

THE OPPORTUNITY

Tom Kalinske had a secret.

For years he had managed to keep it to himself, covering it up with a combination of white lies, noncommittal nods, and uneven smiles, but as he lay on a magnificent beach in sunny Maui with his loving wife and three energetic daughters, he could no longer keep it inside. He had to tell someone.

The right person to tell was Karen, of course. She had always been there for him, and most important, she seemed to possess a magical ability for making his anxieties disappear. She really was his voice of reason, the love of his life, and apparently very much asleep. "Hey, Karen," he said, nudging her now-tanned shoulder. "Karen?" He lifted up her sunglasses, which confirmed that the blazing sun had knocked her into a slumber. Kalinske considered some other subtle means to poke or prod his wife into waking up until, upon further inspection, he noticed that their newborn daughter, Kelly, cradled in her arms, was also asleep. "Sorry about that, ladies," Kalinske said, honoring the unwritten rule that all fathers should abide by: do not, under any circumstances, wake an infant, especially when her fragile little nap allows Mom a rare moment of uninterrupted sun-filled dreams. Karen, it appeared, was off the hook, and wouldn't have to play the role of secret-keeper this time.

Kalinske thought about telling one of his other daughters, Ashley (five years old) or Nicole (three), but they were ankle deep in the ocean capturing hermit crabs in a bright yellow bucket and making the kinds of memories that they'd inevitably forget and only their father would remember. So he returned to reading a day-old *New York Times*, but as he scrolled through previews of baseball games that had already been played, a slender silhouette covered his newspaper. "Hello, Tom," a cheerful voice said. "You are a difficult man to track down."

Kalinske looked up to find a Japanese man with piercing brown eyes and a messy, wind-mangled combover: Hayao Nakayama. "What are you up to?" Nakayama asked, trying to emit a friendly smile, which came out looking more like an ominous smirk. Kalinske would soon learn that Nakayama was incapable of producing a genuine smile. His round face always held too much mystery, making simple, genuine emotions too hard to pull off.

"Well, I was trying to have a nice, relaxing moment with the sun here, until you got between us," Kalinske gracefully shot back. He never allowed himself to appear off guard in conversations and was committed to masking any discomfort or unease with a Jamesdeanian coolness. Nakayama suddenly realized that he was casting a shadow over Kalinske and took a couple of steps to the side. As the sun hit his face, Kalinske smiled and greeted his unexpected guest. "Great to see you, Nakayama-san. What brings you out to Hawaii?"

"I came here to find you. As I just said, you are a difficult man to track down." Nakayama spoke nearly perfect English, albeit with a strange slight Brooklyn accent. It was smooth and seamless, except for the occasional broken phrase. His errors, however, seemed to have less to do with grammatical difficulties and more to do with the rhythm of his conversation. It was almost as if he threw in a few "mistakes" from time to time as camouflage; allowing himself to hide behind the language barrier and play the role of clueless foreigner if need be. "When I heard about your departure from Matchbox I left many messages at your home." Nakayama tried another smile, this one coming out like a spooky grin.

Kalinske bowed his head a bit. After leaving Matchbox, he had tried his best to hide from the universe. He screened all his calls, turned off the fax machine, and rarely left the house. He had been feeling small in the world and he dealt with it by making his world as small as possible. Karen had been good about dealing with his reclusive behavior. She knew he was down and didn't press the issue. Her husband had bounced back from many things over the years, and there was no doubt in her mind that he would soon return to the world more spectacular than ever before. In the meantime, she didn't mind having him around the house. For a temporary recluse, he was quite friendly, good with the dishes, and only sometimes a liability when doing the laundry.

"Yeah, I got your messages. Sorry I hadn't gotten back to you yet," Kalinske said. "I've just been trying to take some time to myself and figure things out a little."

"Ah, yes," Nakayama said. "But don't you know that this is why I called?" Then, right there in the middle of Kalinske's family vacation, Nakayama made him an offer to become the next president and CEO of Sega of America. As unusual as it was to be offered a top-level executive job on the beach, Kalinske wasn't completely shocked. After all, Nakayama was the president of Sega Enterprises, and there had been rumors for some time that he was looking to replace Michael Katz, the current head of Sega's American operation and a man Kalinske knew personally and considered a friend. Nakayama cleared his throat. "What do you say, Tom? I am certain you are the man for the job. We have a wonderful new videogame console."

Kalinske looked at Nakayama, studying the weathered version of a face he knew well. They had first met in the late seventies, when Kalinske was still the golden boy at Mattel. At the time, there were two companies that were the envy of all others: Apple and Atari. Though it was unrealistic for Mattel to jump into the high-tech, high-risk world of personal computers, it could certainly afford to try its hand at mimicking Atari. Though Kalinske mostly dealt with dolls and action figures, he couldn't help but see how ridiculously popular and profitable videogames were becoming, and he decided that Mattel's Toy Division would release handheld electronic games like football and racing. The gameplay was repetitive and the graphics were mediocre at best, but the handheld games were a smashing success. As Kalinske looked to quickly expand this line of business beyond sports, he sought out Nakayama to talk about turning some of Sega's popular arcade hits into a line of portable games. The handheld technology turned out to be too simplistic to handle Sega's games, though, so no deal could be reached.

Despite not moving forward together, Nakayama found Kalinske endearing and was impressed with his encyclopedic knowledge of the toy industry. The two had remained on good terms ever since.

Kalinske's adventures in the videogame industry, however, turned out to be short-lived. After releasing this series of primitive yet highly successful handheld devices, Mattel had decided that videogames were the future. As a result, the company formed an Electronics Division, hired a bunch of brainiacs, and took away anything with batteries from Kalinske's team in favor of this brand-new department. Kalinske was forced to watch from the sidelines as Mattel made a big push to redefine itself behind these handheld devices and their dazzling home console: Intellivision. Kalinske was irked by the move; he had helped create this future and thought he deserved the chance to decide how it should turn out. But ultimately, he didn't care all that much. Videogames were fascinating, but they did all the work for you. No amount of graphics or gameplay could possibly compare to the entertainment value of toys, which, of course, didn't run on batteries but rather were powered by the world's only unlimited resource: imagination.

Besides, the quick end to his relationship with videogames turned out to be for the best. By 1983, it seemed that every company had copied Mattel (who had copied Atari) and gotten into the gaming racket. The market quickly became oversaturated, and the burgeoning videogame industry collapsed. Mattel lost hundreds of millions, Atari lost billions, and Americans from coast to coast lost interest in videogames. After teetering on bankruptcy, Mattel learned their lesson and decided that videogames were no longer their future. Their future would be the same as their past: dolls and action figures.

Kalinske knew that even though Americans had given up on videogames, this wasn't the case overseas. While Atari famously buried three million copies of their notoriously unsuccessful game, E.T. the Extra-Terrestrial, in a New Mexico landfill, the Japanese were flocking to arcades faster than ever. So although Nakayama and Sega seemed to not be welcome in America, the company continued to survive and enjoy success with a generation of Japanese kids and teens who instinctively took to rapidly blinking, brightly colored arcade screens like moths moving toward a glowing light in the dark.

Raising an eyebrow, Kalinske turned to Nakayama. "This new thing you've got is like the Nintendo, right?" Kalinske had never played Nintendo's 8-bit system, dubbed the Famicom in Japan and the Nintendo Entertainment System (NES) in the United States, but he was certainly aware of its massive success. Everyone was. Nintendo was a small but ambitious Japanese company that, in 1985, dared to try to resuscitate the videogame industry in the United States where it had been dead since the failures of Atari and Mattel. Against immense resistance, the NES finally knocked down the fickle walls of pop culture and proved that videogames were not a fad: they were big business. Now, by 1990, less than five years later, Nintendo owned 90 percent of a $3 billion industry. The other 10 percent of the market was made up of wannabes who had seen Nintendo's success and wanted in on the action. Among this group was Sega.

Nakayama rolled his eyes. "No, it is nothing like the Nintendo. Our system is much better. The Nintendo is a toy, but what we have, it is like . . ." He trailed off, struggling to find the perfect words. "Tom, I need for you to come with me to Japan. You must see this for yourself."

Before Kalinske could find a respectful way to protest, he was saved by his five-year-old daughter, Ashley, calling his name, "Daddy!" In that ghostly way that kids can appear out of thin air, she stood before both men and raised her clasped hands to show him something. After doing so, she noticed Nakayama and took a small step backward. "Who is he?"

Nakayama introduced himself and offered the young girl a smile.

"Sweetie," Kalinske said tenderly to his daughter, "Daddy needs some advice. Would you be willing to give your father some advice?"

Ashley loved telling people what to do. She nodded.

"Great," Kalinske said, and paused to figure out how to phrase his question. "So my friend here wants me to go with him on a little vacation to Japan. He wants to show me something there. But I don't think that's a good idea because, you know, I'm already on a vacation with you, your sisters, and Mommy. What do you think I should do?"

Ashley bit her lip and gave this question tremendous thought. As her eyes darted between her father and the man with the funny hair, Kalinske was struck by how fast his daughter was maturing. He felt a pang of pride followed by a pang of sadness. All this time she'd been growing up and becoming a person, and he'd been busy chasing his tail at Mattel or at Matchbox. It was all going on right before his eyes, and he was missing everything. Ashley interrupted his thoughts. "You should go to Japan with your friend."

"What? No."

"Listen to her, Tom," Nakayama said. "She is a wise one."

Kalinske looked into his daughter's eyes. "You don't want me here?"

"Of course I want you to stay. But he just wants to show you something, Daddy," she said, exasperated. "Jeez!"

Kalinske was struck by the wisdom of her kid logic. "Well, if that's what you think . . . then I'll go."

"Okay, okay, whatever," Ashley said. "Now can I show you my surprise?"

Amid all the potential life-changing decision making, Kalinske had forgotten that his daughter had come over with a surprise in her hands. "Oh, yes, please."

She opened her hands and revealed a clump of sand.

"What is it, sweetie?"

A devilish smile crept across her face. "It's a snowball that's made of sand." She laughed a few times, then threw it at her father's stomach. It left a little mark just above his bathing suit before falling to the ground. This was too funny for Ashley to handle, and she ran off laughing so hard that she practically fell over.

Kalinske turned to Nakayama. "Well, I guess I'm going to Japan."

"You will like what I show you."

"I better. Because my wife is not going to be happy."

"She will not be happy now, but she will be happy later," Nakayama said. "When you become president of Sega."

"You're pretty confident, eh?"

"I do not mean to be so forceful," Nakayama said. "I understand this is your family vacation. If you wish to remain at the beach for the rest of the day, we can leave in the morning."

"Uh-oh, are you getting nervous that I might not be as impressed as you first thought?" Kalinske said, sensing a playfulness within himself that had been absent for some time. He felt youthful, curious, perhaps even excited, and he could hear it in his own words. The world seemed slightly bigger, and he felt the private pride of being the only one who noticed it. He looked at his unexpected guest, wanting to say something to extend the moment. "Nakayama-san, can I tell you secret?"

"Yes, Tom. Of course."

Kalinske looked over his shoulder before leaning in and confessing the dirty little secret to his new friend. "I don't even like the beach." Nakayama didn't seem to react to the words, but Kalinske had gotten them off his chest. "I mean, I understand why someone would like it. The sun, the sand, the water; I guess that's relaxing. But I just don't feel that way. All that stuff, I think it's . . ."

Nakayama jumped in to finish his sentence. "Boring."

"Yes!" Kalinske said. "Exactly. It's nice and all, but it's boring."

"Of course, Tom. Of course," Nakayama echoed. "It is boring for people like us."

Suddenly, strangely, Kalinske didn't feel so alone.

Nakayama put his arm on Kalinske's shoulder. "Okay, then, let us now go on a real vacation."

Kalinske smiled. "Let me just ask my wife if it's okay."

He turned to Karen, who continued to lie motionless on her back.

"You have my permission, honey. Go conquer the world and all that," she said.

Kalinske was caught off guard. "You're awake? You heard everything?" He asked, before quickly converting his clumsy surprise into a sly sureness. "Sneaky move, Karen. I'm impressed."

"Don't be. You're not as quiet as you think you are," she said as she lifted up her sunglasses, revealing her shining brown eyes. "Oh, and honey, everyone knows you don't like the beach. It's not exactly a well-guarded secret."

Karen winked at Tom, and with her blessing, he was on his way to Japan.

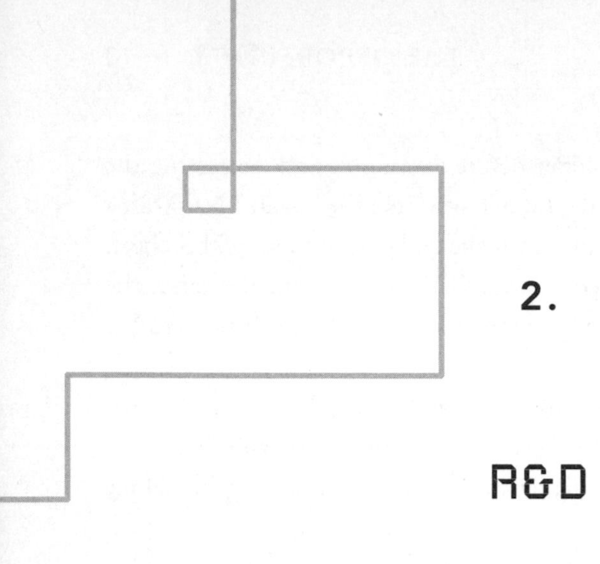

2.

R&D

Like a shark swimming by a school of fish, a hulking yellow Cadillac de Ville crept past awestruck motorists on Tokyo's busy streets. Kalinske and Nakayama sat in the back of the chauffeured car, watching the people they passed instinctively try to peer through their tinted windows. Guppies.

Unlike most wealthy Japanese businessmen, who favored the comfort and grace of a limo, town car, or even an upscale Nissan, Nakayama had an entirely different opinion when it came to transportation. He had imported a big, bulky left-hand-drive Cadillac that was so large and exotic, it made other cars on the street appear to shrink in comparison. Wherever he went and whatever he did, Nakayama always stood out.

As the automotive extension of Nakayama's personality cruised through narrow streets, Nakayama and Kalinske drank whiskey and argued the merits of karaoke. Like most Japanese, Nakayama held enormous respect for the activity and considered it an art, and like most Americans, Kalinske thought it was a cheesy thing to do that seemed like a good idea when drunk but usually ended with regret. These men may have been from two different worlds, but at least they could both appreciate fine whiskey. They refilled their glasses, agreed to table the karaoke discussion, and downed their drinks.

Kalinske finished a nice long sip. "I have a question."

"Perhaps I have an answer."

Kalinske leaned forward. "What about Katz?"

"What about him?"

"What happens to him?"

"To Katz?"

Kalinske shook his head. He had been waiting for the right time to ask about Katz, which was starting to feel like never, but if he was even going to consider taking the job and replacing a friend, they needed to have the conversation. "Oh, don't play it like that. You know we have a history. You had to know I was going to ask about him at some point. So, yes: what about Katz?"

Katz, of course, was Michael Katz, a no-nonsense pragmatist who was already something of a journeyman in the nascent videogame industry. His experience stretched all the way back to 1977, when he had served as the marketing director for Mattel's brand-new line of LED handheld games. After growing the new product line into a $500 million business, Katz moved on to Coleco for their short-lived console venture. After that, he became president of a small, unprofitable computer game company called Epyx, before moving on to Atari in 1985, which by that time was a shell of its former self. Katz had experienced the highs and the lows, and his year as Sega of America's president could probably most accurately be described as somewhere right in the middle.

After a series of gimmicky electronic game systems (like the frigidly named SG-1000, which was cheaply built with spare, off-the-shelf parts), Sega's first real foray into the wide world of home videogames came with the release of the Master System. This was their own NES-like 8-bit console, intended to rival the staggering success of Nintendo. But it was not exactly the triumph they had in mind. The Master System was released first in Japan in 1985, and then in North America in 1986, but in less than two years it was clear that Sega wasn't going to be able to make a dent on the force that was Nintendo. Nakayama decided that if they couldn't win the battle on 8-bit terrain, then they would move the location of the war, and this time at least have the advantage of being there first. They quickly and unceremoniously retired their Master System and shifted their attention to the next generation of videogames: 16 bits, twice as powerful as the NES. Again, they created and released their cutting-edge 16-bit system first in Japan, where it was called the Mega Drive, and next in North America, where they dubbed their technological dynamo Genesis.

In October 1989, Katz was hired to make the Genesis a smash hit in America. In Japan, the Mega Drive had achieved mild success with its initial release, and this gave the powers that be high hopes for their American counterparts. So high, in fact, that Nakayama came up with the rallying cry "Hyakumandai!" (one million units). Despite the shadow of Nintendo, Nakayama fully expected that Katz would be able to sell more than a million Genesis systems by the end of his first year on the job. Katz had tried his best to reach this goal and make a name for the Genesis, but after that year was up he had sold only about 350,000 units and Sega still lacked an identity. Not great results, but not terrible, either. The problem was that Nakayama just didn't think he was the right man for the job. He had good ideas, but he lacked follow-through and the grace to get things done. Big talk, that's what he was, and nothing personified this better than the ad campaign he had chosen: "Genesis Does What Nintendon't." Not only did this bother Nakayama-san because competitive advertising was frowned upon in Japan, but even more so because it was just an empty promise. Genesis does what Nintendon't? What, not make money? Katz knew what Sega was not, but Nakayama believed that Kalinske knew what Sega could, should, and would be.

"Katz had his chance."

Kalinske raised an eyebrow. "His chance? A year?"

"He was only hired for the job until the right person came along."

"Still . . . one year?"

"He thinks he's running a movie studio, not a videogame company. He spends my money like a madman and then calls it an investment. He thinks everything is an investment." Katz had spent a lot of money to secure deals with celebrities, highlighted by $1.7 million for Joe Montana and, most recently, boxer James "Buster" Douglas, the current heavyweight champion of the world. "He has no vision for the company. No identity. So all he does is go out and try to buy one."

Kalinske considered this. "Well, Nintendo has Mario. So naturally you guys should have your own mascot character, someone to crush that little plumber."

"See! You get it, Tom!" Nakayama said, so thrilled that someone else saw the world the same way. "I have tasked our most loyal employees with coming up with our own Mario. You will be astounded by their work, I assure you," he said, his voice vibrating with excitement. "Katz doesn't understand. He just likes to spend money."

While Nakayama's accusation that Katz had spent a lot of money was true, the insinuation wasn't exactly fair. Katz knew that the secret formula to selling a million units lay in making popular games. If people craved the software, then they would undoubtedly buy the hardware. This had been Nintendo's approach: dazzle the market with much-talked-about hits like Super Mario Bros., The Legend of Zelda, and Teenage Mutant Ninja Turtles, and as a result, seduce an entire generation into buying the NES. Unfortunately, this approach presented a problem for Katz because Nintendo held an iron grip on software developers. If game designers wanted their game on the NES, then Nintendo had them sign an exclusive agreement with a stringent noncompete clause. So if Nintendo got a game, there was no way that Sega could offer it on their system, and given Nintendo's monstrous success, why would anyone choose Sega over Nintendo? Katz's solution was to hitch Sega's wagon to household names, believing the association between Sega and the likes of Joe Montana and Buster Douglas would bring about a certain level of respect and legitimacy.

"You know Katz, though. He's a builder," Kalinske said. As much as he understood where Nakayama was coming from, he had a soft spot for Katz. They had become buddies at Mattel, they played tennis together, and their wives got along. "Katz is slow and steady. I thought that's what wins the race, no?"

"This is not a fable, Tom." Nakayama shook his head. "I want you to take his job because you will be able to do it better."

"Again, I appreciate the flattery, but I don't know the first thing about videogames. I know toys. I'm a toy guy."

"No, Tom. You are a salesman."

Kalinske thought about this as Nakayama refilled his glass and the Cadillac de Ville continued along, turning heads on the streets of Tokyo.

They were dropped off in front of Sega headquarters, which Kalinske was surprised to find was bland and innocuous. It looked almost like a college dorm complex, openly nondescript and painted in a crusty, faded-looking yellow-white. The only difference here was that at the top of this humble ten-story building, the name SEGA was emblazoned in blue capital letters.

Nakayama led Kalinske into the building, which somehow turned out to be even more unspectacular on the inside: drab lighting, crowded workstations, and uninspiring windowless conference rooms. As Nakayama introduced his great white hope to the most senior of the hundred or so employees, Kalinske was already having second thoughts.

They made their way into a boxy gray elevator, where Nakayama attempted to reassure his now-skeptical guest. "It gets better."

"No, no, it's fine," Kalinske said.

The elevator stopped on the third floor and Nakayama led Kalinske into what he considered the crown jewel of Sega's operation: the top-secret R&D lab, where long tables were overrun by large computers, unrecognizable mechanical tools, and several televisions that had been taken apart. Kalinske felt like he had entered an evil scientist's lair—that is, if the evil scientist in question had planned to take over the world by means of videogame domination.

Nakayama proudly escorted Kalinske around the room, introducing him to all sorts of gadgets and gizmos that seemed too small and sophisticated to actually exist. With marvelous graphics moving at racecar speed, these seemed less like games than they did playable dreams. This stuff was light-years ahead of what Kalinske remembered from his days at Mattel.

Nakayama pulled him over to a small station and handed him a little black device. "This is called the Game Gear. It will come out here in October and then sometime next year in America." The Game Gear felt great in Kalinske's hands, and as Nakayama turned it on, the screen flooded with graphics that seemed too good to be true. Kalinske didn't know much about Nintendo's home console, but he was familiar with their handheld device, the Game Boy. Like most everyone else, he had been caught up in the whirlwind of Tetris, the addictive puzzle game that came with the Game Boy. The Game Gear featured a similar game called Columns, which seemed to Kalinske to be equally as addictive, yet instead of staring at the dull-yellow coloring of the Game Boy, he was playing a game with vivid, glorious colors. Nakayama wanted to show him more, but Kalinske couldn't stop fixating on the Game Gear. "You take it," Nakayama said with a nudge. "You take it and show it to your daughters. They will love it so much."

Eventually Nakayama pried him away and dazzled him with more glimpses of tomorrow: a CD-based device that played games with near movie-quality graphics, a pair of 3-D glasses that could be worn to bring certain games to life, and some kind of hefty virtual-reality headset. Finally, the tour concluded in front of Nakayama's crown jewel: the Sega Genesis. Kalinske stared at this beautiful black beast. It was sleek and seductive, with graphics and gameplay that blew what little he knew about Nintendo out of the water. He wondered how the hell Katz had struggled to sell this.

Nakayama watched Kalinske's eyes widen like a kid who not only was inside a wondrous candy shop but had just been told that he now owned it. "You like?"

Kalinske took a moment to compose himself. "It's okay."

"Ah, yes, sure it is," Nakayama said. "Shall we go somewhere more private to continue our discussion?"

Kalinske put down the controller he had been inspecting. He couldn't get over how comfortably everything fit into his hands; it was as if they were designed specifically for him. He left with Nakayama, trying his best to hide his boyish enthusiasm. He didn't know exactly where they were headed, but for the first time in a while he felt excited about whatever would come next.

3.

THE STORY OF TOM KALINSKE

Nakayama took Kalinske to a popular hostess bar downtown. Despite the sizable crowd of businessmen, their sporadic drunken chuckles, and the constant flirtatious giggling of the pale-faced women dressed as naive schoolgirls, the place offered a certain level of solitude. Perhaps it was the dim lighting, or maybe it was the collective sentiment that everyone there seemed to want nothing more than a moment of privacy and had no interest in getting entangled in the business of anyone else.

"What exactly are your concerns?" Nakayama asked, as one of the bar's geisha girls strolled over to him and Kalinske with small glasses of sake.

"For starters, I'm not too thrilled about the idea of uprooting my family." Sega of America's headquarters were in San Francisco, and Kalinske would have to move his family there from Los Angeles.

"Northern California is the place to be. That's where things happen. What else?"

Kalinske sipped his sake. "What else? A lot of things."

Nakayama skeptically squinted his eyes. "I think that beneath 'a lot of things' is just one thing. So tell me, what is the problem?"

Perhaps Nakayama was right. Perhaps the stresses that came to mind were just tiny planets of anxiety that were all orbiting a single sun. "Okay," Kalinske said, gathering his concerns. "I don't want to put in everything I have only to watch the carpet get pulled out from under me. I want to be able to try things. I want to be able to fail. I want to be able to make this exactly what I think it ought to be and not have to explain myself every step of the way. Basically, I don't want Mattel to happen again."

Kalinske abruptly finished, realizing that he'd struck a nerve that had been buried for years. Nakayama finished his glass of sake. "Okay, fine," he said. "You come work for me and I let you do things your way. This is the deal. No tricks."

These were the magic words that Kalinske had been waiting for, but when they finally came he was momentarily distracted by something unusual across the bar.

"Do you have an answer?" Nakayama asked.

Kalinske heard the words, but his mind was locked on a well-dressed man sitting at a table about twenty feet away. This man, whose elegant outfit screamed success, was surrounded by beautiful women, scheming friends, and copious amounts of alcohol. Yet despite these temptations, the well-dressed man was completely entranced by only one thing: a Game Boy. As his fingers jabbed the buttons of Nintendo's handheld console, nothing else in the world mattered.

"Tom?" Nakayama asked, trying not to sound too curious.

"I need to think about it," Kalinske replied, and then he pulled out the Game Gear he had received earlier as if this might somehow hold the answer. Could he seriously see himself jumping into the videogame industry? Did he actually believe he had what it took to topple Nintendo? And, for that matter, did anyone? As he considered these questions and thought more about the well-dressed man, the machine came alive in his hands and Tom Kalinske's life flashed before his eyes.

Suddenly his mind was flooded with the sights, sounds, and feelings of being a little boy, racing a red die-cast toy car up and down his legs in the backseat of the family's station wagon, wedged uncomfortably between his brother and sister, as his family moved from Iowa to Chicago. The family had moved a lot during his childhood, which was difficult on a small boy, but he remembered it being a little easier because he always had that toy car by his side. He loved it not only because it was always there for him but because he had built it from a kit and painted it the exact color he wanted. He'd savored so much being the creator of something.

He remembered liking Chicago, mostly because he was only five years old, liked life, and thought that life was Chicago. But then his father got a new job at a water treatment facility in Tucson, Arizona, and the family moved again. Tom had been nervous, but he had his toy car, which he felt kept him safe from everything changing around him.

His family stayed in Tucson for good, and that allowed Tom to dive headfirst into the thick desert heat and become many things: a Boy Scout, an athlete, a baseball card collector. When he was twelve years old, his mother convinced him that he had a beautiful singing voice and dragged him down to the so-called Temple of Music and Art. There he auditioned for and was accepted into the prestigious Tucson Boys Chorus, whose past members included George Chakiris and John Denver. He soon discovered that he really was good at singing, which filled him with red-hot ambition. Within nine months, Tom was promoted to the chorus's traveling group. Over the next five years, he toured the entire country, sang on *The Ed Sullivan Show*, performed at the White House, and traveled through Australia, Mexico, and Canada while cutting albums for Capitol Records.

That period of his life had moved fast—it was still a whirl in his mind all these years later—but the velocity had served him well. He returned home full-time to Arizona for his junior year of high school and made heads turn with his blazing speed. He joined the track team and earned a scholarship to the University of Wisconsin but lost it junior year when he got injured in a car accident. Even now, so many years later, he could still viscerally feel not just fragments of the pain, emotional as well as physical, but also how his now-or-nothing persona had emerged at that time. Without the scholarship, he needed to make money in order to pay for classes and graduate, so at twenty-two, with his back up against the wall, that red-hot ambition returned, steering him toward marketing.

Tom Kalinske and his friend Jonathan Pelligrin had both recently taken an advertising class and decided that male students were a very hard demographic to reach. So Kalinske and Pelligrin decided to start a magazine, called *Wisconsin Man*, that was specifically designed to reach male students. The magazine would feature stories about sports, cars, women, and how to do things like ski, barbecue, or interview for jobs. Local and national advertisers recognized the value in targeting the readers of *Wisconsin Man* and paid handsomely for space in the magazine.

This experience proved to Tom that he was capable of bigger things, so he enrolled in business school at the University of Arizona to study marketing. This time, to pay for school, he wrote and sold ads for a local company that owned radio and television stations. In 1968, his writing, résumé, and colorful life experiences earned him a job with J. Walter Thompson, the renowned New York advertising firm. His responsibility was to come up with new product lines for existing customers. Within a couple of months, Kalinske made a name for himself with the work he did on the Miles Laboratory account.

Miles Laboratory was a health products company that had risen to prominence in the 1940s with their One A Day multivitamins. In the 1960s, they wanted to expand their business to children, and developed Chocks, the first chewable vitamin. Though parents liked the idea of providing supplemental nutrition to their children, kids avoided the vitamins because they seemed to be too much like medicine. To change the minds of this fickle demographic, Kalinske suggested that the vitamins be shaped like characters that kids liked, and arranged for the licensing rights to a recently syndicated cartoon from the animation company Hanna-Barbera. This deal resulted in the creation of a successful new product called Flintstones Chewable Vitamins.

As all these memories flooded back to Kalinske, he couldn't help but feel a certain retroactive naturalness to the trajectory his life had taken, and recognize the confidence that had been a constant presence throughout his life. That confidence had never been more on display than in 1970, when Senator Margaret Chase Smith arranged for a series of subcommittee hearings to investigate the advertising tactics used to sell high-sugar products that were heavily fortified with vitamins and minerals. The allegation was that advertisers were attempting to create a false perception that the health hazards of such products (like cereal, juices, and candy-like chewable vitamins) were offset by the added nutrients. During the hearings, Kalinske took the stand, where he was reprimanded by Senator Smith for essentially selling candy wrapped in a flimsy excuse of good health. "So, Mr. Kalinske," she said, pointing a finger at him. "Do you really think selling drugs to children is a good idea?"

Kalinske knew that he was supposed to sit up there and transform himself into a fountain of apologies, but with the finger pointed in his face, he decided to opt for the truth. "I think it's a great idea! Fifty percent of

America's children are malnourished, and, frankly, I don't care how they get their vitamins as long as they get them. We're helping kids stay healthy." The room went silent. After he was dismissed from the stand, he was approached by executives from Mattel who had been watching the proceeding and were impressed with his performance. They offered him a position as product manager of their preschool business.

Two years later, he was invited to speak with the founder and president of Mattel, and experienced his first career-defining moment. "People are saying that Barbie's done, finished, kaput," Ruth Handler proclaimed in her usual raspy, exasperated, yet somehow optimistic voice. "Barbie just had her first-ever down year last year. And you know what that means. In this business, once you dip, you drop and you don't stop." She finished her rant with a powerful but gentle nod of the head. "People are saying it's time to kill Barbie and devote our resources to other things. What do you say?"

An idealistic but unpolished twenty-seven-year-old Tom Kalinske stood in front of her desk, wearing a pleasant smile, as he tried to make sense of what he'd just heard. He desperately wanted to impress his boss, the living legend responsible for making Barbie the most famous plastic doll ever. To avoid saying the wrong thing, Tom continued to hide behind the shield of his smile.

"Nope," Ruth said, her eyes unflinching. "You don't get to where I am without becoming fluent in the language of smiles. And that one you've got slapped across your face right now says, 'I have no idea how to answer her question, so I should try to remind her that I'm handsome and charming.' Am I right?"

Tom chuckled, and this time he gave her a very different kind of smile.

"Okay then, that's much better. But just because I called you charming and handsome doesn't mean you're off the hot seat. Now answer the question, mister."

"Well, Ruth," he said, almost surprised to hear himself sound so calm, "that's the craziest thing I ever heard. Barbie, done? No way." Kalinske shook his head profoundly, now controlling the room with his every word, gesture, and expression, a gift that would reveal itself to him more and more over the years. "Look, I think it's fair to say that both you and I are in good health and seem destined to live nice long lives. And let me tell you something: Barbie will be around long after you and I are gone."

"Oh, is that a fact?" she asked.

"It is," he said confidently.

Ruth's eyes zeroed in on his. "What makes you so sure that people won't get bored of a doll, albeit a fetching blond one?" A tiny smile delicately bent half of Ruth's face. She wasn't the only one fluent in the language of smiles, and Tom knew what it meant: genuine curiosity, with the potential for an impulsive decision.

"They won't get bored," he said, "because Barbie's not just a doll. She's an idea, a promise to girls of all ages that you can fulfill any dream or fantasy out there. Through Barbie, a girl can be anything she wants to be!" Kalinske nodded slowly. "And yeah, it doesn't hurt that she's not bad-looking."

Ruth emphatically slapped her desk. "Great, that's the right answer. You're promoted. You're now the marketing director on Barbie." Without missing a beat or giving him a moment to gloat, she shooed him away. "You've convinced me. Now get out of my office and go convince the rest of the world."

And that's exactly what he did. He revived the Barbie line with the novel idea of segmenting the market. Instead of simply selling one doll per season, Mattel would offer a multitude of Barbies for differing interests, at diverse price points, and targeted at girls of different ages along the spectrum. There would be a Twist N' Turn Barbie, Ballerina Barbie, Hawaiian Barbie, and even President Barbie. In addition, Mattel would vigorously expand the line for her family and friends with the likes of Big Business Ken and Growing Up Skipper, a version of Barbie's younger sister whose breasts would get larger and waist would shrink with the rotation of her left arm. Looking to fill every possible segment of the market, Kalinske even started a line of high-priced collectible Barbie dolls, which came with limited-edition fashions by famous designers like Oscar de la Renta and Bob Mackie. As a result of this new approach, annual sales soon skyrocketed from $42 million per year to $550 million by the end of the decade.

Kalinske's ability to sell particularly came in handy when he met a woman who made his mind feel both empty and infinite. At the 1979 Toy Fair, there she was: a striking young woman hired to dress as Barbie and present the doll's latest accessories at the Mattel booth. Karen Panitz was her name, a New York actress who'd recently had a bit part in *Saturday Night Fever*. She tried to resist his charm, but that didn't last long, because it was obvious that he understood her and she very much understood him. It wasn't quite love at first sight, but what they had was much better than that: a romance built to last through the many splendors and tragedies of life, and in 1983 they married.

Even though he had it all, there was always a need for more. More ideas, more discoveries, and more to do—the instability was always his favorite part. So when Mattel needed a new toy that could replicate for boys what Barbie did for girls, Kalinske rose to the challenge. He commissioned the development of a heroic male action figure, testing out spacemen, military heroes, firefighters, and superheroes—everything under the testosterone-infused sun. The concept that tested strongest was a muscular, sword-wielding, brown-haired conqueror. Kalinske asked the designer to make the hero's hair blond, and then he and his team worked on digging deeper into their character and coming up with his personality, backstory, and supporting cast. The end result was a unique universe for its master, their new character, He-man. The action figure became one of the year's best-selling toys and quickly rose to the top of the character popularity charts. This led to the creation of a comic series, collectible trading cards, and the hugely popular animated TV show *He-Man and the Masters of the Universe*.

Between Barbie, He-Man, and everything in between, people would say that Kalinske had the "magic touch." He liked when people said it, even though he knew it wasn't true. There was no such thing as a magic touch, and it wouldn't have mattered if there were, because the only thing it takes to sell toys, vitamins, or magazines is the power of story. That was the secret. That was the whole trick: to recognize that the world is nothing but chaos, and the only thing holding it (and us) together are stories. And Kalinske realized this in a way that only people who have been there and done that possibly can: that when you tell memorable, universal, intricate, and heartbreaking stories, anything is possible.

"More?" interrupted the geisha girl from earlier, appearing beside Kalinske with a warm jug of sake. "Yes, maybe?" she asked hopefully, pointing to his glass.

Kalinske nodded, returning to the moment. But before the girl could fix him a drink, she suddenly became transfixed by the Game Gear and, as with the well-dressed man, the world suddenly shrunk around her. Well, would you look at that, Kalinske mused, while having a revelation that would shape

Sega, the videogame industry, and the face of entertainment as a whole. Videogames weren't just for kids; they were for anyone who wanted to feel like a kid. Anyone who missed the freedom and innocence that comes with endless wonder. Videogames were for everyone; they just didn't realize it yet.

"What is this?" the geisha girl asked, finally refilling Kalinske's glass.

As he considered the question, Nakayama noticed that a grin had grown across Kalinske's face, and he seemed to know that it had nothing to do with the sake. This was the kind of expression that you remember someone by for the rest of his life. The kind of expression that either starts or ends a story.

"Can't you see?" Kalinske said to the girl as if it were obvious. "It's the future."

4.

RUDE AWAKENING

As Kalinske settled into his new office in the one-story warehouse where Sega rented space, he couldn't help but think about how different this was from every other place he'd ever worked. It was a far cry from Mattel's eight-story tower in Hawthorne, California, not even in the same league as J. Walter Thompson's high-rise on Manhattan's Madison Avenue, and barely a step up from the college apartment where he'd founded *Wisconsin Man* magazine. Well, he thought, at least this new office came with a view, and then he looked out the window at the small company parking lot. It was his first day as president and CEO of Sega of America and he had only met a few people thus far, but he couldn't resist trying to guess which car belonged to which person.

"You're a colossally outstanding idiot!" someone said, interrupting his thoughts.

Kalinske turned around, looked up, and muttered a barely audible noise that could best be described as the sound a question mark might make. It was Michael Katz, standing by the door, slowly shaking his head, with an unfinished smile on his face.

"I'm sorry, Michael," Kalinske said. "I didn't actively pursue this job."

"You stealing my job isn't what makes you an idiot. That makes you kind of an asshole. But I've always known you were secretly kind of an asshole.

What I didn't know was that you were also secretly a colossally outstanding idiot."

Kalinske offered Katz a seat, but he declined. "How so?"

"For taking this job," Katz said.

"What makes you say that?"

"Oh, I don't know, about a million things, but at the top of the list is the glaring fact that you don't know a thing about videogames!"

Kalinske considered this. "I'll learn," he said at last.

"Yes, and what you'll soon learn is that you shouldn't have taken this job. Do you know what Sega really is? Sega is a joke. Sega is a punch line," Katz ranted. He was good at ranting; at times it appeared to be his best quality. "Sega is a ticking time bomb, and you just signed up to strap it to your chest—"

Kalinske cut him off. "I'm glad you came in here, actually. I've been doing some research, and I wanted to say that you did a very strong job, given the hand you were dealt."

"I know," Katz said, nodding. "I did. And look how far away we—sorry, *you* are from even being a blip on Nintendo's radar."

"Then I guess I'll be the one to go down with the ship," Kalinske said with a dignity he felt distinguished him from most other leaders. "Look, I appreciate the advice, Michael. I know it's going to be an uphill battle."

Katz shook his head. "This just isn't like you," he said. "What happened? Did he take you to the secret lab? Or was it the hostess bar?"

Kalinske tried not to let his eyes reveal the answer.

"Oh, my God. It was both, wasn't it?" Katz snickered.

Kalinske stood up and walked Katz to the door. "Listen, I appreciate the . . . whatever you would call this, but . . ."

"Wait," Katz protested. "I just want to ask you a question. Do you really think he's not going to do the same thing to you that he did to me? Think about it."

Kalinske tried his best not to. This was a question that had scraped the top of his mind on numerous occasions since his Japan trip, but he'd been trying his best to avoid it.

With a sheen of sincerity, Katz met Kalinske's eyes and offered some parting words. "Just remember: you may think you're in charge and you may think he's your friend, but watch your back." And then, before leaving, Katz locked eyes with Kalinske, and it dawned on both men that their future success and failure would forever be strangely intertwined.

"I really do think you did a good job, Michael."

"Thanks, Tom. Thanks."

They shook hands and, for a moment, let mutual respect trump awkwardness.

Kalinske shut the door and walked back to the window, where he spent a few moments staring at the unspectacular parking lot. This was the view, this was his new life. Get used to it, he thought.

There was a gentle knock on the door, and in came Shinobu Toyoda, a thin, soft-spoken Japanese man who wore fine Italian suits, fetching neckties, and a pair of thick glasses that always seemed to be trying to escape down his nose. He served as the executive vice president with the primary responsibility of acting as liaison between California's Sega of America and Tokyo's Sega of Japan. After Kalinske accepted the job, Toyoda had graciously been meeting with the new CEO to give him the lay of the land. Within moments of their first meeting, Kalinske could tell that behind Toyoda's reserved demeanor and perpetual smile-for-the-camera grin was an incredibly shrewd and resourceful man. What he couldn't tell, however, was the true source of this man's devotion. Kalinske had heard varying opinions on Toyoda's role at Sega of America, ranging from "the straw that stirs the drink" to "a Japanese spy who was ready at the drop of a hat to tattle on his colleagues to Nakayama." Thus far, his actions had caused Kalinske to believe the former but, then again, wasn't that exactly how the latter would behave? Kalinske happily greeted his guest. "Toyoda-san, come in!"

"Please, just Shinobu."

Kalinske nodded, convincing himself that Toyoda's desire to be addressed informally should count as a point in the not-a-spy column. Though, once again, perhaps that was just a clever ruse.

Toyoda punctured Kalinske's pulpy thought bubble by advising him that the executive meeting was about to begin. "It will be a great chance for you to meet everyone and understand what they do."

"Perfect," Kalinske said, and then followed Toyoda through the wide hallways of the warehouse. Though the building was small, it had an open feel, which made the many boxes stacked against the wall feel less like clutter and more like the foundation of something to come. "Did I hear you mention the other day that you had recently been in Dallas? What's out there?"

"Ah, yes, my family is there," Toyoda said. After a moment, he seemed to realize that his answer could benefit from elaboration, and he continued, "When I left Japan to work for Mitsubishi, my wife and I made a life in Dallas. So she stays there with the kids full-time and I return for the weekends so we can be together."

"Wait, you fly back every weekend? So you basically commute from Dallas?"

"Yes, exactly," Toyoda said softly.

"That's—" Kalinske was about to say "crazy" until he realized that he would be doing a similar (though shorter) commute for the foreseeable future. With the school year about to begin for Kalinske's daughters, he and Karen had decided that until next summer it would be best if she remained in Los Angeles with the girls, and he rent a small place in the Bay Area. On weekends, he would drive down to Los Angeles to spend time with his family and then drive back north early Monday morning. It was not at all an ideal situation (for Kalinske nor his car's odometer), but it wouldn't have been fair to his daughters to move so abruptly, especially when there was a decent chance Sega might not even be around a year from now. And hey, at least he wasn't going all the way to Dallas each weekend. "That's really nice, Shinobu."

Toyoda led him into a shadowy conference room with a large table and dark wooden walls. The room was filled nearly to capacity with just over a dozen employees. Kalinske took a brief moment to introduce himself and explain that today he was there merely as an observer. This led to a few minutes of glad-handing, ass-kissing, and proclamations of future greatness.

After the compulsory pleasantries came to an end, the meeting resumed—if you could even call it a meeting. In Kalinske's experience, meetings were places where employees could share ideas—some good, some bad, some unclear—and then isolate the best ones for implementation. Meetings were places where status updates were given, strategies were discussed, and, most important, employees left feeling slightly better about what they were doing. This was nothing like that. Here, all the voices blurred into a cacophony of discontent.

"What's the status with Atomic Robo-Kid?"

"Who cares? The game is garbage."

"Well, whose fault is that?"

"Some idiot at UPL for coming up with a crap game, some other idiot in Japan for porting it to the Genesis, and then another idiot here for ordering too many copies!"

"Are you talking about me?"

"Well, now that you mention it . . ."

"Hey, screw that. You're lucky you weren't fired for that Babbage's bullshit!"

Finally there was a respite to the chaos when the verbal daggers were momentarily replaced with collective giggles. "Sgt. Kabukiman," Sega's director of licensing, Diane Drosnes, repeated over laughter, "Yup, that's right, he's back." *Sgt. Kabukiman N.Y.P.D.* was a 1990 comedy about a clumsy New York cop-turned-superhero with powers like heat-seaking chopsticks and fatal sushi. Only a few Sega employees had actually seen the movie, but those who had all agreed it had to be among the worst ever made. Yet despite its seemingly obvious horridness, Sega's game developers in Tokyo thought it was a wonderful film and the Americans needed to obtain the license to make a game based on it. Every month, Drosnes and her colleagues would send faxes explaining why this was a bad idea. But without fail, the suggestion kept coming back from Japan. This perpetual cultural difference was a source of great levity, but after everyone had a quick laugh the bickering resumed in full force.

Kalinske tried to hold his tongue, but it was tough. There were, however, three employees whom Kalinske immediately deemed as highly promising. One of them was the enigmatic Shinobu Toyoda, and the other two, Paul Rioux and Al Nilsen, he knew from previous career stops.

"Both of you, settle down," Rioux said, controlling the room with poise, authority, and a hint of threat. Rioux was a stocky Vietnam vet with hollow eyes and a deep, silvery voice. "I'll take care of giving us a little wiggle room, and then we'll ship through Chicago to make up for some of the lost time." Kalinske had briefly worked with Rioux at Mattel, where Rioux had held a management role in the electronics division and been hailed as the nuts-and-bolts guy of the operation. It was immediately clear that Rioux was the go-to guy at Sega of America, the one who made sure that everyone had at least a vague idea of what they were supposed to be doing. Throughout his life, from fighting in Vietnam to fighting for shelf space, he was an unapologetic gladiator who always got things done.

"Hey, maybe we should create some big event to generate interest in the game," said Nilsen, a large-bodied, larger-than-life marketing dynamo. Beneath round-rimmed spectacles, Nilsen's face rarely depicted emotions. But when he spoke, a wonderfully boundless, kid-trapped-in-an-adult-body enthusiasm could not be contained. "Like maybe, I don't know . . . a National Kid's Day! Yeah, that might work. You know, like Mother's Day and Father's Day, but for kids. And we'll make Atomic Robo-Kid the perfect gift for the occasion. What do you say, guys? Do we have a winner?" Nobody responded, and Nilsen sat back down. From their time at Mattel, Kalinske remembered Nilsen as someone who could and would lay golden eggs from time to time. Seeing Nilsen excitedly churning with ideas put Kalinske slightly at ease.

After a lingering silence, Toyoda meekly spoke up. "If we delay the shipment as Al suggested, but not for such a reason, can we depreciate the R&D costs for the second quarter?" Nobody responded, but it was clear the financial guys in the room loved this idea, and they scribbled it down. Toyoda only spoke up a few times, each time asking a clever question or clarifying an important remark that had gotten lost in the perpetual shuffle.

Kalinske gazed out the window of the conference room, once again attempting to accept that this was now his life. At least he had Rioux, Nilsen, and Toyoda. Kalinske took a deep breath. Everything would be okay. But just as Kalinske was starting to feel a little better about this motley crew, the meeting once again turned into a name-calling shouting match.

"Who cares about depreciating the R&D costs? You can't expect a Band-Aid to cover up a hole the size of the goddamn Grand Canyon."

"Yeah? Well we wouldn't be in this hole if you didn't convince us to overbid on trash like Dick Tracy and Spider-Man."

"What do you want from me? These are our only options. Everyone else is developing for Nintendo!"

"So why don't we get them to develop for us?" Kalinske said, before he even realized that he was speaking. His own words caught him off guard, but he'd already started, so he couldn't stop now. "If they've already made the games for Nintendo, why don't we offer them some money to put them on our system too?" Kalinske said, shining a spotlight of reason in a room full of nonsense. "Maybe we could even get them to make enhanced versions for the Genesis so they'll stand out." Kalinske looked around, expecting enthusiastic nods. Instead, he found a combination of discomfort, dismay, and maybe even some pity.

Toyoda spoke up to fill the void. "Those are all strong ideas, Tom. Thank you for sharing. But unfortunately, at this moment, they will not be most efficient. Nintendo has taken measures to prevent anything like that. They are prepared."

Kalinske nodded, getting the message: Shut up until you know what you're talking about. He thought about trying to compensate for his blunder by making some bold promise or guarantee of success, but realized that might make him appear even more out of his element. For now, there was nothing to do but sit there, listen, and wonder if he'd made the biggest mistake of his life.

When the meeting ended, everyone quickly exited the conference room, except for Paul Rioux, who lumbered toward Kalinske with a thick, Hemingwayesque grin. "I know how you're feeling," he said. "It's how I felt my first day on the job. To come from a place like Mattel to . . . this?"

"Yeah, it's a little different from what I expected," Kalinske said.

"It is. And it took me a little while to figure out that different is good," Rioux said, reflecting. "Anyway, I wanted to come over and say that I was very, very pleased when I heard that you were coming here. You're going to do a hell of a job."

"We shall see . . ."

"We shall. And to help you get there I've put together a dossier of information so that you can familiarize yourself with Nintendo. They're a real beast, and their financials are just . . . I don't even know how to describe it. I've got all kinds of articles, reports, presentations, and things of that nature. It's yours, if you want it."

Kalinske was so grateful that for a moment he considered enveloping Rioux with a giant bear hug. Instead, he took a moment to gather himself and simply say thank you, before setting out to learn everything there was to know about Nintendo.

5.

THE HISTORY OF NOA

(A STORY TOLD IN 8 BITS)

On September 23, 1889, just weeks before his thirtieth birthday, an entrepreneur named Fusajiro Yamauchi opened a small, rickety-looking shop in the heart of Kyoto. To attract the attention of passing rickshaws and wealthy denizens, he inscribed the name of his new enterprise on the storefront window: Nintendo, which had been selected by combining the *kanji* characters *nin*, *ten*, and *do*. Taken together, they meant roughly "leave luck to heaven"—though, like most successful entrepreneurs, Yamauchi found success by making his own luck. In an era where most businessmen were content to survive off the modest returns of regional mainstays such as sake, silk, and tea, he decided it was time to try something new. So instead of selling a conventional product, Fusajiro Yamauchi opted for a controversial one, a product that the Japanese government had legalized only five years earlier: playing cards.

The history of Japan's relationship with playing cards is as noble as it is bizarre. It began in the late sixteenth century, when visiting Portuguese sailors first introduced card games. This popular Western pastime quickly spread throughout Japan. At the same time as this frenzy was overtaking the nation, Japan's military commanders were growing increasingly incensed by the rapid influx of European missionaries. To ward off this flood of Christianity, the Japanese government issued a series of edicts that effectively closed the country's borders and banned many Western items, including clocks, eyeglasses, and, of course, playing cards.

As a means to circumvent the ban on cards, the Japanese obediently stopped producing Western cards with four suits and twelve numbers and instead constructed cards with the seasons (also four) and months of the year (also twelve). These new playing cards, which also featured altered illustrations and were intended for games with modified rules, became known as *hanafuda* cards and lent themselves to more complex games like bridge and mah-jongg. Eventually the government caught on and banned these as well, but even that wasn't enough to deter the popularity of *hanafuda* cards, which survived into the nineteenth century by being played behind closed doors.

That all changed in 1885, when a more open-minded Japanese government withdrew several restrictions that had been imposed on gambling and the manufacturing of *hanafuda* cards. For the first time in centuries, people who had been playing this game illegally, like twenty-six-year-old Fusajiro Yamauchi, were finally able to do so out in the open. This led Yamauchi to spend more time playing the game, which led to a series of entrepreneurial ideas and eventually to the opening of that fateful Nintendo shop on September 23, 1889.

There, in the heart of Kyoto, he and a small team of employees crafted paper from the bark of mulberry trees, mixed these thin slits with soft clay, and then added the *hanafuda* card designs with inks made from berries and flower petals. Nintendo's cards, particularly a series called Daitoryo (which featured an outline of Napoleon Bonaparte on the package), became the most successful in all of Kyoto. The fate of Nintendo looked very promising, until future success was threatened by that of the past. Because those early years had gone so well, all the region's households seemed to already have a deck of *hanafuda* cards, and demand came to a halt. To overcome this issue, he set his sights on the one place where the demand never diminished: casinos.

In the smoky dens of Japan's gambling parlors, the highest-stakes tables would open a new pack of cards for each game. Yamauchi recognized the incredible potential and signed contracts with nearly seventy gambling parlors. With each one going through hundreds of packs every week, Nintendo's profits soared. Up until this point, Yamauchi had been selling his cards exclusively in Nintendo shops, but if he ever wanted to see Nintendo become a household name, he needed a way to reach more households. That path to pervasiveness came when he struck a deal with the Japan Tobacco and Salt

Public Corporation, the state-run monopoly operated by the Japanese Ministry of Finance, which agreed to sell Nintendo cards in their many cigarette shops scattered throughout the nation.

After several decades of staggering success, Fusajiro Yamauchi retired in 1929 and was succeeded by his son-in-law Sekiryo Yamauchi, who ran Nintendo efficiently for nineteen years, but in 1948 he had a stroke and was forced to retire. With no male children, he offered Nintendo's presidency to his grandson, Hiroshi, who was twenty-one years old and studying law at Waseda University. It didn't take long for Hiroshi Yamauchi to make his presence known, and he earned a reputation for quick thinking and a quicker temper. Unsurprisingly, he soon fired every manager that had been appointed by his grandfather and replaced them with young go-getters who he believed could usher Nintendo beyond its conservative past. Also unsurprisingly, his keen insights and rabid efforts to modernize Nintendo accomplished their goal. In 1951, he consolidated all Nintendo's manufacturing plants in Kyoto to greatly speed up the production process. In 1953 he introduced the first plastic-coated playing cards in Japan. And in 1959 he led Nintendo into its first licensing agreement, a potentially game-changing deal with the Walt Disney Company. The Disney playing cards were a stunning success and helped Nintendo reach a new generation of Japanese boys and girls.

Emboldened by this triumph, Yamauchi branched out into a number of other, less lucrative endeavors, including an instant-rice company and a pay-by-the-hour "love hotel." These disappointments led Yamauchi to the conclusion that Nintendo's greatest asset was the meticulous distribution system that it had built over decades of selling playing cards. With such an intricate and expansive pipeline already in place, he narrowed his entrepreneurial scope to products that could be sold in toy and department stores and settled upon a new product called "videogames."

Yamauchi wanted Nintendo to aggressively get into the videogame business, which was really two separate businesses: home consoles and coin-operated arcade games. Yamauchi saw the potential in these industries and took the necessary steps for Nintendo to enter both. In 1973 he established Nintendo Leisure System, a subsidiary devoted to making arcade games. Despite middling results from titles like Wild Gunman and Battle Shark, Yamauchi remained committed to his new vision of Nintendo and continued to allocate a vast amount of resources toward videogames. In 1977 Nintendo released a shoebox-sized orange console called the Color TV-Game 6, which played six slightly different versions of electronic tennis and was met with a mixed reception. Though the console managed to sell one million units, it ultimately lost money for Nintendo because of the exorbitant R&D costs. Nevertheless, Yamauchi remained undeterred. Nintendo continued to put out arcade games (striking out with duds like Monkey Magic and Block Fever) and also continued to release home consoles (like the Color TV-Game 15, which offered fifteen slightly different versions of electronic tennis).

In the same way that, over seventy years earlier, Fusajiro Yamauchi had changed Nintendo's fortunes by expanding its distribution network, Hiroshi Yamauchi saw the potential in a similar idea. By this point, Nintendo already had penetrated most of Japan, so the company would have to look overseas. With this logic, Yamauchi set his sights on the place where the videogame frenzy had started: America.

1. Arakawa

Yamauchi had already dipped a toe into this red, white, and blue pool of water a few years earlier and was encouraged by the results. In the late 1970s Nintendo had begun working with a trading company that would export arcade cabinets to American distributors, who would in turn sell these games to vendors in the United States. Though the profits from this arrangement were minimal, Yamauchi believed that if he could cut out the trading companies and send over someone he trusted to grow Nintendo's business organically, then there was a lot of money to be made in the land of opportunity.

The American market would be risky, tricky, and perpetually persnickety. There appeared to be only one man properly equipped for the challenge: Minoru Arakawa, a frustratingly shy but brilliant thirty-four-year-old MIT graduate. Not only did Arakawa possess the insight and intellect to open a U.S. division of Nintendo, but he came from good stock (his family had been running a prominent Kyoto-based textile company since 1886), he was already living in North America (he was selling condominiums for the Marubeni Corporation in Vancouver), and had an incalculable fondness for America (some of his most cherished memories had come from a post-college cross-country trip he'd taken in a used Volkswagen bus).

In every way, Minoru Arakawa appeared to be the perfect candidate . . . except that he happened to be married to Yamauchi's daughter, Yoko, who blamed Nintendo for turning her father callous. She simply refused to let her husband join Nintendo, as she did not want to watch history repeat itself.

Yamauchi initially proposed the idea to Arakawa in early 1980. Following a pleasant family dinner, Yamauchi spent two hours discussing his plans for the expansion of Nintendo and concluded by stating that the success of his plan hinged on Arakawa. The unusual plea intrigued Arakawa, as did the unique potential of Nintendo's upcoming products (including small, calculator-sized handheld games, and a new console that would play interchangeable cartridges). Lastly, anticipating his daughter's reluctance, Yamauchi explained that this American division would be a completely independent subsidiary. Arakawa wrestled with the decision as well as with the objections of his wife, who cautioned him that no matter what he accomplished, he would always be perceived as nothing more than the son-in-law. Perhaps that would be the case, but Arakawa decided that the opportunity was too good to pass up, and in May 1980 he and his wife left Vancouver to start Nintendo of America (NOA).

Despite her concerns, Yoko greatly loved and supported her husband. So after the family moved to New Jersey, she formally became NOA's first employee and helped select a location for the new company's office. Arakawa and his wife settled on a small space on the seventeenth floor of a Manhattan high-rise located in the center of the toy district at 25th Street and Broadway. They spent their days at the office and their nights observing games and players at local arcades. They learned a lot this way, but no amount of knowledge could make up for the fact that for Nintendo to gain a foothold in America, they needed to build a strong sales network. So Arakawa set up a meeting with a couple of guys who he thought might be able to help: Al Stone and Ron Judy.

2. Stone and Judy

Al Stone and Ron Judy were old friends from the University of Washington, where they had once lived in the same frat house and had been known to embark on promising get-rich-quick schemes together (like buying soon-to-be-discarded local wine cheaply and then reselling it to their college brethren with less sophisticated palates). After graduating, they each moved to separate coasts (Stone to the West, Judy to the East), but eventually the alchemy of their entrepreneurial relationship pulled them back together. Tired of working for other people, they started a trucking business in Seattle, because they liked the changes that were occurring in the transportation industry as a result of deregulation. After founding a company, which they called Chase Express, Stone and Judy proceeded to buy a handful of small trucking companies with the goal of rolling them into one midsized trucking company. Shortly thereafter, they learned the true meaning of "easier said than done." The trucking industry turned out to be too political, too insular, and not quite what these two guys wanted to do for the rest of their lives. So while they continued to invest in Chase Express and hope for a turnaround, they also began looking for alternative business opportunities, ones that did not include five axles and eighteen wheels.

They eventually found their answer, though it still involved big rigs. But unlike their previous venture, this wasn't about the trucks themselves, but rather about what they carried inside. Through a friend in Hawaii, Ron Judy had been informed that a Japanese trading company was seeking a distributor to sell some arcade games made by Nintendo Company Limited (NCL). Intrigued, he agreed to test the waters and received a crate containing a few arcade cabinets of Nintendo's Space Wars. Though the game was little more than a shameless rip-off of Taito's Space Invaders, Judy got his brother-in-law to place the games in some of the taverns he owned in south Seattle. Much to his delight, the machines were quickly overrun with quarters, which convinced him and Stone that this was their future. They formed a distribution company called Far East Video and used their assets from the trucking industry to travel around the country and sell Nintendo games to bars, arcades, hotel lounges, and pizza parlors.

After experiencing the nation's unquenchable thirst for videogames firsthand, Stone and Judy believed they had caught lightning in a bottle. But despite the bottle full of lightning, their bank accounts weren't roaring with thunder. The outrageous profits they were making from the arcade cabinets were being canceled out by the trucking company's continuing losses. They decided that it was time to unwind the trucking company and go all in on videogames, but to do so they needed additional financing, and at the time banks were highly skeptical of this newfangled industry. This left Stone and Judy in an awkward position, not knowing where to turn next, until they got a call from a man named Minoru Arakawa and things came into focus.

From Arakawa's perspective, Stone and Judy were a godsend: a pair of hard-nosed entrepreneurs who had already developed a network of sales contacts. He wanted them to stop working for the company that was selling Nintendo games and come work directly for him. For Stone and Judy, this was a no-brainer on a variety of levels: though their margins would be slightly reduced, this solved their financing dilemma, secured a steady stream of product, and would even get their travel expenses reimbursed. With so much upside for all involved, the only possible way this arrangement wouldn't work out was if Nintendo's upcoming games weren't any good. And, unfortunately, that's exactly what happened next.

Space Fever was followed by Space Launcher (underwhelming), which was followed by Space Firebird (disappointing), which was followed by a slew of unsuccessful non-space-themed games. After this string of mediocre misfires, Stone and Judy were ready to quit, and Arakawa couldn't help but reconsider his new vocation. More than ever, Nintendo of America needed a megahit, like Pong or Pac-Man, to keep the dream alive. And just when it appeared time was running out, Arakawa believed that he'd found what he needed: Radarscope.

At first glance, Radarscope may have appeared to be just another shoot-'em-up space game, but it distinguished itself with incredibly sharp graphics and an innovative 3-D perspective. After receiving positive feedback from test locations around the Seattle area, Arakawa invested much of NOA's remaining resources and ordered three thousand units. But a few weeks later, before the rest of the arcade cabinets even arrived, Arakawa felt an ominous chill upon revisiting the test locations, where he noticed that nobody was playing Radarscope. That foreboding feeling was validated after the three thousand units finally arrived and Stone and Judy found that operators had little interest in the title. Radarscope was fun at first, the consensus appeared to be, but it lacked replay value.

3. Miyamoto

With so much invested in this game, it would be too expensive to send all the bulky arcade cabinets back to Japan and then import something else. The last remaining hope was for a designer in Japan to quickly create a game that would be compatible with Radarscope's infrastructure (and, when finished, send over processors with that new game to America, where NOA employees could swap out the motherboard and then repaint the arcade cabinets to reflect this new game). This task was given to Shigeru Miyamoto, a floppy-haired first-time game designer who idealistically believed that videogames should be treated with the same respect given to books, movies, and television shows. His efforts to elevate the art form were immediately given a boost when he was informed that Nintendo was close to finalizing a licensing deal with King Features, enabling him to develop his game around the popular cartoon series Popeye the Sailor Man. Using those characters, he began crafting a game where Popeye must rescue his beloved Olive Oyl by hopping over obstacles tossed in his way by his obese archenemy, Bluto.

Meanwhile, as Miyamoto set out to save NOA, Arakawa wanted to take precautions against something like this happening again. The major takeaway, of course, was to be more attuned to the fickle nature of arcade players, but there was another lesson here as well. Part of the problem with Radarscope could be attributed to the long shipping time (about four months) and the high cost of transporting the arcade cabinets. One way to cut back on both was to find an office closer to Japan, which prompted Arakawa to move Nintendo of America from New York to a warehouse with three small offices located at Seattle's Segali Business Park.

Shortly after the cross-country move, shipments containing the code for Miyamoto's new game began to arrive. Due to last-minute negotiation issues with King Features, Nintendo had lost the rights to Popeye, which forced Miyamoto to come up with something else. As a result, Arakawa, Stone, Judy, and a handful of warehouse employees didn't know what to expect. They inserted the new processor into one of the thousands of unsold Radarscope machines and then watched the lights flicker as the words "Donkey Kong" came to life on the arcade screen. The initial impression was that this was a silly game with an even sillier name. Who would possibly want to play a game where a tiny red plumber must rescue his beloved princess by hopping over obstacles tossed in his way by an obese gorilla? Yet, with no remaining options, Stone and Judy set out across the country to sell this wacky thing called Donkey Kong.

Seemingly overnight, it turned into the hottest game of the year, and eventually it became the most popular arcade game of all time. Never before had there been a quarter magnet quite like Donkey Kong. It was so successful, in fact, that it eventually attracted the attention of a major Hollywood studio, whose high-priced legal team believed that the game violated copyrights, and they threatened to crush Nintendo. To avoid this potentially crippling blow, Arakawa turned to the only lawyer he knew in Seattle: Howard Lincoln, an elegant, imposing former naval attorney with the claim to fame of having modeled for Norman Rockwell's painting *The Scoutmaster* when he was a child.

4. Lincoln

Lincoln had first crossed paths with Arakawa about one year earlier when his clients Al Stone and Ron Judy needed him to review their contract with Nintendo of America. After that, Lincoln slowly but surely took on the role of Arakawa's consigliore, weighing in on any matter with legal ramifications, which in corporate America was just about everything. As Nintendo of America grew, Lincoln drew up new employment agreements (Stone became NOA's VP of sales and Judy NOA's VP of marketing), looked at various business deals (Arakawa was interested in buying the franchise rights to Chuck E. Cheese Pizza Time Theater), and handled some tough matters (like siccing the U.S. marshals on bootleggers to take down rings of Donkey Kong counterfeiters). Through it all, Lincoln and Arakawa had forged an unshakable lifelong friendship. Which is why Lincoln was the first person Arakawa contacted when, in April 1982, MCA Universal sent a telex to NCL explaining that Nintendo had forty-eight hours to hand over all profits from Donkey Kong due to the game's copyright infringement on their 1933 classic movie *King Kong*.

To sort this out, Lincoln and Arakawa flew out to Los Angeles for a meeting with the movie studio. It didn't take long for them to realize that this was a high-stakes shakedown. Though never explicit, Universal's ultimatum was simple: Settle this, or we'll make life at tiny Nintendo so difficult that the company will have to fold. With Nintendo's fate once again on the line, the prudent thing to do was pay the ransom. But Lincoln believed he could win this thing—and not just win, but actually get Universal to pay damages to Nintendo. It was a risky call, but Arakawa was forever willing to gamble on Howard Lincoln. As a result, they decided to take on the movie moguls at MCA Universal and, in true Nintendo spirit, leave luck to heaven (and Howard).

As a trial neared, Universal didn't just go after Nintendo but also went after those who had licensed the game. Unlike Nintendo, however, these licensees (Atari, Coleco, and Ruby-Spears) weren't willing to take the same gamble, and all opted for settlement. Even with these threats plus a barrage of cease-and-desist letters from Universal, Lincoln remained confident. Part of this was personality, but a larger part was the ace tucked away up his sleeve: in all of his research, there didn't appear to be a single document indicating that Universal had trademarked King Kong. There was no doubt, of course, that they had made the movie, but Lincoln believed that they had failed to take the necessary measures to own what they thought they owned, and that the famous gorilla belonged to public domain. And in early 1983, when both parties revealed their cards, Judge Robert W. Sweet sided with Nintendo. He concluded that they had not infringed and, as Lincoln had predicted earlier, he awarded Nintendo over $1 million in legal fees and damages.

The Donkey Kong fiasco-turned-feat caused many ripples, but three waves in particular were instrumental in creating the eventual tsunami that would be Nintendo. First, Lincoln became NOA's senior vice president, officially making him the ying to Arakawa's yang. Second these events foreshadowed an aggressive, litigious nature that many would say later defined the company. And third, the verdict kept the Donkey Kong cash flowing, which provided Nintendo with a war chest of funds at what would soon prove to be a very crucial time.

5. Borofsky and Associates

By the early 1980s, the videogame bonanza had become so lucrative that everyone wanted in on the action. This included companies that had no business entering the market (like Purina, whose Chase the Chuck Wagon was designed to help promote their Chuck Wagon brand of dog food), companies that didn't quite understand the market (like Dunhill Electronics, whose Tax Avoiders allowed players to jockey past a maze of evil accountants and onerous IRS agents), and lowbrow companies that set out to polarize the market (like Mystique, whose flair for pornographic titles was highlighted by their 1982 anticlassic Custer's Revenge, which follows a naked cowboy on his quest to rape Native American women). With games like these becoming more and more common, the marketplace was overrun by a glut of smut, muck, and mediocrity.

What once was hot now was cold, and just like that, the North American videogame industry ground to a halt. Hardware companies (like Atari) went bankrupt, software companies (like Sega) were sold for pennies on the dollar, and retailers (like Sears) vowed never to make the same mistake again. Meanwhile, as the gods of this golden age were thrown overboard, Nintendo quietly glided through the bloody waters on a gorilla-shaped raft. The continuing cash flow from Donkey Kong enabled Arakawa, Stone, Judy, and Lincoln to dream of a new world order, one where NOA miraculously resurrected the industry and Nintendo reigned supreme. Not now, perhaps, but one day soon.

In Japan, however, that time had already come. Yamauchi's large investment in R&D had paid off once again, this time resulting in the Family Computer. The Famicom, as it was commonly called, was an 8-bit console that stood head and shoulders above anything that had ever come before. Since Japan's conservative markets had avoided any sort of gaming crash, the Famicom was released in July 1983 along with three games: Donkey Kong, Donkey Kong Jr., and Popeye, which Miyamoto ended up designing after licensing negotiations got back on track. The Famicom stumbled out of the gate but was soon rescued by heavy advertising and the release that September of Super Mario Bros. (another Miyamoto brainchild). Things looked promising until it was discovered that some consoles contained a bad chip set, causing certain games to freeze. Instead of simply issuing fixes to customers with faulty systems, Nintendo issued a full product recall. Tens of millions of dollars were lost due to this approach, but Yamauchi believed it was a small price to pay to retain a peerless, quality-centric reputation. His gamble paid off, and soon Nintendo's factory couldn't keep up with the demand.

As sales soared to staggering heights, Yamauchi pressured his son-in-law to introduce the Famicom in America. Arakawa resisted, exercising patience. The U.S. market was still licking its wounds from the videogame crash, and releasing the right console at the wrong time would be a recipe for disaster. For this reason, he continued to rebuff the suggestion until 1984, when he was finally willing to consider the notion—but only if the console Nintendo of America sold looked nothing like a console at all.

This wolf-in-sheep's-clothing logic led to the Advanced Video System (AVS). Though the guts of this machine were nearly identical to the Famicom's, the AVS hardly resembled its foreign relative. Functionally, it came with a computer keyboard, a musical keyboard, and a cassette recorder; aesthetically, it was slim and sleek, with a subdued gray coloring that contrasted sharply with the Famicom's peppy red and white palette. Nintendo's AVS, the nonconsole console, was first introduced at the 1984 Winter Consumer Electronics Show, where it was accompanied by a brochure that proclaimed, "The evolution of a species is now complete." Of the thousands of companies there who had rented booths to showcase upcoming products, Nintendo was the only one who wasn't trying to sell theirs. Arakawa simply wanted to gauge the market reaction, which was as distressing as he had feared: nothing but scoffs, sighs, and sob stories. Nobody there wanted anything to do with Nintendo, except for a tanned man with piercing blue eyes who stared at the Advanced Video System as if it were the legendary sword in the stone. He then introduced himself with an understated sureness that would have made even King Arthur jealous. His name was Sam Borofsky.

Borofsky ran Sam Borofsky Associates, a marketing and sales representative firm based in Manhattan. The purpose of such a firm is, essentially, to serve as the go-between for suppliers and retailers, with the logic being that the commission they take is offset by the additional opportunities they create. And when it came to doing that for consumer electronics, Sam Borofsky Associates was one of the best in the business. Back in the late seventies, they became one of the first firms to represent videogames, and at the height of the boom they had been responsible for over 30 percent of Atari's sales. If Nintendo of America ever wanted retailers to reopen their doors, then these were the guys who ought to do the knocking. From Borofsky's end, the attraction was equally strong. Ever since Atari had imploded, he'd been scouring the country in search of the next big thing, and as he reviewed what Nintendo had to offer, he believed he had found it.

Arakawa, however, still needed convincing, which Borofsky was happy to provide. He spent months detailing Atari's pitfalls (like oversaturation), coming up with solutions for those problems (always, no matter what, underdeliver on orders), and outlining plans for a proposed launch. Meanwhile, Nintendo of America put another new costume on the Famicom, this time dressing it up as an all-in-one entertainment center for kids. The result of the rebranding effort was a clunky gray lunchbox-like contraption and, along with that, a new lexicon to differentiate it from its predecessors: cartridges were now dubbed Game Paks, the hardware was dubbed the Control Deck, and the entire videogame console was rechristened the Nintendo Entertainment System (NES). And to round out the renovation, the NES came with a pair of groundbreaking peripherals: a slick light-zapper gun and an amiable interactive robot named R.O.B.

With all the pieces in place, Borofsky finally persuaded Arakawa that now was the time to strike and that he was the man to lead the charge. Upon receiving the green light, Borofsky begged, pleaded, and haggled with retailers all over New York. From Crazy Eddie and The Wiz to Macy's and Gimbels, he reached out to them all. If and when a retailer was finally willing to consider Nintendo, Borofsky would head over there with his trusted associate Randy Peretzman, the man with the twenty-six-inch suitcase. Peretzman was Borofsky's VP of sales, a gritty but graceful Bronx-born salesman who operated with an unwavering tell-it-how-it-is attitude. As the go-to guy for demonstrations and presentations, he was entrusted with Nintendo of America's first prototype of the NES, which he gently packed into the soft foam cutouts of a twenty-six-inch hard-sided gray suitcase. With the suitcase in hand, he would pinball around the city visiting distrustful retailers, intent upon proving that his luggage would trump their baggage.

Eventually Peretzman, Borofsky, and his tireless associates persuaded retailers to gamble on Nintendo and to stock the NES in time for the coveted Christmas season. The orders came in dribs and drabs at first, but by late 1985 the numbers were slowly beginning to add up. Then it was time for Nintendo's ultimate test: the product launch. If things went well, the NES would be rolled out nationwide the following year. But if things went poorly, then Nintendo of America would forget about consoles once and for all.

6. The Bruces

As the test launch approached, things did not look promising. Focus groups suggested that the NES would be a colossal flop, R.O.B. kept malfunctioning during Peretzman's pitches, and the press showed no interest in covering Nintendo (leading one employee to suggest a publicity stunt that involved throwing a fleet of R.O.B.'s off the Brooklyn Bridge). Arakawa, however, remained undeterred. He temporarily relocated a handful of employees to the East Coast after leasing a warehouse in Hackensack, New Jersey, where Nintendo could house inventory, build in-store displays, and, most important, resemble a legitimate company to still-skeptical retailers. To keep tabs on the progress, Ron Judy made frequent visits to New York, often accompanied by Bruce Lowry, NOA's bright and blustery VP of sales. After making a name for himself at Pioneer Electronics, Lowry had joined NOA in April 1981 to launch a new consumer division. Though arcade games were and would be Nintendo's bread and butter, Arakawa wanted to supplement that income with Game & Watch, a line of wallet-sized handheld games that played on tiny LCD screens. Unlike the electronics Lowry had sold at Pioneer, the Game & Watch titles were targeted at children, which prompted Lowry to become familiar with the toy industry. Because of his entrée into that world and his understanding of its nuances (such as the fact that regardless of when toy buyers ordered products, they didn't have to pay until December 10 of that year), Lowry was occasionally able to help Borofsky persuade the big toy chains. At the top of their wish list was Toys "R" Us, whose eventual decision to stock the NES provided Nintendo with much-needed momentum going into the launch.

On the morning of the big day, the Nintendo of America team gathered at FAO Schwarz, where Nintendo had paid for an elaborate window display and an attractive floor space that featured a small mountain of televisions with game footage playing. The moment of truth had finally arrived, and within moments of the store's opening an excited customer eagerly approached the display, grabbing an NES and all fifteen of its games. The NOA team looked on, watching everything they had been working toward so suddenly come to fruition. It was a dream come true—until they were snapped back to reality upon learning that Customer #1 was actually just a competitor doing his due diligence. Nevertheless, that invincible feeling would soon come again, many times over.

That Christmas, the NES was available in over five hundred stores. Though no staggering success, Nintendo managed to sell half of the 100,000 units they'd stocked in stores, which effectively proved to the world that the videogame industry was not dead but had simply been hibernating. Nintendo of America was so overcome by the results that Lincoln paid Borofsky the ultimate compliment: For what you've done for Nintendo, you will be with us forever unless you commit a crime or we go out of business. Borofsky let this sink in, and then prepared to roll out the NES nationwide. Next up: Los Angeles. But that's when things got complicated.

With Nintendo expanding, the company wanted to beef up its sales force and hire a handful of regional sales managers. Because New York was the center of the toy universe, the East Coast would be the most important region. Naturally, Nintendo wanted Sam Borofsky to take the role and come in-house, following in the footsteps of Stone, Judy, and Lincoln. Borofsky, however, wished to remain independent; he didn't want to restrict himself to representing a single region or company. This came as no big surprise, but Nintendo still wanted someone who could handle the task, so they targeted Peretzman, who gladly accepted the opportunity. Borofsky understood Peretzman's decision to leave his company for NOA, but it led to an awkward situation in which his former employee was now his client. The situation became even more tenuous when Arakawa hired Worlds of Wonder, the toymaker famous for Teddy Ruxpin and Lazer Tag, to launch the NES around the country. Borofsky was under the assumption that after New York his firm would get the full business, but Arakawa had interpreted the relationship differently. Feeling bad about the misunderstanding and trying to mend fences, Arakawa gave Borofsky much of the tri-state area. Though the situation was not ideal, this compromise was enough to satisfy both parties and prevent anyone from jumping ship.

The same could not be said, however, about Bruce Lowry. The spotlight of success in New York caught the attention of other Japanese videogame companies with an interest in following Nintendo's lead. One of these would-be competitors was Sega, who had just released their own console on the heels of the Famicom's success. After Nintendo had shown that the videogame market was still viable in America, Sega hired away Lowry to launch their 8-bit Master System and go head-to-head with the NES. To replace Lowry, NOA brought in a new Bruce, one they believed to be of equal or greater value: Bruce Donaldson.

Donaldson was a relentlessly affable former Mattel electronics VP who immediately provided a sage-like certainty and serenity to a young company that had been learning by trial and error. Having survived the boom and bust of the Atari age, he reveled in the chance to do it all again, but he vowed to get it right this time. He arrived at Nintendo in early 1986, when the company was enduring a fair amount of growing pains.

The New York launch had been a success and everyone agreed that expansion was in order, but the where, when, and how of pulling it off were constantly in flux. Originally the plan had been to leave New York and conduct similar "tests" in Los Angeles, San Francisco, and Chicago. The problem, however, was that some of the New York retailers (Toys "R" Us in particular) wanted to go nationwide right away. This was fantastic news, except that Nintendo didn't have enough units to supply all their stores. As a compromise, Nintendo and Toys "R" Us settled on seven regions, but this new equation quickly became unbalanced when transporting the units to those specific regions got needlessly complicated. Situations like this caused political unrest among retailers, who viewed Nintendo as playing favorites. To compound matters, as NOA was tinkering with expansion at home, they were simultaneously looking abroad. Al Stone moved to Germany and initiated plans for European expansion; meanwhile, Ron Judy was beginning to lose his entrepreneurial itch and wasn't sure if he should move to Germany as well or quit the videogame industry entirely. And to make everything more complicated, Nintendo now had competition on the horizon (like Lowry's Sega) at a time when they were still fending off the ghost of Atari. Donaldson tried to fill in where he could, but as the company grew, so did the gaps (in personnel, logistics, marketing, etc.).

Heading into 1987, what NOA really needed was someone to ensure that at the end of this roller-coaster ride, Nintendo would wind up on top. Someone to prove that the NES was more than just this year's Christmas fad. Someone who could exploit the potential for expansion and transform Nintendo from a niche sensation into a global juggernaut.

7. Main

That someone turned out to be Peter Main, though at the time he was dealing with matters much more pressing than corporate expansion: beef dip sandwiches and garlic butter buns. As the president of White Spot, a Canadian fast-food chain, Main was used to eating, sleeping, and breathing burgers, but in the summer of 1985 his mind went into overdrive when an outbreak of botulism swept through Vancouver. Health officials alleged that improper refrigeration of garlic oil concentrate was the likely cause and that Main's restaurants were responsible for the epidemic. Following this horrifying news, he spent much of the next year doing damage control, ensuring that the issue had been resolved and defending the integrity of his beef dip sandwich. When the public outcry finally died down and White Spot's reputation was restored, he stepped down from his post and took a long vacation to decide what he'd like to do next. That's when Arakawa called and asked Main to join Nintendo of America.

Before Peter Main and Minoru Arakawa were ever colleagues they were friends, and before that they were neighbors, back when Main marketed toothpaste at Colgate and Arakawa sold real estate for Marubeni. They first met in 1977 when the Arakawas moved in next door to the Mains. For years Main had wanted the previous tenant to cut down a tree that was obstructing his view, but the owner had declined, so as soon as Arakawa unpacked, Main seized the opportunity to convince his new neighbor that the tree must come down. They were friends from that point on, and ever since Arakawa had left Vancouver to start NOA, he had been trying to recruit Main, claiming that if Main could persuade him to chop down the tree, then surely he could convince kids around the country to play videogames.

For years Main had declined these job offers. Videogames were a far cry from hamburgers, and a part of him feared that despite being a big fish in Canada, he might flounder in America's big pond. But even though he turned down Arakawa, he often provided friendly advice on Nintendo of America's strange yet profitable forays into the restaurant business (Arakawa had gone ahead and bought the British Columbia franchise rights to Chuck E. Cheese as well as a pair of seafood bistros in Vancouver). Main's expertise as a restauranteur only fueled Arakawa's desire to rope him in, but time after time Main declined the overtures—until that fateful call in late 1986. This time Main was open to a major life change, and it didn't hurt that Ron Judy was planning to relocate to Europe, which would effectively make Main NOA's number three. It seemed like a good opportunity, but there was still a lot that could go wrong, so Main decided to leave the decision to luck. For the holidays, he and his wife were headed to Asia for a much-needed vacation. Just before the trip, Main told Arakawa that if the U.S. embassy approved his application for an H1-B visa to work in the United States, then he would head to Nintendo; if not, then he would open a restaurant of his own in Canada. With the odds of securing such a visa being only about 10 percent, Main didn't expect to be hocking cartridges anytime soon. But on the second night of his trip, Arakawa and Lincoln called his hotel room in Hong Kong and happily announced that the visa had been approved. And so in April 1987, Peter Main became Nintendo of America's VP of marketing and sales.

Though Main lacked any videogame experience, his outsider mentality allowed him to look at the business not as an offshoot of the toy, arcade, or electronics industry but as something novel and spectacular. To spread this new gospel, he choreographed what he would later describe as Nintendo's "storming of Normandy," a full-out advertising, promotion, and distribution blitz that accompanied the rollout of the NES into stores nationwide. Meanwhile, Main provided a trustworthy-looking corporate (and Caucasian) face to a company that many in the outside world still viewed as nothing more than a foreign curiosity. Changing people's perceptions of both videogames and Japanese business presented a variety of challenges, but Main always found a way, because above all else he was an expert charmer. And that charm, that talent for cultivating friendships, gained the company credibility with Wall Street, trust from retailers, and respect from parents wanting to know what they were buying.

Month after month, Nintendo of America grew stronger. They sold 2.3 million consoles in 1987 and 6.1 million in 1988. As staggering as these numbers were, sales of the hardware were nothing compared to the software: the company unloaded 10 million games in 1987, and 33 million more in 1988. With numbers like these, it didn't take Main long to realize that, at the end of the day, it was the software that drove the hardware; the console was just the movie theater, but it was the movies that kept people coming back for more. This personal revelation led to a Hollywood-like title-driven business strategy, and his coining of the phrase "the name of the game is the game."

Main's approach to sales and marketing coincided with Arakawa's overarching philosophy of "quality over quantity." As Nintendo exploded, there were plenty of opportunities to make a quick buck (hardware upgrades, unnecessary peripherals), exploit the company's beloved characters (movies, theme parks), or dilute the brand by trying attract an audience older than Nintendo's six-to-fourteen-year-olds. But these kinds of things didn't interest Arakawa. He wasn't driven by making money, at least not in the short term. What propelled him, what kept him up at night, was a desire to continually provide Nintendo's customers with a unique and flawless user experience. As proof of this never-ending obsession, he set up a toll-free telephone line where Nintendo "Game Counselors" were available all day to help players get through difficult levels, and he initiated the Nintendo Fun Club, which sent a free newsletter to any customer who had sent in a warranty card. Both programs were very costly and could have been offset by charging small fees or obtaining sponsorship, but Arakawa believed that doing so would compromise Nintendo's mission. And to further safeguard Nintendo from the dangers of impurity, he and his team put into place a series of controversial measures:

1. The Nintendo Seal of Quality: Ron Judy had the novel idea of mandating that all games pass a stringent series of tests to be deemed Nintendo-worthy, ensuring high-caliber product and making software developers beholden to Nintendo's approval.

2. Third-party licensing program: Howard Lincoln's strict licensing agreement enabled software designers to make games for the NES but restricted the quantity they could make (five titles per year), required full payment up front (months before revenue from a game would be seen), and charged a hefty royalty (around 10 percent). In addition to these stringent terms, all game makers needed to purchase their cartridges directly from Nintendo. This ensured peerless quality but also allowed NOA to dictate price, schedule, and production allocation, which became a particularly touchy matter during the notorious microchip shortage in May 1988.

3. Inventory management: Heeding Sam Borofsky's suggestion, Peter Main devised an incredibly rigid distribution strategy that purposefully provided licensees and retailers with only a fraction of the products they requested. The goal of this technique was twofold: to create a frenzy for whatever products were available, and to protect overeager industry players from themselves.

Though NOA's methods drew ire from retailers, anger from software developers, and eventually allegations of antitrust violations from the U.S. government, there was no denying that whatever Nintendo was doing was working—so well, in fact, that Peter Main needed additional reinforcements as he inflicted Nintendo-mania on his adopted homeland.

8. Nintendo Power

Help came in the form of Bill White, a straitlaced marketing whiz whose smallish eyes and oversized, round-rimmed glasses emitted an eternally boyish vibe. Though he was only thirty years old (and, after a haircut, could have passed for thirteen), he spoke about brand recognition, market analysis, and strategic alliances with the expertise of someone twice his age. Part of that precocious nature was due to an almost religious belief in the power of marketing, part was due to his father's history as a Madison Avenue ad man, and part was due to a chronic insecurity that could only be quieted by winning at everything he did. Peter Main saw the potential in White and hired him in April 1988 to become Nintendo's first director of advertising and public relations.

When White joined NOA, the marketing department consisted of just three people: himself, Main, and Gail Tilden, an exceptionally smart brunette with an encyclopedic memory. The lack of manpower forced White to wear many hats (commercial producer, press secretary, Peter Main's whipping boy), but his most important responsibility was to forge corporate partnerships. Though Nintendo continued the take the videogame world by storm, the rest of the world still didn't know what a Nintendo was. To build the brand, White courted Fortune 500 companies, resulting in pivotal promotions, like Pepsi placing a Nintendo ad on over 2 billion cans of soda and

Tide featuring Mario on the detergent maker's giant in-store displays. His coup came with the release of Super Mario 3, when he negotiated for McDonald's to not only make a Mario-themed Happy Meal but also produce a series of commercials centered around the game. By virtue of his efforts, White became Main's right-hand man, something of a protégé. But as Main fed White's ambitions and the young marketer swallowed up more and more responsibility, this left Tilden, the other member of the marketing equation, with less and less to do. This displeased Arakawa, who set out to find a better way to utilize one of NOA's most dynamic employees.

If Bill White wore many hats, then Gail Tilden owned the hat shop. Tilden had joined the company in July 1983, back when the videogame fad was supposed to be on its very last legs. As Nintendo of America's ad manager, working under Ron Judy and Bruce Lowry, she had to find new and exciting ways to promote Nintendo's latest arcade games. Her creativity, resourcefulness, and nothing-is-beneath-me attitude impressed Arakawa, who later entrusted her to handle the marketing for New York's test launch. She spent the summer living in the city, hiring an ad agency, picking a PR firm, and introducing the paradigm-shifting "Now You're Playing with Power" campaign with Nintendo's very first commercials. After shepherding the company through its formative years, she was disheartened to see her voice diminished, but didn't see any logical way to raise the volume back up. So when she became pregnant with her first child in 1987, she began to consider converting her maternity leave into a permanent one. Arakawa, however, didn't want to see her go. He had seen what Tilden could do and how quickly she could do it. But as much as she appreciated his desire to keep her there, the situation was what it was, and it wasn't likely to change. Nevertheless, Arakawa continued to brainstorm after her departure, and a couple of months later finally found a solution.

Tilden was at home, nursing her six-week-old son, when Arakawa called and asked her to come into the office the next day for an important meeting. She was caught off guard but knew that Arakawa was not the kind of man to ever waste anyone's time, so the following day she and the baby headed to NOA headquarters. After dropping off her son with some trusted coworkers, she went into a meeting with Arakawa and some Japanese colleagues from NCL in which they discussed the possibility of expanding the Nintendo Fun Club's newsletter into something bigger. By 1988, the Nintendo

Fun Club had over a million members receiving the monthly newsletter, and the company needed to hire over five hundred Game Counselors just to keep up with the more than 150,000 calls per week. The appetite for Nintendo tips, hints, and supplemental information was insatiable, so Arakawa decided that a full-length magazine would be a better way to deliver exactly what his players wanted.

Tilden was put in charge of bringing this idea to life. She didn't know much about creating, launching, and distributing a magazine, but, as with everything that had come before, she would figure it out. What she was unlikely to figure out, however, was how to become an inside-and-out expert on Nintendo's games. She played, yes, but she couldn't close her eyes and tell you which bush to burn in The Legend of Zelda or King Hippo's fatal flaw in Mike Tyson's Punch-Out!! For this kind of intel, there was no one better than Nintendo's resident expert gamer, Howard Phillips, an always-smiling, freckle-faced videogame prodigy.

Technically, Phillips was NOA's warehouse manager. He had held that job since February 1981, but along the way he revealed a preternatural talent for playing, testing, and evaluating games. After earning Arakawa's trust as a tastemaker, he would scour the arcade scene and write detailed assessments that would go to Japan. Sometimes his advice was implemented, sometimes it was ignored, but in the best-case scenarios he would find something hot, such as the 1982 hit Joust, alert Japan's R&D to it, and watch it result in a similar Nintendo title—in this case a 1983 Joust-like game called Mario Bros. As Nintendo grew, Phillips's ill-defined role continued to expand, though he continued to remain the warehouse manager. That all changed, however, when he was selected to be the lieutenant for Tilden's new endeavor.

To name something is to make it real, so ideas for the magazine's title were immediately tossed around. The leading candidate was *Power Player* (inspired by the "Now You're Playing with Power" slogan), but that trademark had already been taken, and besides, Arakawa wanted the name Nintendo to be used. The brainstorming continued—*Nintendo Now*? Or maybe *Playing with Nintendo*?—until they finally settled on one that felt just right: *Nintendo Power*. From there, Tilden and Phillips began generating column ideas, and coming up with sections such as "Pak Watch" (which provided previews of upcoming games) and "Classified Information" (which revealed top-secret tips, tricks, and codes). They were pleased with what this was turning into, but agreed that something was missing: a more direct way to connect with their players, a way that said, We may be making a magazine, but we love these games as much as you do. They couldn't quite figure out how to accomplish this yet, but they didn't have time to chase their tails. They needed to keep moving and define the look, layout, and feel of *Nintendo Power*. To do so, they flew out to Japan and met with Work House, a small design company in Tokyo, who could help create a cross-cultural aesthetic that would please the parent company and appeal to kids from every country in the world.

Unsurprisingly, it was no cinch to sync up the cultural stylings of East and West. Work House liked ostentatious headlines, but Tilden liked understated. She wanted bright, sunny layouts, but they preferred pale, cloudy ones. It was hard to find common ground, but Tilden wouldn't relent. This wasn't about ego; it was about giving kids a reason to run to their mailbox each month and hide a flashlight under the bed so they could stay up late reading under the covers. To defuse the growing tension, Phillips jokingly suggested that it was no use arguing with Tilden because she was NOA's "dragon lady." His quip failed to lighten the mood at the time, but the nickname for Tilden stuck.

As Phillips was hard at work playing through games in order to rate them for the magazine, the Dragon Lady was struck by the fact that in a way, Phillips played videogames for a living. That was every child's fantasy, wasn't it? At that point she realized that Phillips was the key to building a bridge between Nintendo and its players. She approached Phillips with this vision, leading them to the creation of a comic strip called "Howard & Nester," which featured a professorial version of Howard Phillips finding subtle ways to give game hints to Nester, a kid who needed advice but was too much of a know-it-all to ever ask. After fleshing it out, they loved the comic idea but felt it was still missing one thing. The two-dimensional Phillips character needed a telltale trait, something iconic like Superman's S or Popeye's pipe. The answer turned out to be rather easy. At the request of his wife, Phillips always wore a bow tie on special occasions, so it was decided that cartoon Howard would always wear one as well.

In July 1988, Nintendo of America shipped out the first issue of *Nintendo Power* to the 3.4 million members of the Nintendo Fun Club. Over 30 percent of the recipients immediately bought an annual subscription, marking the fastest that a magazine had ever reached one million paid subscribers. And as the magazine's audience grew, so did the influence of Howard Phillips. If *Nintendo Power* gave kids a chance to step inside the candy factory, then he was their Willy Wonka, magically and eccentrically showing them how the sweets were made. Though Mario was Nintendo's mascot, Phillips became the face of Nintendo. Peter Main took advantage of this, sending Phillips all over the country for press events and on-camera interviews. This miffed Bill White, who had been working so hard to present Nintendo as a royal kingdom, not the Magic Kingdom. After all, it was Michael Eisner who did Disney's press interviews, not someone in a Mickey Mouse costume. It wasn't that White wanted to stop exploiting Phillips—the guy was pure gold—but they needed some way to present him as Nintendo's jester and not as the king. Main agreed, and found an elegant solution by turning to the beer industry. Ever since producing its first ale in 1759, Guinness has always appointed a spokesman called the Master of Brew, who supposedly inspects all aspects of the brewing process, from the purchase of barley to the experimental work conducted in the brewery's lab. Following in that tradition, Main appointed Phillips Nintendo's first Game Master. Shortly thereafter, Howard Phillips became a national celebrity, boasting a Q score higher than Madonna, Pee-wee Herman, and the Incredible Hulk.

The rise of the Game Master was just the latest sign of Nintendo's unprecedented success. By 1990, Nintendo of America had sold nearly thirty million consoles, resulting in an NES in one out of every three homes. Videogames were now a $5 billion industry, and Nintendo owned at least 90 percent of that. The numbers were astounding, but Nintendo's triumph went beyond that. Arakawa had proven that he was more than just the son-in-law, Lincoln had proven that he could take on anyone, and Main had proven that he could swim with the sharks.

Although Arakawa, Lincoln, and Main were at the top of the ladder, the unlikely triumph of NOA was a team effort from top to bottom. There was John Sakaley, the wild and crazy renegade who fought for every inch of retail space as if his life depended on it. He pioneered the industry's first store-within-a-store by transforming ordinary retail spaces into magical, snow globe–like areas called the World of Nintendo. So mesmerizing were these interactive displays that kids couldn't help but fantasize about accidentally getting locked in these stores overnight and staying awake to play Nintendo's games. Sakaley also developed the Nintendo Fun Center, a mobile entertainment gaming kiosk that was popular with patients in childrens' hospitals.

Then there was Don Coyner, the man behind Nintendo's happy-go-lucky, game-footage-fueled commercials. Previously he had been an account director at Foote, Cone & Belding, where he created the famous slogan for Kraft's macaroni and cheese, "It's the Cheesiest!" He brought that same gift for innocent, well-crafted cheesiness to Nintendo, overseeing dozens of commercials for games like Dr. Mario and Metroid, as well as other products like Game Boy and the Power Pad.

There was also Lance Barr, who designed the iconic look of the NES and the feel of its heaven-in-your-hands controllers, and Don James, the product development guru who helped Barr with the design. (James, along with Arakawa, also named Nintendo's pixelated plumber, Mario, after Mario Segale, the company's mysterious landlord whom nobody had ever met.) There were hundreds of others at NOA whose unseen efforts backstage helped define and then refine the Nintendo experience.

Together, they had single-handedly resurrected an industry. And they did it all with only 8 bits. Imagine what they could do with 16 . . .

6.

THE NAME OF THE GAME

"There you are!" Sega's plucky product manager, Madeline Schroeder, said as she sidled into the office's tiny kitchen and found Tom Kalinske standing in front of the coffee machine with a cup of joe in his hand.

Although she felt certain that he had heard her exclamation, her new boss did not look up from the papers in his other hand. She thought this odd (and rather rude), until a split-second forensic analysis of the situation revealed the following: the cup in his hand was empty and upside down, the coffee machine wasn't even turned on, and the dazed look on Kalinske's face was symptomatic of one who'd just seen a ghost. Either Tom Kalinske didn't know how to use a coffee machine or whatever he was reading had made him catatonic. "My, oh my," Schroeder said, settling upon the latter. "That must be some compelling literature right there."

In a flash, the pale expression on his face became an a-okay smile. "Oh hey there, Madeline," Kalinske replied. He was good like that, with names, only needing to hear them once to remember them forever. "You've probably seen this already, but as you suspected I found it to be particularly engrossing."

He handed her an article that he had been reading as part of his crash-course on all things Nintendo. It was a piece written by Anthony Gonzalez that had appeared in the *New York Times* the previous year, titled "The

Games Played for Nintendo's Sales." This is what Kalinske had been reading when his coffee craving suddenly disappeared:

> **THURSDAY, DECEMBER 21, 1989**
>
> SEATTLE – Meet the man behind Nintendo, the video game maker that is the talk of America three Christmases running.
>
> To his admirers, Peter Main, vice president of marketing for Nintendo of America, is a master seller of children's entertainment. They say he is a skilled businessman who has learned lessons from the unhappy history of the video game business and helped revive it into a $3.4 billion industry in only three years.
>
> To his critics, however, he is an aspiring monopolist, squeezing supply and jacking up prices. Charging monopoly, a competitor has sued and a Congressman has called for a Justice Department investigation.

Of all the pieces that Kalinske had recently read about Nintendo, the beginning of this one seemed to perfectly sum up his competitor: they were either heroes or villains, and the truth was entirely a matter of perspective. Unlike other companies obsessed with the facade of political correctness, Nintendo made no effort to hide its obsession with control (the article goes on to say "By design, the company does not fill all of a retailer's order and keeps half or more of its video cartridge library inactive"), nor did the company worry about alienating other developers ("But far more in dispute is the company's other tactic: building the hardware system in its games with a special 'lockout' computer chip").

When Kalinske first began to investigate what lay at the heart of Nintendo, their controlling, our-way-or-the-highway philosophy scared the hell out of him, mostly because it made sense. Although Nintendo didn't always act with the best bedside manner, their tactics generally tended to benefit the industry. They had just resurrected the videogame business from a terrible crash, and they had taken it upon themselves to put in safeguards to prevent such a thing from happening again. So retailers might complain that orders weren't being filled, or developers might grumble about being locked out, but this was all Nintendo's way of avoiding an Atari-like glut of bad games. Nintendo, in many ways, really did know best.

Although this realization had indeed petrified Kalinske after he stopped asking himself what the hell he had gotten into, he began to wonder if this might actually be Sega's greatest advantage. Maybe Nintendo really did know best, but if there was one thing Kalinske had learned about consumers throughout his career, it was this: the only thing they valued more than making the right decision was making their own decision. So if Nintendo represented control, Sega would represent freedom, and this cornerstone of choice would be the foundation of Kalinske's plan to reboot, rebuild, and rebrand Sega. He was in the midst of this mini-epiphany when Schroeder entered the kitchen, hence the dazed look on his face. He had seen a ghost all right, but it wasn't anything malevolent—no, it was the alluring ghost of Sega's Christmas Future.

"You have to hand it to Nintendo," Schroeder said, finishing the article. "Like them or not, everything those guys touch turns to gold."

"You're right," Kalinske replied, turning on the coffee machine. "So I guess we'll just have to make sure everything we touch turns to silver. And, you know, while we're doing that, we'll find a way to convince the world that silver is more valuable than gold."

"Count me in," Schroeder said and then smiled, a Cheshire grin through and through. For weeks prior to Kalinske's arrival, she had been hearing how great things would be with him in charge and how everything would turn around after he took over. At the time, she thought it was just false hope for the hopeless. Now, while that skepticism still persisted, she couldn't deny there was a whole lot more hope.

"But enough pontificating," Kalinske said. "I vaguely recall your entering with the words 'There you are.' So what can I help you with?"

"Right," Schroeder said, quickly rewinding her mind. "I had wanted to ask you about your trip to Japan. Was there anything new on the mascot front?"

"I'm not sure, but Nakayama-san assured me that he'd get us a Mario-killer sooner rather than later."

Schroeder eyed Kalinske as if trying to somehow conduct a telepathic polygraph test. "And how much do you trust Nakayama's assurances?"

Kalinske thought about this for a moment. "I'm from the innocent-until-proven-guilty school of thought. I have no reason not to trust him . . . at this point."

"Okay, just curious," Schroeder said. "Did he at least show you the hedgehog?"

"What hedgehog?"

"That freak from the mascot contest," she said. Kalinske had no idea what she was talking about, so Schroeder filled him in. Prior to Kalinske's arrival, Sega had held an internal mascot contest, encouraging employees to come up with a new face for the company (which would supplant the current face of the company: Alex Kidd, a disappointing rip-off of Mario). The Japanese programmers submitted a host of diverse entries, including an armadillo (later developed into Mighty the Armadillo), a dog, a cat, a cheetah, a Theodore Roosevelt look-alike in pajamas, and a peppy rabbit that could use his extendable ears to collect objects. The top two choices, however, were an anime-inspired egg and a teal hedgehog with red shoes created by Naoto Oshima that he called Mr. Needlemouse. Nakayama had presented these two finalists to Katz, who quickly declared that they both sucked. He thought the egg was absurd and the hedgehog just didn't make any sense; nobody even knew what a hedgehog was, so how could anyone ever possibly care about one? Despite the vote of no confidence from Katz, Nakayama forged ahead with Mr. Needlemouse and asked Oshima to explore what kind of a game would best suit his character. Oshima partnered up with Yuji Naka, a brilliant hothead in the programming department who was responsible for one of Sega's most popular series: Phantasy Star, a sci-fi role-playing game (RPG) about a resilient young female warrior bent on galactic revenge with the help of a muskrat named Myau and a wizard named Noah. Oshima and Naka worked together to build a game around Sega's new mascot, and it would fall onto Schroeder's shoulders to whip the game into shape and introduce it to the world. "I can't believe the always-transparent Nakayama failed to mention all this. Curious," she said lightheartedly. "So basically our entire jobs, careers, and livelihoods depend on this hedgehog guy."

Kalinske realized the truth in her words. "Well, I can't think of any animal I'd rather pin my hopes and dreams on than a good old-fashioned hedgehog," he said.

After hedging his way through the hedgehog conversation, Kalinske entered Nilsen's office.

"Well, if it isn't Mr. Kalinske!" Nilsen announced as his boss closed the door and sat down on the chair in front of his desk.

"Hey, Al," Kalinske said, jumping right in. "What the hell is a hedgehog?"

"You mean, aside from the demise of Sega?" Nilsen joked. "According to one Mr. Michael Katz, at least."

"He wasn't a fan?"

"He authored a scathing multipage letter to Nakayama saying why it wouldn't succeed in the U.S. and other assorted doom and gloom."

Kalinske was taken aback, feeling as if there were trapdoors hidden around every corner in this company. "That's a tad disconcerting."

"Seriously, though, don't worry about it . . . yet," Nilsen reassured him. "We haven't even seen any gameplay. And in this business, it could look like a duck and talk like a duck, but in the end nobody cares if it's a duck, or even a neon-green wolverine, as long as it makes for a fun gaming experience."

His words worked, and Kalinske eased up. "Okay, that makes sense."

"There's only one thing you need to know to survive in this world."

"And what's that?"

"The name of the game is the game," Nilsen said, with the bouncy, up-lifting rhythm of a prayer worn into the soul by repetition. If Nilsen had known that the originator of the phrase was none other than Nintendo's Peter Main, he likely would have washed his mouth out with soap. But with the bliss of quotational ignorance, Nilsen repeated the mantra and then pointed to a copy of Atari's game E.T., which was framed on his wall. "I keep this here as a reminder. Most consider it to be the worst game ever made." Nilsen pressed his finger up against the glass. "Look at this thing: based on a blockbuster movie, blessed by none other than Steven Spielberg, and had more marketing money pumped into it than any other game."

"And still it failed?"

"Miserably! You can still see the markdown stickers on the game," Nilsen said, pointing to the tiny stickers showing its various price points. "It went from $49.95 to $34.95 then, ouch, $12.99, $3.99, and finally I became a proud owner of the worst videogame ever at $1.99."

"You really know this business, don't you?"

"On occasion."

"Maybe you should be the one in charge?"

Nilsen waved him off. "Oh, stop it. We both know that you're going to turn this company around."

"I am?"

Nilsen looked at him strangely, as if he were genuinely surprised that Kalinske had no idea that this was going to happen. "Never been more sure of anything in my life. Truly." Kalinske graciously nodded, and then the two of them sat in Nilsen's office for about an hour, talking about everything from videogames and toys to their families and sports. As Kalinske spoke with Nilsen, the company started to feel more like a home. This was his opportunity, and he would find a way to make it work. "I think you're right, Al," Kalinske said, sitting up. "I think that things might actually work out pretty nice."

"Fantastic," Nilsen replied. "Then I shall eagerly await the next senior staff meeting to see what you've got cooked up."

Kalinske looked across the long table, sizing up his troops. He felt ready to shoulder the responsibilities as president and CEO, ready to turn these guys into *his* guys, and ready to lead them into battle. In a couple of months he would go to Japan to meet with the board of directors, outline everything he had learned, and propose the changes that needed to be made to turn Sega into a household name. In the meantime, however, there was only one thing on his mind: distribution.

If distribution was the lifeblood of a company, then Sega had just undergone a blood transfusion. In 1988, when Sega's Master System proved unable to make a dent in NES sales, Sega had struck a deal with the toy manufacturer Tonka to allow them to handle distribution. But despite the weight that Tonka's name carried in the toy world, the company had no idea how to market and sell videogames. If the Master System wasn't already dead in the United States, Tonka's failed attempts at distribution fired the final shots. Nakayama blamed Tonka for much of the Master System failures and wouldn't allow such a thing to transpire again. As a result, Paul Rioux had worked tirelessly to unwind the deal with Tonka so that Sega of America would be solely responsible for their own distribution. In Nakayama's eyes, nothing should hold back the company now. Kalinske, however, was savvy enough to know that many obstacles stood in his way: lack of brand identity, poor sales history, and, most important, Nintendo's unbending grip over the retailers.

"The fact is, you can't sell something if there's no place for anyone to buy it." Kalinske looked at his employees, who likely had to try their best to think of any response besides *Well, yeah*, and *Duh.* "Yes, I know, it's a pretty obvious concept, but unfortunately that's our biggest problem at the moment." Kalinske pointed to a map on the wall, indicating which retailers were carrying the Genesis and in which regions the game systems were available. "We need to convince more stores to jump on the bandwagon. I know, I know, easier said than done, but I think instead of approaching each of them piecemeal, we're better off using a top-down approach. If we sign up the big boys, the rest will fall into line."

"What exactly did you have in mind?" Nilsen asked.

Kalinske narrowed his eyes and answered, "Wal-Mart."

Getting the Genesis into Wal-Mart wasn't as easy as sending them a free system and letting them see how much better it was than the Nintendo, though Kalinske believed it should have worked like this. Unfortunately, Wal-Mart sold Nintendo products, and they didn't just sell, they flew off the shelves. Nintendo single-handedly accounted for about 10 percent of Wal-Mart's profits, and the giant retailer felt an obligation to keep the game maker happy. Kalinske, however, was willing to rock the boat.

"Maybe it is best to look elsewhere first?" Toyoda suggested. "Take some time?"

"Time is a luxury we don't have," Kalinske explained. Nintendo had recently announced plans to release their 16-bit machine, the Super Famicom, in Japan later that year (which meant it would probably hit America one year later). Sega's days had always felt numbered, but with this recent news the end felt even nearer. "Not with the Super Famicom knocking on our door. Can someone tell me where Katz was with Wal-Mart?"

Rioux explained that Katz had made a trip down to Wal-Mart headquarters in Bentonville, Arkansas, to pitch them on Sega and said that things went okay. Kalinske nodded. In his mind, okay was the worst possible result. Okay was worse than bad. At least bad was memorable. "Well, then," Kalinske said, "I think it's time that I go back down and show them that Sega is no longer in the business of being okay."

The team vigorously prepared for weeks, leading to a combination of stress, brainstorming, and hijinks that, when laminated by Kalinske's perpetual sheen of professionalism, created a dare-to-be-bold corporate culture. At some point, Sega started to feel less like the American outpost of a Japanese company and more like the cast of an exciting new Broadway musical with defined roles, a choreography of ideas, and ensemble numbers that brought everyone together. The pieces all seemed to be in place, and with the curtain now rising, it was finally showtime.

7.

POSTCARDS FROM ARKANSAS

Wal-Mart headquarters felt like a military compound during peacetime. It was huge, impressively segmented, and delivered the impression that things were not nearly as calm as they seemed. Kalinske entered the complex and was escorted into the office of Wal-Mart's electronics merchant, a man whose every move called to mind the word "veteran." The men shook hands, enjoyed some small talk about college football, and eventually meandered toward the conversation that had brought Kalinske here. "How much do you know about videogames?" Kalinske asked.

"Pretty much just whatever Nintendo tells me," the electronics merchant replied.

"Well, then, allow me to introduce you to the future," Kalinske said, and presented his Sega gadgets, along with the product analysis reports and market data that his ragtag team had worked so hard to prepare. He leaned forward, dropped his businessman persona, and spoke in a just-between-us tone. "I understand that Wal-Mart feels an obligation to keep Nintendo happy. I get it. But this isn't a Tengen situation."

Kalinske was referring to Tengen, a videogame publisher and developer created by Atari Games. Tengen owned the rights to most of Atari's popular games from the early 1980s and wanted to license some of the company's more popular titles for the NES. Hideyuki Nakajima, Tengen's manager, approached Nintendo to work out the details, but he quickly found out that there was nothing to be worked out; Nintendo had a standard licensing agreement that dictated the same terms for all of its licensees. If Tengen wanted their games on the NES, they'd have to sign a very one-sided contract with Nintendo that would prohibit them from releasing their game on any other console and compel them to give Nintendo 30 percent of their revenue as a royalty. It also stipulated that they'd have to buy physical cartridges directly from Nintendo, which was not only expensive (about $10 each) but often highly frustrating, as it gave Nintendo the leverage to pick and choose which orders to fill.

Nakajima thought this was absurd, so in 1986 he arranged for a special meeting with Nintendo of America's president, Minoru Arakawa, and senior vice president Howard Lincoln to discuss renegotiating these standard terms for what he believed was an exceptional situation. Nakajima kindly reminded Nintendo that Atari, Tengen's parent company, had essentially created the videogame industry and deserved some special privileges, particularly the ability to publish more than the five titles per year that Nintendo's standard agreement specified. But Arakawa and Lincoln stood their ground and reiterated that no special treatment would be offered.

After thinking over the terms of the deal, Nakajima resiliently decided to find a way around this lousy licensing agreement. So he had Tengen engineers begin by trying to find a way around the security device inside the NES. Nintendo's console had a lockout chip containing a protocol called 10NES programming, which detected unlicensed cartridges and prevented them from working. Tengen engineers tried everything to break this code, even chemically peeling layers from the NES chips to allow for microscopic examination. Yet despite these efforts, Tengen was unable to crack the code, and in 1987 it signed a contract with Nintendo.

But after releasing a slew of successful games like Pac-Man and RBI Baseball, Nakajima was infuriated by how much Nintendo's royalty cut into his company's profit margins. He rationalized that the only way to get around the lockout chip was to figure out exactly what made up its 10NES programming. He needed to obtain a copy, but there were only two places where this code could be found: Nintendo headquarters and the U.S. Copyright Office. Since breaking into Nintendo was impossible, Tengen would approach the U.S. Copyright Office and claim that they were filing a copyright infringement suit against Nintendo. Despite this being entirely fictional, they went the whole nine yards to make it appear true, and even signed an affidavit legally testifying about the accuracy and urgency of this lawsuit. The Copyright Office handed over the code, and Tengen worked backward to create a program called the Rabbit, which could unlock the NES. Now, not only did Tengen possess the ability to produce as many games as they saw fit, but by learning about Nintendo's distribution techniques (through the three games they had legally released as third-party licensees), they were equipped to contact the retailers directly. Essentially, Tengen had relegated Nintendo to middleman status, and then cut them out completely.

This was a brilliant plan in theory, but in practice the problem came when Nintendo gave the retailers an ultimatum: us or them. Though Nintendo couldn't legally threaten to stop supplying retailers, they had enough strength to use the power of what-if ("What if our trucks got lost going to your stores? What if we stopped filling your orders in full?") to make this an easy decision. The retailers bit the bullet, got rid of all Tengen products, and wrote off the losses. To further flex their muscle, Nintendo eventually took Tengen to court and got an injunction to prevent them from producing their illegally created games. Nakajima and Tengen had no choice but to bow out of the business, and soon the name Tengen served as nothing but a cautionary tale.

Now, as Kalinske sat before Wal-Mart's electronics merchant, he tried to make it perfectly clear that this was an entirely different situation. Sega hadn't done anything illegal, nor had they screwed over Nintendo; they were just a competitor who had the better product. "To be perfectly honest," Kalinske said, "I think carrying our products will boost Nintendo sales. The money we spend on print and television is only going to help the industry overall, and you and I both know exactly who the industry is."

The merchant looked through some of Sega's materials again and smiled.

Kalinske scooted up to the edge of his seat. He didn't need a big order, just an order. A single order. That would be enough to motivate his employees, give Sega the credibility it sorely needed, and confirm to Kalinske that maybe he wasn't in completely over his head. Please, just a single order, especially with the Super Famicom coming.

"A little while ago I was interested in stocking an electronic handheld game," the electronic merchant explained. "A stupid little football game, not even close to anything on the Game Boy. For fifteen bucks, maybe Mom and Dad buy it for Sonny if he makes all A's, or hits the game-winning shot. But then a pal of mine, someone in the same line of work, he passes along a rumor to me: there was a small store struggling to keep up with the big boys, and they decided they were going to lower the price of the NES by five cents just so they had some small advantage. Well, they advertise this in the Sunday newspaper, their five-cent advantage, and another small store sees this and calls Nintendo to let them know. A week goes by, Nintendo sends out the trucks to deliver the product, and lo and behold, there's nothing left for the store with the five-cent discount, and by sheer coincidence, the guy who passed along the tip gets a bigger allocation than normal." The merchant tapped his fingers on the desk. "But like I said, this was just a rumor I heard. It's probably not true. After all, that would be illegal."

Kalinske shook his head. "Not just illegal, it'd be un-American."

The merchant gave a gummy smile. "Perhaps one day we'll return to a place where the streets are paved with gold and all you need to succeed is a good idea, a strong work ethic, and some kind of bootstraps. Or perhaps we'll continue to move in the opposite direction." The merchant pondered this for a second and then stood. "Personally, I like the place with golden streets. And believe it or not, I like you, Mr. Kalinske. But my answer is no."

"I understand," Kalinske said, standing to leave. "And I appreciate your ode to better, simpler times. But you know what the sad thing is? The man in your story, the one who tipped off Nintendo, I don't really blame that guy. He was just trying to find an angle. If you ask me, the people really killing this country are the ones who realize the American dream is being crushed but don't bother to do anything about it." Kalinske thanked the merchant for his time, and then flew back to Sega with a 16-bit chip on his shoulder.

8.

THE BIRTH OF AN ICON

Back in the office, Kalinske stared at his phone with Wal-Mart on his mind. He knew that he had laid it on pretty thick with the electronics merchant and wasn't sure whether he ought to call back and apologize. Before he could persuade himself in one direction or another, however, the phone rang. Kalinske quickly answered, so certain that it must be the man on his mind that he even took half a second to try to hide the excitement in his voice.

It wasn't Wal-Mart. Of course it wasn't Wal-Mart; the man Kalinske had met with didn't even have his direct line. It was Nakayama, whose ominously chipper voice boomed through the phone. "Tom! How is everything? How are you adjusting?" Nakayama and Kalinske spoke just about every day, but still their conversations often began with this open-ended question.

"I had a great meeting with Wal-Mart," Kalinske said. "I think they're close."

Nakayama was an intelligent man who understood many of the intricacies of the industry, but he didn't quite grasp the distribution difficulties in the United States. "What is the holdup?" he asked. In Japan, where Nintendo also reigned supreme, Sega had still managed to get their products into all the biggest retailers. "I thought everything would be smooth after we moved on from Tonka. I was told this was the plan," Nakayama stated. "But I have not called to discuss distribution. I am calling with good news."

"Wonderful," Kalinske said. "Let's hear it."

"The new company mascot is ready, and he is sure to be a success."

"This is the hedgehog named Mr. Needlemouse?"

"Ah, you have heard," Nakayama said, surprised. "We have made some changes, and his name is now Sonic."

"Okay," Kalinske said. "Well, when can I see him?"

"I will send him over now," Nakayama said, and then barked orders in Japanese to someone on the other end. "He will enter through the fax. I will stay on the line to hear your reaction. You will be very pleased." Kalinske made his way over to the fax machine as it buzzed and huffed, printing out lines of what would be the company's savior. "My guys here have already begun work on the game engine. They showed me an early version, and it is fast like nothing else."

The fax machine stopped sputtering, and Kalinske picked up the sketch. "Ah," he said, trying not to sound repulsed. "Very interesting." Kalinske stared at the drawing, trying to see in it what Nakayama saw, but it was no use. The hedgehog looked villainous and crude, complete with sharp fangs, a spiked collar, an electric guitar, and a human girlfriend whose cleavage made Barbie's chest look flat. "I assume this is his girlfriend?"

"Yes," Nakayama said. "That is Madonna."

"Kind of racy, no?"

"Tom," Nakayama said, and sighed. "This is not the reaction I expected."

Kalinske continued to stare at the drawing. "Sorry, Nakayama-san, sometimes it just takes a little while for things to sink in for me," he said, still shocked that this bruiser was supposed to be his messiah. "I'll tell you one thing, though—if Sonic and Mario were alone in an alley, I have no doubt who I'd put my money on." He had been expecting a Mario-killer, but not one that literally looked like a serial killer. Maybe this Sonic could sell in Japan, but in America he belonged inside a nightmare.

Kalinske got off the phone with Nakayama and took the fax to Madeline Schroeder's office. "I have good news and I have scary news. Which do you want first?"

"This doesn't sound promising."

He handed her the artwork. "What do you think?"

She looked it over. "I think we'll be the first videogame company whose core demographic is goths."

"Nakayama loves it."

"Of course he does," she said. "It's so weirdly Japanese. I'm surprised the girlfriend's boobs aren't hanging out of a schoolgirl outfit."

Despite his sour mood, Kalinske laughed. "Her name is Madonna."

"Of course it is," she said. "What kind of leeway did he say we had?"

"We didn't exactly have a Q&A session."

Schroeder put the drawing on the desk. After a long silent inspection they both spoke at the same time, saying the exact same thing: "Can you fix it?"

Schroeder sighed. "You know, I expected something pretty terrible. I mean, the second-place winner in the contest was an egg, for God's sake. This is certainly not ideal, but it's actually not as bad as I was bracing for. We can make this work."

Her optimism was contagious. "Great," Kalinske said, standing up. "Then let's turn this punk into a global icon."

"And how exactly do you propose we begin?"

"Oh, I know of a little place where the icons all hang out together. Why don't we grab Al and go check it out?"

Kalinske, Schroeder, and Nilsen went on a field trip to Toys "R" Us to pay a visit to some famous friends: Mickey Mouse, GI Joe, He-Man, Mr. Potato Head, and the newly popular and ever-rowdy Teenage Mutant Ninja Turtles. Kalinske walked them through the store, pointing out one billion-dollar property after another and explaining what made each character unique, likable, and timeless. There didn't appear to be a single toy in the store that Kalinske was unfamiliar with; he knew which company had developed each toy, why they had done so, and how they had gone about marketing it. There was just no place that Kalinske felt more in his element than inside a toy store.

Toy stores were more than just a comfort zone or realm of inspiration to him. They were like a library of cultural mythology. His biggest takeaway from the toy industry had been the importance of story. A toy might be just a piece of plastic, but if you added a compelling narrative and a character mythology, you could transform that piece of plastic into the next big thing. He had proved it with Barbie and with He-Man and the Masters of the Universe, and he was starting to feel more and more confident that he could do it with Sonic as well.

They stopped in front of a Mickey and Minnie dollhouse. "He's the ultimate friend," Kalinske said. "No matter what, Mickey remains upbeat and encouraging. It's like he lives to put a smile on the face of others."

"Sounds kind of pathetic, if you ask me," Schroeder said. "I prefer my friends to be a little more selective."

"Well, not everyone can be as popular as you, Madeline. There are a lot of kids out there who just want someone to like them. Enter Mickey Mouse."

Kalinske continued his tour and stopped in front of a large display of Teenage Mutant Ninja Turtles, the most recent plastic sensation. "I've been thinking that these guys embody the tone that we should be trying to strike. Playful, but edgy—cool, but no leather jackets. You know what I mean?" Nilsen and Schroeder nodded, taking it all in. "And I've seen a few episodes of the cartoon. They do a great job of establishing the universe."

They passed through the boys' action hero section into the pink and purple world of girls' dolls. Kalinske didn't notice until he was face-to-face with Bathtime Fun Barbie, dressed up like a mermaid. Schroeder and Nilsen noticed his subtle flinch.

"Don't like running into her, do you?" she asked.

Kalinske rolled the question over. "It can be a little strange sometimes."

"I'm sure it doesn't help that she's just about everywhere," Nilsen said.

Schroeder could tell that the sight of Barbie really did strike a nerve. "Would it make you feel any better if I told you she was just a piece of plastic?"

"If only that were true," he said with a sigh. He glared at the doll and moved on, his mind racing with ideas to try to put Barbie in her place.

Day after day, Kalinske, Schroeder and Nilsen worked to turn this critter into something more than lines on a page. At first their primary focus was subtraction, removing the fangs, the collar, the guitar, and the girlfriend. Then as he began to look more and more like a lost little hedgehog, they worked to add back some of that attitude, focusing less on gimmicks like a guitar or a girlfriend and more on his backstory and character. To better understand this speedy blue hedgehog, Kalinske had Schroeder write a thirteen-page bible that detailed the who, what, where, when, and why of his personality. He had grown up in Nebraska, lost his father at a young age, trained hard to develop world-class speed, and befriended a brilliant scientist who acted as a father figure until an experiment gone awry turned him into an evil villain.

Eventually, the creative forces at Sega of America got to the point where they no longer felt like they were making up the hedgehog's story on the fly, but actually learning more about a character who truly existed. As they continued to redefine this character from a marketing standpoint, the designers and engineers at Sega of Japan were busy working on a "game like no other" in which the hedgehog would star. During this time, Nilsen did what he was often known to do and took things further by rechristening him Sonic The Hedgehog (with his middle name literally being "The," on the reasoning that it would make a cool story one day).

Sonic wouldn't just become the face of the company but also would represent their spirit: the tiny underdog moved with manic speed, and no matter what obstacles stood in his way, he never ever stopped going. Sonic embodied not only the spirit of Sega of America's employees but also the cultural zeitgeist of the early nineties. He had captured Kurt Cobain's "whatever" attitude, Michael Jordan's graceful arrogance, and Bill Clinton's get-it-done demeanor.

When the newly refined hedgehog was ready, Kalinske called up Nakayama. "We made some changes. I want you to take a look."

"Okay," Nakayama said. "I will call you back."

"No, I'd like to stay on the line and hear your reaction," Kalinske said as he faxed over a copy of Sega of America's revised hedgehog.

Nakayama chuckled, but his good mood quickly devolved into a cold neutrality. "Oh," he said. "This is not even the same hedgehog that we gave you! Where is his lady friend? And those sharp teeth of his?"

"This is not the reaction I was expecting," Kalinske said, echoing not only Nakayama's earlier words but also his distinctly disappointed tone.

Nakayama thought for a moment. He was a man who chose his words wisely, so it was significant whenever he took an extra moment to do so. "It doesn't matter what I think. It only matters what will sell." But over the following days, tempers at Sega of Japan began to flare. The games designers believed they should be in charge of every aspect of Sonic. In normal circumstances, this would likely be the case, but since the character of Sonic had initially been created for the goal of success in the United States, Sega of America believed that they knew best when it came to the tastes and preferences of their audience.

Days later, Nakayama called Kalinske back, sounding less open-minded. "My people do not like what you have done to their creation. It no longer resembles what they had in mind. We must revert to the original."

Kalinske realized for the first time that despite being under the same umbrella, Sega was essentially two companies: Sega of Japan (SOJ) and Sega of America (SOA). It didn't matter to SOJ that the new hedgehog might be better; all that mattered to them was that it wasn't theirs. Although the friction between parent company and subsidiary was subtle, it certainly did exist and was real in a way that Sgt. Kabukiman was not.

Kalinske knew this was the moment that could make or break the company. He had to put it all on the line and urge Nakayama to reconsider. "I've been in the videogame business for about five minutes," he began, "but I've been in the toy business for over twenty years. You know what the toy business really is? It's not about size, shape, color, or price; it's about character. You want to play with characters you like. You want to become a part of their world and let them become a part of yours," Kalinske said, overwhelmed with passion. "I can only speak for myself, but there's not a character out there that I'd rather spend some time with than our new Sonic The Hedgehog. And if I feel this way, I think there are a lot of others who will feel exactly the same." Kalinske stopped and took a deep breath. He thought for a moment about reminding Nakayama about his promise to let Kalinske do things his way, and he also considered suggesting they conduct some market tests to see which hedgehog was more popular, but at the end of the day none of that mattered. This was about a vision, and if Nakayama couldn't see that, then he didn't deserve Sonic.

Nakayama finally broke the silence. "Tom, maybe I agree, but you must understand that there are people here of premium integrity who think differently."

"I understand," Kalinske replied. "So how about we try and change their minds?"

To share Sega of America's vision for Sonic, Schroeder was sent to Japan with the unenviable task of convincing the programmers that although they may know how to develop great games, she and her colleagues knew how to develop great characters. This fateful meeting at SOJ began friendly enough, but when it became clear that Schroeder wasn't interested in revising her vision, tempers began to flare. As a compromise, they suggested that each faction of the company simply have their own Sonic: you use your Sonic, and we'll use ours. To support this multi-Sonic worldview, they cited how Mickey Mouse wasn't exactly the same all over the world.

First off, Schroeder thought, I don't even think that's true. And secondly, even if Mickey does get localized in certain regions, she felt fairly confident that there wasn't a territory in the world where Mickey had fangs (or Minnie had double-Ds). Thirdly, and most important, she didn't want two Sonics. This wasn't about Sega of America getting their way, but about creating something immortal that existed in the world's collective imagination. And to do that, there could be no S(OA)onic and S(OJ)onic. Schroeder tried to make this point, but soon enough everyone had left the room. Although this impromptu boycott seemed to point toward a Sonic schism, whatever she had said in Japan appeared to have done the trick. When Kalinske next spoke with Nakayama, he and his team were given the green light to proceed as they saw fit.

With this mandate, Sonic sprinted toward the finish line, hoping to one day race past Mario and declare war on Nintendo. But in the coming months, those David-vs.-Goliath dreams were too often dashed by skirmishes between Sega of America and Sega of Japan. This cultural clash would lead to a standoff in which every decision, big or small, escalated into a battle of pride, principle, or sometimes just pure pettiness. This growing divide would be hard on everyone, but it would undoubtedly be hardest on Shinobu Toyoda, the liaison between the two factions. If Schroeder fought for a change, and Naka fought against it, it was Toyoda who was thrust in the middle and played the role of peacemaker. Kalinske knew that when it comes to war, everyone ultimately has to pick a side, and in the process of watching Toyoda constantly trying to broker peace between SOA and SOJ, he finally saw this man's true colors. Toyoda looked like a Japanese man and sounded like a Japanese man, but when push came to shove, he remained loyal to SOA. It was the small things that Kalinske noticed: the way he translated fighting words into diplomatic terms, the way he might claim Nakayama had approved something that he'd never even seen. Most important were the subtle ways he moved the emotional chess pieces to get what SOA wanted, for example, by doing something like adding a ridiculous character detail and then gaining leverage over SOJ by offering to remove it.

But before any of those unnecessary battles in the Sonic wars would be fought, Kalinske was informed about a more pressing conflict: Sega's negotiations with Electronic Arts.

"What negotiations?" Kalinske asked.

"I'm sure I must have mentioned this earlier," Nakayama replied.

"Nope, doesn't ring a bell."

"Well, the situation at present is that Trip Hawkins of Electronic Arts has found a way to reverse-engineer the Genesis and now they have decided to make games without our approval."

Kalinske was floored. "Didn't we just release some of their games?"

"Yes," Nakayama said. "You must go to Electronic Arts and show them that we mean business."

Kalinske sighed deeply, looking at the fax machine. At least he had Sonic. At least there was that.

9.

TRIPPED UP

"Don't act so surprised," snapped Trip Hawkins, the brilliant but mercurial founder of Electronic Arts. "You had to imagine it would only be a matter of time."

On Nakayama's instructions, Kalinske had traveled with Rioux, Toyoda, and Sega's legal counsel, Riley Russell, to the offices of Electronic Arts, where they met with Hawkins, marketing wizard Bing Gordon, CEO Larry Probst, and EA's legal counsel. Over the past several months, negotiations had been going on between the companies about how to proceed now that EA had reverse-engineered the Genesis. It was a potentially fatal situation that could knock Sega out before they ever really entered the ring.

Kalinske wondered what other problems were out there, ones that he didn't yet know about. After all, he had only just recently been informed of the fiasco involving Sega's planned football game. As part of Michael Katz's plan to hitch Sega's wagon to a constellation of superstars, the company had signed Joe Montana to a $1.7 million licensing agreement. Since Sonic wouldn't be released until next year, late 1991, Joe Montana Football was intended to be the flagship title for the Genesis—the game that branded Sega, the game that sold consoles, the game that kids begged Santa for this Christmas.

The problem, though, was that Sega's games were designed in Japan, where they didn't know the first thing about football, so Katz had decided to find a software developer in the West who could create a quality product in a short period of time. Luckily, a local software company called Mediagenic already had a football game in development that was about 30 percent complete. It was far enough along that it could be ready in time for Christmas, but unfinished enough that there was still room for it to be built around Montana. A few weeks before Kalinske joined Sega, however, Katz had discovered that the game was nowhere near complete. Ironically, when Katz found out about this, his first call was to Hawkins in the hopes that EA would sell them their new John Madden Football game and allow Sega to tweak it and rename it after Joe Montana. But Hawkins refused, believing he had a hot franchise in the making.

"Come now. You must have something to say," Hawkins asked. "Comments? Questions? Perhaps a more eloquent form of flabbergast?"

"I guess the only question that comes to mind is how the hell you pulled it off, but I suspect that I don't truly want to know the answer to that." Hawkins didn't respond, but his eyes shone with the glimmer of a shooting star swiftly slashing through the sky—one about to crash down to earth and destroy everything in its path.

It had all started about a year ago when Hawkins had a significant change of heart. Since founding Electronic Arts, he had been viciously resistant to the idea of creating software for videogame consoles. He saw them as pesky toys, nothing at all compared to the future of personal computers. This mind-set had made him look like a genius when Atari blew up in 1983, but by 1987, when the Nintendo phenomenon was in full effect, it made him look like some combination of foolish and pretentious. Even as Nintendo continued to rise, Hawkins vehemently defended his position, believing that the NES was no more than a Cabbage Patch Kid–like fad, and reminding his employees that the computer was the future. Besides, the graphics on the NES were mediocre at best and couldn't handle EA's enormous talents. But this all changed with Skate or Die!

With pressure mounting from his EA cohorts, Hawkins finally agreed to a small concession: he would allow EA to license their game Skate or Die! to another software company named Konami, granting them rights to distribute the game on various other systems (including the NES). The decision to forgo directly putting the game on the Nintendo yet allowing it to happen anyway may seem strange, but it stemmed mostly from the fact that Hawkins simply didn't want to deal directly with Nintendo and their strict licensing agreements, which didn't exist in the computer world.

EA received only a fraction of what Konami had made (which was only a fraction of what Nintendo had made), but the royalty from Skate or Die! for the first month alone was more money than EA made on its best-selling computer games. It was then that Hawkins decided while personal computers were still the future, console games apparently were also the future. Yet despite this realization, Hawkins still couldn't stomach the notion of becoming beholden to Nintendo, which caused him to take a closer look at Sega. From a technical standpoint, the Genesis (with its 16 bits and 68000 Motorola processor) was better equipped to handle EA's games, but even though Sega had such a puny market share, their licensing agreement wasn't all that different from Nintendo's. The rates were cheaper, of course, but conceptually they both believed that the hardware companies deserved to be paid a toll by the software makers. So to avoid this, Hawkins arranged for Electronic Arts to reverse-engineer the Genesis. Unlike Tengen, who bent the law to discover a work-around, EA would do this legally by setting up a "clean room" environment, which would create a Chinese wall between the engineers dismantling the machine and the engineers trying to rebuild it with the desired change (in this case, circumventing the console's lock-out system). Now, a year later, Trip Hawkins excitedly stood before the guys from Sega with the grandeur of someone who had just knocked over a little boy's sand castle.

"All right, Trip," Kalinske said, while replaying the connect-the-dots of crap that had made up his day, "tell me how you did it."

"Eh, who cares?" Hawkins mused. "A to B to C to D, and here we are. The question is, now what? What do you think is fair?"

"You're getting nothing!" Rioux bellowed. "I'll bet you they didn't legally reverse-engineer the thing! Those guys are too dirty to ever bother with a clean room."

"Would you really like to gamble away your company on a hunch?" Hawkins asked. He turned to Kalinske. "You're new to these conversations. What say you?"

Kalinske snickered. "Who cares what I think? You're holding all the cards."

Hawkins struck a defensive tone. "You can drop the high-and-mighty routine. You think I wanted to do this? You think I like driving off the course?"

"How is this my fault?"

"You, Katz, Nakayama, and those zombies at Nintendo. You guys just don't get it. I spend years making a game, hundreds of thousands of dollars in development costs, and then when it comes time to recoup, I need to buy the cartridges from you, get your saintly blessing, and then after all that, pay you ten bucks for every game I sell."

When Hawkins finished, Kalinske remained silent for a moment, this time by choice. Finally he said, "What? No fake tears? I mean, if you're going to lob a sob story at me, at least go the whole nine yards." The Sega team couldn't help but crack smiles.

"I'm not being dramatic. I'm not by nature a dramatic person," Hawkins said, which might have been the falsest statement he had ever spoken.

"It's the price of doing business," Kalinske said. "Those thousands you spent on making games . . . what about the millions we spent making consoles? We barely break even on those systems. We're giving away razors in order to sell the blades."

"But they're my blades."

"Yeah, well, they're our razors!" Kalinske shouted. Toyoda nudged Kalinske in an attempt to calm him down.

Hawkins shot up. "Steve Jobs is an obsessive maniac and even he doesn't make us pay him money to put our games on his computers." This was, after all, the basis for Hawkins's stubbornness. He came from the computer world, where anybody could make any game for any system. In some cases, the computer manufacturer even paid you money to develop for their system.

Kalinske sighed. "Oh, Trip, didn't your mother ever teach you the difference between right and wrong?"

Hawkins sighed too. "Oh, Tom, the funny things a man will say when he's down in the fourth quarter."

Though the banter briefly lifted Kalinske's mood, that comment reminded him of the Joe Montana Football problems. Even if Sega and EA found a way to play nice, there was still that to deal with. These battles with EA, Japan, and Wal-Mart only served to take Sega's eyes off Nintendo and weren't worth fighting. Kalinske briefly discussed the situation with Rioux,

Toyoda, and Russell. They all agreed that having a close relationship with Electronic Arts, even if it wasn't profitable, would be an incredible win for Sega. EA made great games, which was exactly what Sega needed at the moment. To make that happen and also save face, Toyoda came up with a strategy that he whispered to the others.

"Okay," Rioux said. "How about we play *Let's Make a Deal*?"

"I don't like game shows," Hawkins replied. "The hosts freak me out."

"Well, despite your strange phobia," Kalinske said, "here's what we propose: Sega will grant you permission to make authorized games for the Genesis, and instead of the ten bucks for the cartridges, we'll only charge you four."

"And . . . ?"

Kalinske looked into his eyes. This really was a shakedown. In addition to the 60 percent cost reduction, Sega would also allow EA to publish up to sixteen games per year and self-manufacture their cartridges.

Hawkins took this in. "That all sounds well and swell, but you're forgetting that we can do all of this for free, no?"

"I'm forgetting nothing. I'm sure you realize that if you go through with this, you're gonna bankrupt both our companies with lawsuits back and forth."

Hawkins thought this over, warming to the idea. "So that's your final offer?"

"Actually, there's something else that we want from you," Kalinske said. "In exchange for keeping us out of court and saving us both from ever having to have a conversation about razors and razor blades again, I want you to give us Madden."

"No way!" Hawkins yelped.

"What about for some cash and promotion?"

"It's not about the money," Hawkins said.

Kalinske understood and respected that. He knew from Mattel that these weren't stocks, bonds, and commodities that they were selling; these were emotions, experiences, and ideas. "Why don't you use EA's Madden engine to make our Montana game? Same gameplay, but switch everything up so that it looks and feels different. If you guys can help us out with this, we'd even be willing to offer EA a nice royalty on Montana sales."

Hawkins stroked his chin, playing out scenarios in his always-whirling mind. Essentially, EA would take their own game and tweak it enough so that customers didn't realize that John Madden Football and Joe Montana Football were actually the same game. Hawkins thought this was a crazy idea. Insane, even. But it was a crazy idea that would leave Kalinske and company holding the bag if people ever figured it out (and, in the meantime, EA had managed to get the Sega royalty offer up to 24 percent of this proposed Madden-turned-Montana game).

Hawkins extended his hand across the desk and Kalinske did the same. It was a perfect moment . . . until Hawkins felt compelled to speak again. "I just want to say, that there really is a way of looking at this where I'm the good guy and you're the bad guy." He nodded rabidly, seemingly trying to persuade not only Kalinske but himself. "There's a way of looking at this where I'm the hero who chops through the bullshit and oppression in order to come out the other end with a spoonful of freedom."

"Please don't say another word," Kalinske said, trying to put an end to the long, arduous day. "Every time you open your mouth it makes me feel like I've just made a huge mistake."

10.

EXTREMELY DANGEROUS

(DON'T TRY THIS AT HOME)

Another day, another plane. Kalinske's go-go-go life was starting to feel eerily familiar to that of his Matchbox days, where he spent most of his time waiting in airport lines, flying through the sky, and landlocked in places where his watch didn't match up with the local time. At least now he had company.

"Are you ready to have some fun?" Nilsen asked as he and Kalinske walked up the steps to the New Orleans Convention Center.

"Honestly, I'm ready for anything that does not involve reverse-engineering or defanging hedgehogs," Kalinske said, as they were about to enter the 1991 Amusement and Music Operators Association (AMOA) expo.

The AMOA trade show was as exciting, confusing, and carnivalesque as its name made it sound. The organization had been created in 1948 when sixty-eight angry jukebox owners banded together to fight against paying royalties on the songs their machines played. Out of that alliance came the idea to hold an annual trade show that would bring together various parts of their niche universe, from jukebox designers and equipment distributors to bar owners and music producers—basically anyone who had a stake in customers inserting quarters into a jukebox. Meanwhile, as the prestige of the AMOA grew, it began to attract the attention of companies producing other mechanized products that offered entertainment in exchange for coins, such as air hockey tables, love testers, and eventually arcade games. As jukeboxes and the less interactive diversions of yesteryear waned in popularity, the arcade business became the dominant species of coin-operated entertainment. Kalinske knew little about arcade games beyond what he'd learned watching his daughters play Frogger or Ms. Pac-Man at pizza places; eager to replace his ignorance with competence, he was excited to enter the show.

As soon as they stepped inside, Kalinske felt as if he'd been directly transported into the imagination of a five-year-old: row after row of pinball machines, kiddie rides, and arcade cabinets as far as the eye could see. Ironically, out of the thousands in attendance, there wasn't a single child in sight.

"So how does this work?" Kalinske asked.

"It's great. There are a couple hundred booths here, about half of which are utterly absurd," Nilsen said, pointing to a man in a suit eagerly grasping for swirling dollars inside a tall translucent rectangle. "While we occasionally take a moment out of our busy schedules to mock that half, we'll mostly be trying out arcade games, scoping out trends, and seeing if there's anything we should consider licensing as a Genesis game. Then we ought to sync up with the Sega of Japan arcade folks so they can meet you and see that now we're for real. Sound good?"

"Very. I mean, you basically just said that we get paid to play games."

"That's why we have the best job on earth."

It was a great job. Or at least parts of it were great. As a lifelong sports fan, Kalinske had a tendency to think of things in terms of athletic analogies. Normally, he played the role of coach (motivating and instructing Team Sega) or general manager (wheeling and dealing, like with Wal-Mart and EA), but today he got to play the role of scout: purely an observer, looking for diamonds in the rough. When teams won, coaches and general managers were the gods responsible for every speck of greatness, but when they lost, they were incompetent morons who didn't deserve the oxygen they were breathing. Given all that, today's anonymity was a nice respite from those two opposite yet equally irrational perspectives.

"So," Nilsen said, "shall we get down to business?"

Though they spent a pleasant afternoon checking out what was new in the arcade realm, their real excitement was reserved for the pay-per-view boxing match that evening. Prior to Kalinske's arrival, Nilsen and then-president Michael Katz had taken a risk investing in a franchise built around Buster Douglas, a no-name boxer who had recently defied 42-to-1 odds and knocked out the heavyweight champion of the world, Mike Tyson. Part of their gamble was based on an appreciation for the young boxer's scrappy fighting style, but the larger part was based on a desire to stick it to Nintendo, who published the popular game Mike Tyson's Punch-Out!!

To rub it in Nintendo's face, Sega wanted to fast-track their new boxing game so that it would be ready for release shortly after Douglas's next match. But given how long it normally takes to design, create, test, and finalize a game (about a year), it would not be possible to start from scratch in February (when Douglas defeated Mike Tyson) and be ready in time for October (when Douglas would, everyone hoped, knock out Evander Holyfield). So Nilsen asked Hugh Bowen, a product manager he had hired and trusted, to find a good already-complete boxing game that Sega could buy and essentially slap Douglas's face onto. He had some concerns about this type of approach, but after all, that's what Nintendo had done with their famed boxing game. It originally had been called Punch Out!! until Nintendo's Minoru Arakawa signed Mike Tyson, threw him on the cover, and made him the final fighter players had to beat in order to win the game. That had obviously worked out nicely for Nintendo, and Nilsen wanted to replicate the recipe for Sega, though, as always, he'd want to spice it up a little. Instead of fighting against the professional boxer as if he were some kind of villain, he wanted players to enjoy the experience of being a pro boxer. Hope became reality when Bowen found an arcade game called Final Blow, made by Taito. Needing to act fast, Nilsen quickly inspected and approved the game to be "Buster-ized," Katz signed off on the plan, and Nakayama arranged a deal to use the boxing title on the Genesis. Sega's R&D took Taito's game and made the minor but necessary alterations (such as swapping the Detroit Kid, the game's original main character, with Buster Douglas) and then racing against time to have the game ready for release within a month of Douglas's first fight as champion. Which was, finally, tonight.

"By the way," Nilsen said as they played a blood-spattering zombie game, "at some point we should see what the folks at Gottlieb are up to."

"Gottlieb?" Kalinske nodded. "They're the ones who made Q*bert, right?"

"Hey, look at that. You're really getting a firm grasp on the business!"

"Slowly but surely," Kalinske said, almost surprised to realize he was indeed starting to get a handle on things. "Slowly but surely."

Nilsen smiled. "Don't worry. I won't tell anyone. We'll try to keep expectations as low as possible so it's that much more impressive when you shock the world."

That night, Kalinske and Nilsen got to the sports bar early to stake out a prime position for the fight. Neither being a particularly big drinker, they each nursed a beer, shared some appetizers, and reflected on Kalinske's crash course in the arcade industry.

"So, any grand observations?" Nilsen asked.

"Nothing too crazy," Kalinske said, scanning through his memories of the day. It had been as light and enjoyable as he hoped for. They'd spent some time with the Sega of Japan arcade guys, who had some cool things in the pipeline but seemed a bit standoffish. And they had indeed gone to the Taito booth, where they'd seen a game that Nilsen liked called Hit the Ice. It was a hockey game, but unlike typical hockey games, this one featured only three players per team (forward, defense, and goalie); even more unusual was that the players were encouraged to break the rules by slashing, tripping, and kicking opponents in the groin. It was cartoony and ridiculous, but definitely fun. "Still, all things being equal," Kalinske started, trying to find the words to match his mild discomfort, "I could do without all the violence. Lots of those games we saw seemed to be a bit excessive with the blood and gore. Doesn't any of that stuff ever bother you?"

"I don't love it. But it doesn't get to me." Nilsen shrugged. "Besides, the arcade games tend to be a bit racier. They're usually made for an audience a little older than what we're doing with the Genesis."

Kalinske subtly shivered. "I just hate the idea of using that garbage to sell games, you know?"

"I absolutely do," Nilsen said, right there on the same wavelength. "But the good news is that we don't need that stuff to make great games. And at the end of the day, that's really all that matters. Just remember: the name of the game is the game."

"I've heard you say that before."

"And you'll hear me say it again. Because it's true."

"The name of the game is the game?"

"The game has to be good. That's all the matters. That's the only thing."

Kalinske nodded, getting it, and glad to have an excuse to feed his conscience. He looked across the table and tried to find a way to express his appreciation. This had really been a nice day, a great respite from the time he spent in front of a firing squad. But as he sought a way to savor the moment, he was interrupted by an excited hubbub inside the sports bar as the fight was about to start.

Kalinske and Nilsen looked up at the TV, and despite their best efforts to play it cool, they couldn't help but do a double-take. Their fighter—the heavyweight champion of the world, the face of their new game franchise—did not look like the face Sega had signed. Douglas was bloated, at least fifteen pounds overweight, and even as he danced around in his boxing trunks he looked sluggish.

The guys looked at each other.

"Uh-oh," Nilsen said.

The words quickly proved to be prophetic. Holyfield overmatched his chubby opponent and knocked out Douglas in the third round.

After it ended and Douglas was put out of his misery, Kalinske looked at a shell-shocked Nilsen. "Don't worry," he said. "Everything is going to be okay."

"Somehow I doubt that," Nilsen said, unable to think about anything besides how hard he'd worked to get the game ready in such a short amount of time and how the press was going to have a field day with this.

Kalinske put a hand on his friend's shoulder. "We'll be fine."

Instead of spending a celebratory night in New Orleans, victory-dancing their way through the Big Easy, stopping only to tell anyone who appeared willing to listen that Sega was for real and Nintendo had better watch themselves, they went back to their rooms and called it an early night.

Back in his hotel room, Kalinske wondered how so much had gone so wrong in such a short amount of time. He was beginning to worry that the vaunted reputation he'd built over two decades might quickly and irrevocably sink with Sega. He had come to the company with the goal of giving it an identity, but the more he learned, the more uncertain he became of what that identity might be. In a few weeks he was supposed to get in front of Sega's board of directors and tell them his plan for how Sega could become competitive. What was he supposed to say? Maybe he ought to just stand in front of them, shake his head, and succinctly say, "Let's just give up."

To subdue these thoughts, Kalinske turned on the TV. There were only twelve channels (though one of them, the hotel boasted, was HBO), but that would be enough to provide him with the numbing therapy of television. He watched the middle of some Sean Connery movie, trying to figure out if this was a James Bond movie or not.

When the film went to a commercial, he quickly forgot all about the movie. There was something going on with this commercial, and it got his juices flowing.

It began with a quaint, to-grandmother's-house-we-go bridge. Plain. Pretty. Picturesque. And then, after the visual lullaby, it hits you: Boom! A warning appears on the screen: Extremely dangerous. Don't try this.

Two men stand on the bridge, preparing to jump into a tumultuous body of water. The man in Nike Air shoes breathes nervously, while the other bends down to add some air to his Pump shoes, made by Reebok. Just then the audience realizes they are bungee-jumping. In slow motion the two men spread their arms and jump. Downward they go, soaring with immaculate weightlessness as time seems to stand still.

Until suddenly their bungee cords snap them back. The man in the Reebok Pumps dangles upside down, safely above the water. Meanwhile, the other bungee cord can be seen, but the man is gone and all that remains are his flimsy Nike Air shoes.

After it ended, Kalinske stared at the television screen. This was it—this was the message. Edgy, sarcastic, clever, and fun. The ad perfectly summed up Kalinske's vague and disjointed notions of how Sega should define itself and its products. How would he pull this off? He didn't know, and right now he didn't care.

He wanted to play the commercial again, over and over. He wanted to call Karen, or Nilsen, or even Nakayama and say, I've got it! But truly, he didn't know what he had. Luckily, he had the rest of the night to figure that out.

• • •

The following morning, Kalinske met Nilsen downstairs for breakfast at the hotel restaurant. Though he didn't have any tangible reason to feel upbeat, he seemed to be as cheerful as the bright yellow walls.

"How'd you sleep?" Kalinske asked as they took their seats.

Nilsen, normally a man of many words, answered by placing a copy of *USA Today* on the table, which featured a cover story on Buster Douglas's embarrassing loss.

Kalinske slowly shook his head. "How screwed are we?"

Nilsen opened his briefcase and pulled out the documents related to product development budgets and schedules. "Everything's already been manufactured and shipped. The cartridges, game cases, manuals, and everything else are on a boat somewhere in the middle of the Pacific Ocean. So, in short, very screwed."

"Yeah, I figured," Kalinske said. He knew this to be factually correct, but there was something about that commercial last night that made him feel resilient. "But still, come on. There must be *something* we can do. I mean, the fact that an obscure boxer knocked out the champ and then got fat and lazy doesn't change the fact that we have a better product than Nintendo, right? In a way, it's almost comical."

Before Kalinske finished the sentence, Nilsen's mind was already furiously at work; that's how it went with him. Often he'd seem to suddenly zone out, but actually that was when he was most zoned in. He would become bombarded with images, related and unrelated alike: Buster Douglas, game cartridges, the Pacific Ocean, bright yellow walls, refreshing orange juice with its tiny bits of pulp. These seemingly random remnants of memory would collide against an eclectic soundtrack in his head: a chord from Beethoven, a crash from MC Hammer, and the eternal optimism of that "Zestfully clean" commercial—and in a matter of seconds, unlikely connections formed.

Nilsen knew that some creative types worked slowly and deliberately, but that was the complete opposite of how his mind worked. "Wait!" Nilsen proclaimed. "Let's make it a cool inside joke."

"I like it," Kalinske said, "but don't quite understand. What do you mean?"

"Let's not run from this thing. We've got egg all over our face, but instead of running to go wipe it off and pretending like it never happened, let's just not wipe off the egg. Let's stand here, embarrassed, and just laugh at ourselves."

"So we turn into the skid?"

"Exactly! We embrace the failure. Maybe we even turn the game into a collector's edition. But whatever we do, we don't run and hide. After all, even though it's named after an overweight former heavyweight champion, it does happen to be a really fun game. And the name of the game—"

"Is the game."

Nilsen nodded, proud of his boss. "So, what do you say?"

At the time, Kalinske couldn't have known that by going along with this plan, James "Buster" Douglas Knock Out Boxing would go on to be a critical and commercial success. He didn't know that the game would sell so well that Sega would have to order a second shipment and that upon its rerelease it would be packaged with the tongue-in-cheek branding of "Sega Classics." And he didn't know that this game, and the company's laugh-at-us stance in releasing it, would give Sega a certain credibility and coolness with gamers and the press. At the moment Kalinske didn't know any of these things. But he knew that what Nilsen said was the best idea he had heard in a long time, and for them to have any chance of pulling off the impossible, it was exactly the type of thing they needed Sega to become: the fun, rebellious, sarcastic underdog.

"Let's do it," Kalinske said, with the enormous confidence, fearlessness, and appreciation the idea deserved.

11.

LIGHTNING IN A BOTTLE

It is a rare thing for five smart, skeptical adults to be so awed that in unison, their jaws drop in astonishment. But what Kalinske, Rioux, Toyoda, Nilsen, and Schroeder were staring at in the conference room was quite possibly the most extraordinary thing they'd ever seen: an early, incomplete, minute-long demo for Sonic The Hedgehog. The game was so fast, so exhilarating, and so unlike anything they'd ever seen before that their facial responses hardly did it justice.

"This is . . ." Kalinske began. "This is genius."

"A work of art," Schroeder chimed in.

"A home run!" Toyoda declared.

"Consider my expectations exceeded," Rioux added.

They looked to Nilsen, their resident gamer, for his assessment. "Eh, it's okay," Nilsen said, drawing blank stares from his colleagues. "Actually, wait, that didn't sound right. Can I borrow someone's sarcasm? Mine doesn't seem to be up to snuff."

The five of them, like bank robbers who had just completed an unbelievable heist, continued to stare at the game and allow themselves to indulge in a series of what-ifs. What if Sega could change the way videogames were played and perceived? What if they could take off the gloves, pound their chests, and declare war? What if they could actually go after Nintendo?

When Kalinske returned from New Orleans, this was exactly the kind of fearless ambition he wanted his team to harbor as he prepared for his big trip to Japan. The employees now admired him; he was smart, approachable, and inspiring. He was a man obsessed with competitive advantages, and soon enough, that kind of strategic thinking became infectious. Between his vision, Rioux's meticulous day-to-day management of operations, and Toyoda's constant efforts to protect America from Japan, the what-ifs began bubbling inside Sega's employees. But to fully believe, they needed something to see, and that's what Sonic The Hedgehog was: proof of possibility.

"This is it," Kalinske said, fixated on Sonic, their beloved blue hedgehog, who somehow managed to come across as both rebellious and adorable. This was the character who would embody the Sega spirit and give them a puncher's chance against Nintendo's 16-bit machine. "We need everyone to see this," Kalinske proclaimed, and then stuck his head out the door and invited everyone in the office to check it out.

The Sega of America employees gathered around the television set and had the same reaction that had just occurred in the five who'd first seen it.

"This is . . . wow," an astounded employee said.

"Holy crap," said another. "We're going to sell a million of these, aren't we?"

It was, of course, intended to be nothing more than a rhetorical comment, but the idea of having something that a million people would want to buy sent Kalinske's mind into overdrive. The combination of seeing this game, Sega's new Buster-ized attitude, and the fact that more great software was on its way (both internally and from Electronic Arts) led him to what he believed was a game-changing realization. "No, we're not," Kalinske said. "We're not going to sell a single one of these. And that's exactly what I'm going to tell Japan."

12.

THE REVOLUTION WILL BE PIXELATED

Kalinske, with Toyoda by his side, stood in front of eighteen unimpressed Japanese men: Sega's board of directors. The room was stuffy and claustrophobic, and the atmosphere inside it gave the impression that a bout of laughter might be punishable by death.

The only sympathetic face in the crowd belonged to Nakayama, though Kalinske noticed that in this context there was something different about his demeanor. Among the board members, he seemed less like an excitable and affable business visionary and more like a cheerfully cunning Machiavellian politician. Nevertheless, Nakayama's presence was a relief. He was the man who had believed in Kalinske even before Kalinske believed in himself.

"Let us begin," Nakayama said with a nod to Kalinske and then a briefer one to Toyoda.

"Ready?" Kalinske asked Toyoda, who would be translating for him.

"Always and never," his translator answered.

"Excellent," Kalinske said, with a hint of here-goes-nothing in his voice. Kalinske never felt more at ease, more like himself, than when he was speaking to a crowd. As he began, whatever nervousness he had been feeling instantly evaporated, and he possessed the confidence of a man who believed he could convince anyone of anything. "Members of the board, I want to thank you very much for inviting me here today to speak with you."

"私を招待していただきありがとうございます" Toyoda dutifully translated.

"It is both an honor and a privilege to be asked to share my thoughts with you," he continued. When Kalinske had first taken the job, Nakayama advised him that as soon as he had gotten his bearings, he would be invited to speak with the board and provide his assessment of Sega, his opinions on the burgeoning videogame industry, and plans for how the company might be able to carve out a bigger piece of the market. Kalinske understood this was something of a formality, but initially he'd been worried that after only a few months he wouldn't have much to say. Now, however, he was worried he wouldn't have enough time to say everything on his mind.

Simply put, the videogame industry was nothing more than a modern, pixelated version of the Wild West. There were no rules, no code of ethics, and absolutely no law, except for Nintendo's attempt to appoint themselves sheriff. Consequently, Kalinske was in a position to implement his own laws. And today, in front of these unimpressed gentlemen, that was exactly what he planned to do.

Kalinske had trekked to Japan to deliver his "Four-Point Plan," a bold vision that he hoped would turn videogames into an acceptable, cool, mainstream activity. If Nintendo fancied themselves as makers of the ultimate toy, well, then let them have their children. Sega was ready to scoop up everybody else; the hard-core gamers who demanded only the best, the teenagers who were always looking for new ways to procrastinate, and even Nintendo's precious children when they were a few years older and ready for something more sophisticated.

So instead of standing in front of the board of directors with a saccharine smile, appeasing eyes, a double thumbs-up, and an assurance that he'd keep on course, he'd showed up with a proposal that he believed would either flip the gaming landscape upside down or blow it up in spectacular fashion. "I hope that my suggestions, observations, and criticisms do not offend you, but are accepted in the spirit in which they are intended," Kalinske began. "We all want what's best for Sega, and I truly believe that over the next few years, we can go from *wanting* the best to *being* the best." He had organized his thoughts about how to reach that next level into four easily understandable areas of attack, beginning with the most controversial:

1. Games: When customers purchased a Sega Genesis, it currently came bundled with the game Altered Beast. That had to stop. Yes, it was a popular arcade game, but like most arcade games, it was too short, too repetitive, and too unsophisticated when played at home with limitless time. Plus, middle America had complained that the name Altered Beast sounded like devil worship. This was all completely unacceptable. Sega had to put their best foot forward and bundle with its consoles the game that absolutely, positively would differentiate them from Nintendo: Sonic The Hedgehog. Giving away their best game for free would cost the company tens of millions, but it should be considered an investment that would help them make hundreds of millions down the line.

2. Price: As he had mentioned to Trip Hawkins, Kalinske believed in the Gillette philosophy of giving away the razors in order to sell the blades. And with over thirty million American homes owning the NES, Sega had to be willing to take a loss on sales of the Genesis hardware just to get it through the door. Once they had established an installed base, then they'd start making back their money by selling games. Not only that, but a lower retail price for the Genesis—$149, down from $189—would make Nintendo's 16-bit console, whenever they introduced it, appear to be that much more expensive.

3. Marketing: Let Nintendo have the kids. Kalinske wanted everyone else, and he aimed to let them know they were wanted. Sega needed to redefine itself as hip, cool, and in-your-face. Doing so would not only speak to older generations but present videogames as a mainstream form of entertainment, no different from books, movies, and music. Toward that end, Kalinske proposed to increase Sega's advertising budget and create edgy advertisements that mocked Nintendo and appealed to teens and college students rather than younger kids.

4. Development: Sega of Japan developed great games, but many of those great games tended to have a particularly Asian appeal. Unsurprisingly, their

Japanese brethren at Nintendo had a similar tendency. If Sega wanted to plant their flag in America, they needed to make games for that specific demographic. And SOJ needed to accept SOA's input on how to alter Japanese titles for the American audience, similar to how they had revised Sonic The Hedgehog. To do this, Kalinske wanted to expand Sega of America's product development budget and staff.

When Kalinske finished, he inspected the faces of the board members. Their expressions suggested that they were even more unimpressed than they had been before. Any hints of friendliness had now been replaced by shock, confusion, and rage. And moments later, a barrage of angry questions, comments, and concerns flooded the room.

Toyoda nimbly worked to translate the ruckus as Kalinske tried to address their worries one by one. Eventually, however, there were too many criticisms, and Toyoda couldn't keep up with the chorus of condemnation.

"You'd think with the law of averages," Kalinske said to Toyoda in an aside, "that at least one of them would have something nice to say."

"That guy over there called you a 'tall handsome American,'" Toyoda replied.

"Oh, yeah? And was that all he had to say about me?"

"No comment."

Kalinske could hardly believe what he was seeing. These people didn't just disagree with what he had said; his suggestions had actually provoked them to abandon rationality and reply with fury. Somehow he had struck a nerve. And while it is never a good idea to aim at the Achilles' heel of someone you're doing business with, Kalinske knew it was an especially terrible idea to do so in Japanese culture, where form, respect, and honor are valued above all else.

Suddenly Kalinske felt like the walls were closing in around him, and he experienced a powerful sense of déjà vu. It was happening again. Just like at Mattel, where he had given that company the best years of his life and assumed that he would be there forever, until suddenly he wasn't.

He still had trouble making sense of what had happened at Mattel. Everything had been going so well there. His job was his life, and he loved it that way, because at Mattel he was the best version of himself. In 1981, after a decade of rising through the ranks, he was rewarded for his enormous contributions and named president of the Toy Division. He felt like he had reached the pinnacle of his career, and considered it an honor to spend the rest of his life selling the magic of toys to kids around the world.

The promotion enabled him to control most aspects of the production, distribution, and marketing of toys—most, but not all. Ultimately, because Mattel was a public company, the board of directors had the final word. And if there was one thing the board members cared about—more than growth, money, or making great products—it was being right. Knowing this, Kalinske tried his best to agree with the board, but that wasn't always possible, especially with the new path he envisioned for Mattel.

In the early 1980s, "conglomeration" was the business world's buzzword du jour. It was no longer enough for a company to excel at what they had been doing; now they had to buy up stakes in unrelated companies and manage those as well. The benefits of entering into unfamiliar industries were said to be diversification, tax breaks, and the possibility of synergy. Kalinske, however, knew the real reason: power. He could see this in Mattel's board whenever it announced new acquisitions, ranging from the understandable (Ringling Bros. Circus and Western Publishing) to the absurd (Turco Steel and Metaframe Pet Supply). Though Kalinske tended to disagree with this type of corporate expansion, he voiced his doubts to the board only when investment in these new businesses came at the expense of what had gotten Mattel to this point: toys. Whenever he objected, his dissent was noted and not forgotten. The board of directors had the final word, and there were always ways for them to silently retaliate.

By 1983 the board's trigger-happy acquisition strategy, coupled with major losses due to the videogame crash, had led Mattel to the brink of bankruptcy. Miraculously, the company managed to survive—with the last-minute help of junk bond guru Michael Milken. He raised enough money to keep Mattel afloat while simultaneously making the company leaner by selling off its recent, unrelated acquisitions. In the face of failure, the board's power weakened and some members were let go, but those who remained were not happy to have been proven wrong. Nevertheless, they were in no position to retaliate, since they needed Kalinske's help to rebuild the company. Two years later, as Mattel flourished, the board of directors rewarded Kalinske's strong stewardship by offering him the position of CEO. He didn't particularly want the job, as it would pull him away from the Toy

Division, but he interpreted the offer as an olive branch from the board of directors and accepted.

The relationship briefly worked well until Kalinske made a jarring discovery: in the years since Milken had restructured the company, interest rates had dropped by half but Mattel was still paying a much higher rate to its debt holders. He quickly realized what he perceived to be the reason: any change would require approval from the members of the board, many of whom were also the debt holders. Believing this to be a conflict of interest, Kalinske requested this matter be voted on, but only by independent board members who did not hold any of Mattel's debt. Naturally, his resolution passed, but it also fostered anger among many of the board members.

The directors were still not strong enough to replace Kalinske, but they were powerful (and clever) enough to point out that he was only thirty-eight and surely would benefit from having some leadership assistance. So they brought in John Amerman, the fifty-three-year-old head of Mattel's International Division, to become his co-CEO until Kalinske had matured enough to handle the post alone. Unsurprisingly, that never came to pass. In truth, Kalinske and Amerman got along fine, but the perception was that battle lines had been drawn, and the internal politics between Kalinske and the board spilled into every division of the company, with employees feeling obliged to take sides. Eventually the perceived turmoil fractured the company and caused it to grind to a halt. Kalinske stepped down, ceding the position to his "older and wiser" co-CEO.

Following the unceremonious end to his tenure, the board tried to keep Kalinske with Mattel by offering him the COO position. But this time he declined, sick of playing politics and believing that there had to be something better out there. Now, years later, he was in front of an equally angry board of directors, having foolishly believed that Nakayama had offered him that something better.

As the grumbling from Sega's board continued to mount, Kalinske resorted to what he did best: speaking. Foolish as it may have been, he continued to plead his case. "It sounds crazy, I know, but these are the type of calculated risks that Sega needs to take. Please, just trust me. Aren't you sick and tired of being bested by Nintendo?"

It was clear that this comment had struck another nerve. Kalinske turned to Nakayama, who looked almost pleased. Had this been Nakayama's plan all along? Had he hired Kalinske purely for the amusement of watching him fail? Had the entire thing—the wooing in Hawaii, the red-carpet treatment in Japan, the promise of autonomy in San Francisco—all been part of some carefully orchestrated charade by Nakayama to get revenge for something years ago that Kalinske either had never known about, or no longer remembered? These thoughts hit him harder than any of the gripes from the board did. A few months ago he'd had no interest in this job, but now it was the only thing that mattered to him. It was his last chance to make his family proud, to prove that he was as successful as he'd always believed he had been.

Nakayama angrily pounded his fist on the table, silencing the room. He stood up sharply, shook his head, and locked eyes with Kalinske. Nakayama opened his mouth to rule on this matter, and as he did a sly smile graced his face. "Tom, no one agrees with anything you're saying. In fact, everyone in here thinks you are nuts." Nakayama took a deep breath. "But this is why I hired you. You may go ahead with the plan."

For a moment, Kalinske's eyes showed disbelief. Then his lips curled into a thin, almost imperceptible smile. "Thank you, Nakayama-san."

Nakayama nodded. Then all the emotion seemed to vanish suddenly from his face. "Now, whatever you do, don't mess up," he said, and promptly exited, leaving behind eighteen furious Japanese men, a dazed translator, and one tall, handsome American who was now officially allowed to shake things up. And the timing of this could not have been better, as Nintendo was about to release their most deadly weapon yet.

PART TWO

SONIC VS. MARIO

13.

THE WINDS OF CHANGE

A gust of wind rose from the Ujigawa River and reverberated through the night, rattling between the low-lying mountains of the Higashiyama, Kitayama, and Nishiyama ranges. From there, the breeze whispered its way through the tranquil city of Kyoto, weaving between an ornate contrast of Zen gardens and imperial palaces, before crashing into the impenetrable exterior of an unassuming warehouse.

Inside, oblivious to the persistent thumps of wind, an assembly line of Japanese laborers worked through the night. Together they operated with the coordinated chaos of scurrying ants. The laborers had been given strict orders that not a single second could be spared, and as a result, their routine became religion: unload, assemble, test, package, and ship. Over and over they did this, until every single unit of Nintendo's new gaming console had been sent out for delivery. And while all of this was going on in Kyoto, similarly stealthy efforts were taking place at other warehouses throughout Japan as part of Nintendo's master plan: Operation Midnight Shipping.

The clandestine nature of these arrangements had been ordered by Nintendo's president, Hiroshi Yamauchi. While preparing for the launch of the company's 16-bit Super Family Computer (Super Famicom), he had uncovered rumors that the Yakuza, Japan's organized crime syndicate, planned to hijack deliveries. Though the organization typically trafficked in drugs, currency, and women, their sudden interest in electronic goods was not completely surprising. Wherever strong demand existed, the Yakuza took the necessary measures to be ready with supply. And in late 1990, when retailers received word from Nintendo that the 16-bit system would be available later that month, demand soared to incredible heights.

On November 3, Osaka's famous Hankyu department store announced that it would take reservations for the Super Famicom. A week later, they had to stop accepting preorders due to the sheer number of requests. Most retailers didn't even last that long before changing gears. Some stores set up lottery systems to determine who would be lucky enough to purchase Nintendo's new product, while others allowed customers to place preorders only if they agreed to buy other products as well. By the end of it all, 1.5 million people had managed to get their names onto the coveted preorder lists.

However, the majority of these chosen ones would be hugely disappointed. In keeping with their controversial tradition of understocking orders, Nintendo planned to ship only 300,000 units, leaving 80 percent of those with reservations out of luck. If Nintendo had completely had its way, however, they wouldn't have shipped a single unit. Instead, they would have been content to keep selling their 8-bit products. But Sega's 16-bit system, Kalinske's plans for change, and rumblings about a top-secret mascot had forced their hand.

Nevertheless, Nintendo was prepared. They had been working on a 16-bit system of their own since the late eighties, and from a technological standpoint they could have had something ready for market by late 1990. But because they had to move faster than expected, there was one issue that could not be rectified in time: backward compatibility.

When Yamauchi had originally tasked his engineers with building Nintendo's next-generation system, he made several demands related to price, performance, and graphic capabilities. He also insisted that the new hardware should be able to play the old 8-bit software. This was gravely important, because without that compatibility the millions of Nintendo games already purchased would instantly become obsolete, and the parents who had bought those games would become angry and less likely to want to pay for new products from the company that had made them feel that way. The burden of accomplishing all this fell onto the shoulders of engineering wizard Masayuki Uemura and his sixty-five-person team, dubbed R&D 2, who had been responsible for creating the original 8-bit system and the lock-out chip that rejected non-Nintendo games.

Uemura's Super Famicom dazzled on numerous fronts. The new console could generate 32,768 unique colors (the Genesis had 512) and eight channels of audio (the Genesis had six), and it could retail for 25,000 yen (about $250). Yet despite his best efforts, Uemura was unable to incorporate backward compatibility without greatly increasing the price (by about $75). Yamauchi discussed this issue with his son-in-law, Minoru Arakawa, who harbored plans to soon release a U.S. version of the system. Arakawa pointed out that compact discs had recently begun to replace cassette tapes and vinyl records without causing much of a stir. Perhaps modern consumers were becoming savvy enough to realize that new technology tended to make previous iterations obsolete. They concluded that Nintendo was strong enough to deal with the possible backlash and couldn't afford to hold off on a 16-bit system any longer.

To placate the possibility of angry parents, they wanted to make sure they had an "it" game for the new endeavor. Naturally, they decided that their new supersystem deserved a new Super Mario Bros. game. This, however, resulted in another problem. Shigeru Miyamoto, the visionary game designer behind the Mario, Zelda, and Donkey Kong franchises, was still in the process of learning the limits, benefits, and nuances of 16-bit technology when he was asked to rush the completion of his new game, Super Mario 4 (later retitled Super Mario World). He was proud of his latest iteration—the new costumes, the clever foes, and the bright, beautiful new worlds—but the perfectionist within him worried that it felt too similar to the previous Mario games. By this point, however, there was no turning back. Nintendo was moving full steam ahead, ready to enter the battle for 16 bits.

Somewhere in Kyoto, another mighty gust of wind roared past a nondescript warehouse, signaling that sooner or later a storm would be coming.

Two days later, Tom Kalinske sat in his office, reviewing the latest screenshots from Sonic The Hedgehog. The game was coming together beautifully, each level more exciting, energetic, and exotic than the last. It wouldn't be ready for six months, but Kalinske viewed this as a good thing, as Sega could use the time wisely. He and Nilsen had decided that they would not do the expected and hype the hedgehog every step of the way. Instead, they were intent on keeping their secret weapon a secret. There would be no game demos sent out in advance, no early print or television advertisements, and no information given to the press. Then, as the mystery of Sonic spurred curiosity, Sega would "leak" screenshots or character art as the release date approached. Kalinske knew it would be hard to keep Sonic under wraps (especially with Sega of Japan's habit of revealing confidential information toward the end of financial quarters), but this seemed like the best strategy for unleashing the wonder that was Sonic.

"It's here," Toyoda said, poking his head into Kalinske's office.

Kalinske glanced up, noticing that Toyoda looked distinctly excited and nervous. "What's here?" he asked.

"*It,*" Toyoda said, nodding. "It's here."

Seconds later, Kalinske realized what had arrived. "Bring it to the conference room," he said, standing up. "I'll go ahead and gather the troops."

Toyoda nodded briskly and shuffled off. For a moment Kalinske didn't move, embracing the anticipation. After the magnitude of the moment fully hit him, he sprang into action and zigzagged through the office to collect everyone who was interested in seeing Nintendo's Super Famicom. Of the 300,000 consoles that were delivered through Yamauchi's Operation Midnight Shipping, Sega of Japan had somehow managed to get hold of two of them. They kept one for themselves and sent the other to Sega of America so that Kalinske and company could see the face of their enemy.

Sega had been positioning themselves against the NES, but Kalinske had always known that the real war, the one that mattered, would come once Nintendo released their 16-bit system. He didn't yet know when that would occur in America, but if history was any indication, Nintendo of America would probably go to market with its 16-bit system a little less than a year after Nintendo's Japanese parent. If that was Nintendo's strategy here, then their new console would hit stores in the fall of 1991.

Kalinske very much hoped that was the case, as it would play right into his hands. Not only would it give him at least nine months of lead time to counteract Nintendo's haymaker, but it would give Sega enough time to properly tease the public about Sonic, build buzz, and then release him from his cage at a time when he could do the most possible damage. Please, Kalinske thought, please be so foolish as to hold off for a year. Give us more time to learn your weaknesses, more time to acquire space on store shelves, and more time to work our way into the minds of consumers. Of course, all this wishful thinking was predicated on Nintendo's 16-bit system being something Sega could even compete with. Luckily, Kalinske was about to find out once and for all what he was really up against.

"Isn't this exciting?" Schroeder asked, gathering around the table with the rest of her eager coworkers. "I've got butterflies."

"Let's not make too big a deal out of this," Kalinske said, but a moment later he broke into a grin. "Well, now that I've got my cautionary-boss moment out of the way, let's see what we've got."

Almost as if on cue, Toyoda entered the conference room. He was carrying a gray box whose cover featured a colorfully drawn silhouette of the Super Famicom and two controllers.

"Kind of looks like something a kid with chalk might draw at recess," Nilsen said.

Toyoda opened the box and dutifully laid its contents on the table, prompting the Sega employees to breathlessly lean in like doctors carefully observing a cutting-edge surgery. The lively picture on the cover was a sharp contrast to the dull, boxy system itself.

"It is just me," Schroeder asked, "or does anyone else think that if this machine could talk, it would ask to be put out of its misery?"

As everyone chuckled, Kalinske picked up one of the two identical controllers. "These are real nice, though," he said. They were indeed; soft, sleek, and popping with colorful buttons in blue, red, yellow, and green.

"Look, they have a fourth button," Toyoda said, referring to the fact that the Genesis only had three.

Nilsen took the controller from Kalinske. "Wait," he said, feeling grooves atop the controller. "I think they've hid a couple more buttons up here. They've got them hidden everywhere."

"Who cares?" Kalinske said. "Let's get past the superficial and fire this up. Al, will you do the honors?"

"Sure thing," Nilsen said, scooping up the system and bringing it over to the television to hook it up. He was Sega's resident AV guy, often joking that if the whole marketing thing didn't work out, he'd always have a job setting up games, movies, and presentations. "We're all set," Nilsen said, attaching the last wire. "Just need a game."

Kalinske handed Nilsen a gray cartridge containing Super Mario World. But just as Nilsen was about to insert it, he paused dramatically.

"What's wrong?" Kalinske asked. "Another hidden button?"

"No," Nilsen said, holding up the bottom of the cartridge. "Look. There's no way any of the old games will work on this new system."

"Maybe they have some kind of converter," Kalinske suggested.

"Very possible," Rioux said. "Like we have for Master System games."

Nilsen shrugged and inserted the game. "Maybe. But if I had to guess, I'd say that Nintendo just hurt themselves way more than anything we ever could have done to them. Arrogance, after all—" Nilsen cut himself off when the game's title flashed across the screen.

Sega employees silently watched as Nilsen played through the first level. The first thing that struck everyone was that while Sonic was brilliantly fast, Mario was obliviously slow. Sonic was chaotically determined, Mario was peacefully subdued. But, most important, Sonic was something new and Mario was, strangely and amazingly, just more of the same.

As if proving this point, a bubbly jingle played as 16-bit Mario's face turned embarrassingly red and he died onscreen. Nilsen had been unable to guide the titular character away from the wrath of a slow but determined turtle wearing a football helmet.

In the wake of the red plumber's death, Kalinske finally broke the silence. "Hey, Al, do you have any plans for this upcoming weekend?"

"Uh, no. Not yet." Nilsen said, this time successfully avoiding the charging turtle in the football helmet. "Why?"

"How would you like to go on a date?"

Nilsen paused the game. "With . . . ?"

Kalinske pointed to the Super Famicom. "Nintendo, of course. Take this thing home, play through more of the game, and make sure there are no surprises. If the controllers have hidden buttons, who's to say that the game doesn't have a few tricks up its sleeve? What do you say?"

"Sure thing. I'd be happy to put the moves on Mario," Nilsen said, a smile crinkling across his face. "Just make sure this doesn't get back to the princess."

14.

SEGAVILLE

One of Kalinske's greatest assets at Mattel had been a list of industry contacts—a list so extensive that it rivaled the yellow pages. In those days, he didn't just know the top executives at every company along the toy spectrum but also knew which marketing guys they trusted, which salesmen they didn't, and which gabby receptionists didn't realize their gossip revealed vital pieces of information. Simply put, Kalinske had a Rolodex that put other men's to shame. He still had that same Rolodex, but with its majority of contacts being in the toy business, he was now trying to build up a similar resource for videogames.

To create such a network, Kalinske spent his mornings on the phone speaking with industry players, financial analysts, and friends from his past life who might have insight into trends. Normally, he wasn't looking for much, just some game development news, rumors of a hot property, or even a golf date. Today, however, he was expecting to hear the first wave of feedback on Nintendo's new system.

His initial calls went to retailers. They usually had good contacts overseas and, more important, had the most to gain or lose. If Nintendo had put out a great product, they were sure to make another fortune. But if it was lacking, then the retailers would be looking for a plan B, which Sega was more than happy to provide.

Because of the time difference, he started with guys on the East Coast. To his surprise, everybody he called spoke about the Super Famicom as if it were the second coming. Apparently the thing had taken Japan by storm, selling out in a matter of hours. This was an unexpected pie in Kalinske's face, though he was glad to hear that many of the retailers were still secretly rooting for Sega. "Sooner or later, those jerks will get what's coming to them," Tasso Koken said in a huffy New Jersey accent. Koken was a buyer for the Wiz, a Northeast electronics chain that had recently gained traction by virtue of a catchy jingle ("Nobody beats the Wiz!") and betting big on sponsorships with all the New York sports franchises. "Now, I hate to admit that it seems like it'll be later rather than sooner, but trust me. It's going to happen."

"I appreciate the kind words," Kalinske said.

"Not just words," Koken said. "Actions, my friend."

"I like the sound of that. Were you able to get us more shelf space?"

"No. But I got you *better* shelf space. End caps, front-facing. Everything eye level. Cream of the crop."

"Every little bit helps," Kalinske said with a chuckle. "Let me know if you hear anything else about Nintendo."

"Will do. But so far it sounds like the real deal."

Kalinske thanked Koken for the intel and hung up. As he sat at his desk wondering if he was seeing only what he wanted to see, Nilsen tapped on the door and poked his head into Kalinske's office. "Have a second?"

"Only if you played through more of the game," Kalinske said, waving him in.

"More? I played through *all* of the game," Nilsen said with a grim expression on his face. "And so I come bearing bad news."

"How bad?" Kalinske asked, bracing himself.

"Very bad," Nilsen said, his frown morphing into a grin. "For Nintendo."

"Yeah?" Kalinske said, raising an eyebrow.

"Don't get me wrong," Nilsen said. "It's fun, it's finely tuned, and it's expertly leveled. It's just so well crafted from a gameplay standpoint, which is what Nintendo does so well and exactly what I would expect from them."

Kalinske smiled. "But?"

"But I spent the entire weekend playing it, and at no pointed did I feel wowed. Yeah, the graphics were a little better. There were some slight gameplay tweaks, which were nice, but nothing revolutionary."

Kalinske clapped his hands together with excitement.

"Yup, I was doing a lot of that myself," Nilsen said. "The whole time I just kept thinking over and over and over: this is 12-bit Mario. A step in the right direction, but definitely not a leap forward. Nintendo dropped the ball."

"I knew it," Kalinske said. "All morning the retailers have been telling me that they've heard it's the best thing since sliced bread. I admit, I was a little surprised at first. But that's got to be because they haven't seen it firsthand."

Nilsen nodded. "That sounds about right. In the famous words of our esteemed cliché makers, seeing is believing."

"Yes, it is," Kalinske said. "And you, sir, have seen."

"I have. And I'm one hundred percent confident that we have a winner."

Kalinske thought for a minute, letting the golden words echo in his head. Sega had the better product. It was no longer speculation or fantasy but a matter of fact. This was everything Kalinske had hoped for, and now Sega had no excuse not to beat Nintendo. "Okay, we have a winner," he said, but that wasn't enough. There needed to be something more, some way to make this distinction obvious and enticing just like in that Reebok bungee commercial. "I want you to prove it. To me, and to the world."

"How, exactly?"

"I don't know yet. And neither do you," Kalinske said. "But I'm confident that at some point you will and, when that happens, heads will roll."

"All right," Nilsen said, turning to leave. "I'll start strategizing."

"Wait," Kalinske said. "Did you bring it back? The Super Famicom?"

"Yup. It's in my car. You need it?"

"I think so," Kalinske said. "It's time I pay a visit to an old friend."

Now armed with complete confidence in his product and a mandate from Japan to make changes as he saw fit, Kalinske arranged for another meeting with Wal-Mart. This time he traveled down to Arkansas with Toyoda, hoping that a tag team effort would change the electronic merchant's mind once and for all.

"Wipe that smirk off your face, Mr. Kalinske. It's very unbecoming," the man from Wal-Mart said, shaking his head and rolling his eyes as the two men from Sega entered. "I assume this is the esteemed Shinobu Toyoda?"

Toyoda nodded and quietly introduced himself as he and Kalinske took a seat.

"You ought to be taking notes and learning from your colleague," the merchant said to Kalinske. "Guy walks in, doesn't say much, and, most important, no smirk."

"It's not a smirk," Kalinske said. "It's a smile. That's all."

The merchant emitted a guttural sound of skepticism. "I'm dubious. But I invited you back, so I'm willing to play along. What's this about?"

"I came to say that I've seen the future."

"Oh, yeah? Is that right, Nostradamus?" the merchant asked, and then turned to Toyoda, who nodded vigorously in support of Kalinske. "And what exactly did you see?"

"I saw the American release of Nintendo's new system. Sometime just before the Christmas season," Kalinske said, and then donned a look of mock horror. "And people are furious. This new Nintendo costs an arm and a leg and doesn't play any of the old games. That would all be well and good, except that the new system isn't even that much better than the old one. It's all a sham." Kalinske swapped out the look of horror on his face with one of dread. "And Wal-Mart's watching all of this happen. Wishing that there was something that could save their Christmas season. Lo and behold, there is: the Sega Genesis, the world's most advanced videogame system."

The man from Wal-Mart chuckled. "Nice prophecy, Mr. Kalinske. But there's a slight hitch: Nintendo's new system isn't a sham. From what I hear, it's pandemonium in Japan. They sold out in less than twenty-four hours. More than a quarter million units."

"This is true," Toyoda confirmed, leaning forward. "But it is all hype."

"My nonsmirking friend here is correct," Kalinske said. "A big opening weekend for a movie isn't proof that it's any good. It just means they had nice a poster."

"Fair enough. But couldn't we have had this tête-à-tête over the telephone?"

"But then I couldn't have shown you this," Kalinske said, reaching into his travel bag and pulling out the Super Famicom. "You can see for yourself. Is there a television somewhere where we can set this up?"

"Yes, but that's not going to happen."

"What do you mean?" Kalinske said, trying not to sound as surprised as he felt.

"There's nothing you can possibly show me on there that will change the fact that Nintendo has made us a lot of money and will continue to do so. Maybe this 16-bit thing won't be as good as it's cracked up to be. And maybe parents will make a stink. But in the end, they're going to buy it, so it doesn't really matter."

Kalinske and Toyoda didn't believe that, but they realized there was nothing they could do or say to change this man's mind. Seeing *was* believing, but Wal-Mart wasn't even willing to see what was right in front of their face, let alone what was coming down the line. So they finished the meeting on cordial terms and got a cab for the airport.

Moments after getting in the cab, however, Toyoda saw something that caused a jolt from head to toe. "Stop here," he told the driver, only a few blocks from Wal-Mart headquarters. Kalinske was confused, but Toyoda smirked and held up a finger, as if to say, I did not smirk before, but trust me, this will be worth the facial muscles. They paid the driver and got out of the cab.

With a travel bag over his shoulder, Toyoda led Kalinske toward a For Rent sign hanging from the awning of an unoccupied space in a busy strip mall. When they finally reached the storefront, Toyoda didn't need to say a word. Kalinske immediately knew what he was supposed to be looking at, and the possibilities sent his mind into overdrive. The location was perfect: centrally located, close to Wal-Mart headquarters, and right next to a major highway. He stared at the empty store and imagined a barrage of various displays popping up: Sega hardware and software as far as the eye could see. But they wouldn't even be selling these things. No, this would purely be to drum up interest and drive Wal-Mart crazy when potential buyers were unable to get Sega products from their stores. Kalinske and Toyoda stared at the empty store, their smiles widening as their mutual vision expanded in scope and size.

"They will come and play for free," Toyoda said without turning his head.

"Every day of the week," Kalinske added. "For as long as they would like." As if making room for an imagined swarm of visitors, Kalinske backed up a few steps and glanced up. "Look," he said, pointing to a restaurant billboard. Toyoda saw it right away, as Kalinske had before. Sega would cover every inch of the town in ads telling people where they could go to play videogames for free. Billboards, bus stops, park benches.

Kalinske and Toyoda were going to turn Bentonville, Arkansas, into Segaville. They had no idea if the plan would work, but it would certainly force Wal-Mart to take off their blinders and at least look at what Sega wanted them to see.

15.

THE PHYSICIST IS DISPLEASED

"Name?" Nintendo's receptionist asked, looking up at a tall, finely dressed gentleman with short dark blond hair.

"Olaf Olafsson," the man replied. As his copper-colored eyes fixed on the receptionist, a smile slid across one side of his face, while a slight scowl draped across the other. "And I am here to see Mr. Lincoln," Olafsson said in his singsongy style of speech, which bordered on iambic pentameter. "He should, of course, be expecting my visit. Though perhaps I have arrived a tad early." The receptionist confirmed his appointment with Howard Lincoln and asked him to wait in the brightly lit brown and white lobby.

Olafsson walked passed a glass case containing an ornate crystal horse's head and sat down on a brown couch. It had been a long day thus far, and it was nice to get off his feet. He had grown accustomed to spending much of his days in transit, but traveling to Nintendo of America's headquarters in Redmond, Washington, didn't make for the easiest trip. Nevertheless, there was something very endearing about Redmond. Normally, he divided his time between Manhattan, Los Angeles, and the postcard cities of Europe, so it was nice to be somewhere quaint and quiet. Taking a moment to appreciate his surroundings, he leaned back on the couch and reflected on the unusual life trajectory that had led him here.

Olaf Olafsson had been born and raised in Reykjavik, Iceland, where he developed a keen passion for math, science, poetry, and athletics. When he was seventeen years old, he parlayed his exploits as an amateur Icelandic Renaissance man into a scholarship at an American college. Embracing this opportunity to expand his horizons, he left his homeland and emigrated to the United States, where he studied physics at Brandeis University. In 1985, as he neared the end of his undergraduate studies, Olafsson saw his future cleanly divide into two distinct paths: he could continue on this track by getting his master's and PhD and then joust for position in the small but distinguished community of physicists, or he could live an unexpected life filled with many and diverse interests. Though the spirited competition involved in the former option intrigued him, he was leaning toward the latter. His mentor, a physics professor named Stephan Berko, wished to talk him out of abandoning the scientific path, and so set him up to meet a distinguished former student: Michael "Mickey" Schulhof, a high-level executive for Sony's American division. In an unexpected twist, the meeting ended with Olafsson not only retiring from a career in physics but accepting a job at Sony. Schulhof hired the young man from Iceland to introduce a new technology called CD-ROM. Rising to the challenge, Olafsson traveled around the world to demonstrate the amazing audio and visual power of compact discs to tech companies like HP, Apple, and Microsoft.

In early 1991, Olafsson was promoted from introducing hardware to selling software. More specifically, he was named president of a new division called Sony Electronic Publishing, which would be responsible for the production of any digital content for computers, multimedia players, or videogame systems. This included anything from a CD containing the complete works of William Shakespeare to an interactive game based around an upcoming blockbuster film. The primary source of this content would come from a pair of multibillion-dollar entities that Sony had recently acquired: a prestigious film studio (Columbia TriStar Pictures) and a premier record label (CBS Records). As a result, Olafsson's initial job was to familiarize himself with the various branches of Sony's newest assets and figure out how, if at all, they might fit under the new umbrella of electronic publishing. Whether on the set of a new movie or in the recording studio with a popular artist, Olafsson thoroughly enjoyed learning all aspects of the entertainment industry . . . except when it came to videogames.

Personally, he knew very little about videogames. They had not been a part of the social scene back in Iceland, and by the time he got to America he was too old to get sucked into Nintendo-mania. So to understand the industry, he relied on information from Imagesoft, Sony's game publishing imprint, based in Santa Monica. Imagesoft was a third-party licensee of Nintendo and had released two games for the NES: Solstice and Super Dodge Ball. Both had been well regarded by critics and well received by consumers, but neither was particularly profitable. This led Olafsson to believe that for all the talk of videogames becoming big business, there wasn't really much money to be made. To test his hypothesis, he met with the Imagesoft team.

It turned out he was wrong. There was actually a lot of money to be made in videogames; it's just that Nintendo was making most of it. And whatever money they weren't making themselves was, essentially, being allocated to companies they deemed worthy of eating the remaining slices of the pie. Olafsson realized that the secret to Nintendo's power lay in the cunning licensing process they had created, which required would-be business partners to jump through a complicated series of hoops.

Getting into business with Nintendo involved signing their take-it-or-leave-it licensing agreement, which would grant a publisher, like Sony's Imagesoft, the right to produce up to five different games per year. In exchange for this privilege, the publisher would have to purchase the game chips directly from Nintendo, give them the exclusive rights to their titles, pay a substantial royalty on units sold, and agree to a variety of other strict conditions. Olafsson didn't find this agreement to be particularly sporting. Nintendo wanted to exert their leverage? They wanted to gouge developers, producers, and publishers? They wanted to be paid everything in advance, before a single game was ever sold? Fine, Nintendo had earned the right to call the shots. But it was the next few steps in the process that Olafsson found to be particularly startling.

After signing the licensing agreement, a publisher would then invest a great deal of time, money, and energy in producing one of their five titles. When they were finished developing a game, they would submit a completed version to Nintendo's headquarters in Redmond, Washington. At some point in the future (it could take days, weeks, or months) the publisher would find out if their game had been approved. But, as they eventually realized, games were never approved the first time. So Nintendo would fax over a list of changes, and the publisher would redevelop the game to make the necessary alterations. The publisher would once again send the game to Nintendo, who would either approve the revised submission or, more typically, send another fax requesting additional changes. This process was repeated until a game was deemed to be up to Nintendo's standards.

Following the approval, it was finally time for the game to be produced. Naturally, having invested a lot of resources, the publisher would want to order a large quantity of games to increase their profit potential. But because the licensing agreement contractually ensured that Nintendo would be the manufacturer, they would get to determine how many copies would be made. A good rule of thumb was that a publisher would receive about 25 percent of their desired order, though that total fluctuated based entirely on Nintendo's evaluation. Lastly, the publisher would then wait for the game cartridges to be manufactured and delivered. This would generally take a couple of months (a month for production, another for shipping overseas) but, like every other step of the process, there were no guarantees. If dastardly winds knocked a cargo ship off course, delivery could be delayed a week, but if a global chip shortage struck, as Nintendo insisted was the case in 1988, then everything could be delayed indefinitely.

Olafsson was appalled by the absurd dynamics of all this. His company was taking all of the risk, yet Nintendo possessed total control every step of the way and received a large share of the rewards. His dismay turned to disgust upon learning that one of Imagesoft's games, Super Sushi Pinball, had been flat-out rejected and would never be released. He had no idea if this game was any good, nor did he care. His company had spent nearly a million dollars to create this game, and to develop a marketing campaign around the whimsical tagline: "Finally a game that tastes as good as it plays." Given this hefty investment, its success or failure should be determined by consumers, not by Nintendo. That was not right, not by a long shot. So he had made the decision to visit Nintendo's headquarters and try his best to balance the equation.

As Olafsson looked around the lobby, he made sure to extract any emotion from his demeanor. He had not come here to rant, rave, or rage. He was here, quite simply, as a businessman who believed it was in Sony and Nintendo's best interest that a special arrangement be made. After all, the two companies had a rich history together. Sony was the manufacturer of the audio chip for the Super Famicom, and in the coming years that relationship would become even tighter as a result of their joint venture to launch a CD-based game system together. The two companies needed each other, and that should be evident in all facets of their relationships.

"Great to finally meet you," said Howard Lincoln, a tall, serious-looking man with soothing eyes and a long, moon-shaped face. He was Nintendo's senior vice president and had a manner of speaking that reflected the prestige of that title. Powerful, but never pretentious. "Very kind of you to make the trip up here to see us."

Olafsson stood to shake hands. "Ah yes, but of course."

"I know Redmond's a bit out of the way, but it's got a nice charm, no?"

"Absolutely. It's quite a pleasant respite," Olafsson began, deploying just the right amount of untraceable European accent so as to sound worldly but not pretentious, "from the scratching and clawing of busy cities."

"Well put," Lincoln replied. "Why don't I give you a tour of the place and then we can get down to it. How does that sound?"

"Lovely," Olafsson said, waving his hand forward. "You lead, I follow."

Lincoln took Olafsson through the colorful offices, whimsically decorated with game characters. It felt almost like a tiny Nintendo theme park, except for the noise, or lack thereof. In contrast to the vibrant aesthetics of the place, the employees were relatively quiet. They were all certainly quite friendly and accommodating, but there was a silent seriousness to them, both individually and as a group.

After the tour, Lincoln brought Olafsson into a modern, white-walled conference room that was filled with images of Donkey Kong, Nintendo's large, lovable, goofy gorilla. The two men sat down at a long, glossy table and briefly discussed this game, which was one of the few games that Olafsson had played himself.

"I was quite young back then, but I could tell right away that the game would become a hit," Olafsson explained with a nostalgic smile sewn to his face. "I remember being enveloped with this addictive sense of wonder. Which, I must say, is high praise from someone who doesn't much like videogames."

Lincoln chuckled at the memory. "What's this about not liking games?"

"Oh, I don't know," Olafsson said with a shrug. "They're certainly fine. I just think they're a bit of a silly thing. Not so much for me."

"That's understandable," Lincoln said. "But then, I must ask, why be in this business at all?"

"That's a very good and valid question," Olafsson said, pointing playfully at Lincoln. "Though it becomes slightly rhetorical in this audience of two because we both know the answer."

"Because business is business is business?"

"In theory, yes. But under the circumstances that bring me here today, I have to ask if that's really the case. I've read through the licensing agreement, reviewed our contracts, and spoken to my guys about some of their adventures with Nintendo. This business, shall we say, is unlike any other that I am familiar with."

Lincoln nodded, unsurprised and not at all flustered by these concerns. "I'll grant you that. It's a new industry."

"One with a promising road ahead."

"Absolutely," Lincoln replied. "But ten years ago they were saying the very same thing: 'Videogames are a can't-miss!' That, however, didn't turn out to be quite right. A lack of quality control led to chaos and the whole thing fell apart. Look, I know what people say about Nintendo, and what publishers, retailers, and everyone else thinks about our 'strict' agreements." Lincoln paused to look Olafsson in the eye, trying to show him that this wasn't just lip service but sincere understanding from a rational individual. "But there really is a method to the madness. From idea to purchase, it's our job to make sure that a product lives up to Nintendo's standards in order to create a premium entertainment experience. And if we can do that, then consumers, retailers, and everyone else along the food chain gets fed."

Olafsson nodded sympathetically. "You're obviously an intelligent man. That being the case, I think you can imagine where I might attempt to point out a few imbalances in that logic. Nevertheless, I acknowledge that we're on different sides of the table here and, naturally, ought to possess divergent perspectives on issues like these. "*But,*" Olafsson continued, sharpening his tone. "If Nintendo gets to make the rules, then, given the company's long-standing relationship with Sony, I would think that some leeway should be in order."

"In what sense, exactly?"

Olafsson squinted, choosing his words. "In a very general sense, Mr. Lincoln. But I would venture to say that a good start might be for Nintendo to stop treating us like we're nothing but slaves on the plantation."

Lincoln recoiled at the comparison. "That's a bit drastic, wouldn't you say?"

"My metaphor? In tone, perhaps. But in sentiment, it feels spot-on."

Lincoln shook his head. "I think we're getting off course. And in the spirit of avoiding any further derailment, I'll just say the following: Nintendo values its relationship with Sony. But our licensing agreement for making games is the same with every company and there are no exceptions."

"I see," Olafsson said, realizing there was no use to any of this. If this was how Nintendo wanted to play the game, then this was how it would have to be. As he sat there, he had many thoughts about this peculiar new industry, many thoughts indeed. But regardless of the shape, size, and color of these thoughts, he was very much displeased. At this juncture, however, there was no value in showing off his displeasure. Instead, Olafsson asked his host to tell him about some of the great promotions that Nintendo had planned for the coming year.

"Excellent question," Lincoln replied, and then launched into an upbeat monologue that he must have already given many times. As Lincoln talked about Mario's new pet, some kind of green dinosaur with an off-putting name, Olafsson began to zone out and look for new ways to play the hand that he'd been dealt. In the middle of his private brainstorm, his eyes grazed one of the paintings on the wall. It was an image of Donkey Kong readying to throw a barrel that he held over his head. Olafsson returned his eyes to Lincoln, but his mind was stuck on Donkey Kong.

When they had first entered the room, the image had brought back to Olafsson the visceral memory of playing Donkey Kong, but now he was beginning to remember the details of the game. If memory served correctly, Donkey Kong was the story of an arrogant gorilla who enjoyed beating his chest and throwing dangerous obstacles in the way of a courageous little red plumber. Because it was unlike any other game at the time, Donkey Kong was exceptionally difficult to beat. But if a player inserted enough quarters, put in the necessary time, and studied the patterns of the game, it eventually became beatable. And as Lincoln continued to speak about Nintendo's future, Olafsson couldn't help but think about what lay ahead for the eight-hundred-pound gorilla in the room.

16.

ROPE-A-DOPE

"Whenever you're at war, you must hit the other guy in the mouth as hard as you can with the first punch," Rioux explained to Kalinske and Toyoda as they plotted Sega's strategy for the upcoming Consumer Electronics Show. "And if you can't hit 'em hard, you might as well not even fight. That's the attitude in real war, and that ought to be our attitude here as well."

Every year there were two Consumer Electronics Shows (CES): the winter show in Vegas and the summer show in Chicago. In the fragmented, ever-evolving videogame industry, which had more in common with the Wild West than, say, with Silicon Valley or Wall Street, the CES provided one of the few opportunities for all the players to come together. The events tended to be little more than each company pounding its own chest, but they were still a big deal. It was where hype began, reputations were made, and scuffles occasionally took place.

This would be Kalinske's first CES, and he wanted to come out with guns blazing, especially in light of Sega's progress (or lack thereof) with Wal-Mart. It had been nearly a month since Sega rented the retail space in Arkansas, and Wal-Mart hadn't even flinched. Instead of relenting, however, Kalinske doubled down. As per the fantasy he had envisioned with Toyoda, they sent his executive assistant, Deb Hart, down to Bentonville with the mission of bringing Segaville to life. She bought up every available billboard in town, handed out flyers on the streets, and arranged for every seat cushion at the University of Arkansas football team's final home game of the season to bear Sega's logo. Hart had done an incredible job, but all of this was becoming a costly gamble. With no results to show thus far, Kalinske was tempted to compensate by making a splash at CES, but he was compelled by the point that Rioux had made. "If we're being honest with ourselves," Kalinske said, "we just don't have the firepower yet to start going at Nintendo. So perhaps a little rope-a-dope is in order?"

Rioux nodded, but Toyoda didn't understand the reference. So Kalinske explained how in 1974 Muhammad Ali squared off against George Foreman for a title fight in Zaire that had been dubbed "The Rumble in the Jungle." At the time, Foreman was bigger and stronger and packed a more powerful punch than Ali, who realized his only hope was to find a clever solution. Ali's plan was to spend the early rounds getting physically abused until finally Foreman tired himself out. For most of the fight, Ali verbally taunted Foreman and then stood in a protective stance, absorbing punch after punch. For four rounds Foreman dominated, but eventually he started to grow tired. When the fifth round began, Ali took advantage of his weakened opponent and demolished him with a series of jabs. Three rounds later, Ali knocked out Foreman, regained the title belt, and went down in history as a tactical genius.

"Ah," Toyoda said. "Like playing the possum?"

"Exactly," Kalinske confirmed. "I'm not saying we take a dive, but we want to go at Nintendo right before they launch in America. There's no point in building momentum now only to lose it. We should tease with Sonic, but show no gameplay, go heavy on Game Gear, and hold back on the rest until this summer."

Kalinske looked to Toyoda, who nodded, and then to Rioux, whose face always seemed to be made of stone. "Paul?"

"What can I say?" Rioux asked rhetorically, as a grin chiseled across his face. "Float like a butterfly and sting like a bee."

A few weeks later, Kalinske checked into his room at the Alexis Park Hotel in Las Vegas, where he and about twenty Sega employees were staying for CES. Following the decision to play things close to the vest, he had arrived in Vegas with low expectations. Even so, it didn't take long for things to spin out of control.

Sega began the week by hosting a preshow sales meeting for retailers where nothing at all seemed to work. Demos didn't play, art designs got mixed up, and the AV system kept breaking down (not even Nilsen could fix it). Kalinske had known that their strategy for CES would make things challenging, but this felt like amateur hour. He had never experienced anything like this at Mattel, and worst of all, this failure was on his shoulders. If Sega's people weren't properly prepared, then it was because he had failed to prepare them. As president and CEO, he would always get more credit than he deserved, and even more so always get more of the blame. That's just how it worked.

In the face of calamity, he tried to rescue Sega's reputation at the podium. He didn't get the overwhelming reaction he was used to, and on one occasion he was even tempted to whip out the demo of Sonic and talk about how things would be changing, as outlined in his "Four-Point Plan." But as temping as it was to push the big red panic button, he knew that was not really an option.

Following the preshow miscues, the first day of CES didn't provide much of an improvement. While televisions, stereos, and VCRs were magnificently displayed in the dazzling Las Vegas Convention Center, the videogame companies were treated like second-class citizens and relegated to a tent outside. Inside of it, there were about a hundred different videogame companies who had booths to show off their products. They ranged in size from tiny to Nintendo, who clearly dominated the festivities with an upscale display of games and peripherals that was so large there was enough room to comfortably fit a stage in the middle. By the end of the show, attendees had taken to calling the Nintendo display the "Death Star."

Because of the decision to hold back on hyping some of their upcoming games until the summer, Sega's modest booth focused primarily on the Game Gear, a slightly sleeker, Americanized version of the handheld device from Japan. Kalinske, Rioux, and Toyoda all liked putting the spotlight on Game Gear, though each for different reasons. Kalinske figured this would make for a good opportunity to temporarily deemphasize the Genesis but still manage to further define Sega as edgy, unconventional, and technologically advanced. Rioux thought very highly of portable game systems and believed deep down that this business would one day overtake home consoles. And Toyoda knew that the Japan executives loved showing off hardware more than software, so they would be happy to see so much attention paid to Game Gear.

At its core, Game Gear appeared to be an easy sell: it was the color version of Nintendo's Game Boy. Color TV trounced black-and-white broadcasts, so Game Gear should quickly take its place atop the handheld food chain. While on paper that was true, there were a few hitches to this thinking: Atari's recently released color handheld, Lynx, had gotten crushed; the battery life on Game Gear was lousy; and their best game for it was an obvious and inferior Tetris ripoff. Beyond Game Gear, Sega's only other gaudy product was the Joe Montana football game that had been rescued by EA. Although they hadn't been able to get it ready in time for the Christmas season, it had actually turned out to be a pretty great game and would at least give Sega some momentum heading into the new year (that is, as long as consumers didn't realize it was basically just a repackaged version of EA's John Madden Football).

While Sega's employees put the final touches on their booth, Kalinske excused himself to catch an 8:00 a.m. press conference by Nintendo. Though CES wasn't yet open to the public, the press conference was heavily attended, partly because the speech would serve as the unofficial kickoff to the show, but primarily because anything Nintendo said or did was a huge deal. As a result, Kalinske had to stand in the back of the room, behind journalists, financial analysts, and Nintendo's many fans within the electronics industry.

He tried not to admit it to himself, but he missed the whirlwind atmosphere that was so evident in Nintendo's press conference and which he remembered from his time at Mattel: the thrill of anticipation, the curiosity in the air, and of course the applause, which came quickly and reached thunderous levels when Peter Main, Nintendo of America's VP of sales and marketing, stepped to the podium. Main, a balding man with bewitching brown eyes and round, Lennon-like spectacles, nonchalantly quieted the crowd and introduced himself. "There are many great things on the horizon that I'm excited to tell you about," Main said, jumping right in. Like Kalinske, Main was an excellent speaker, but in a very different kind of way. When Kalinske spoke, it was as though he transported audiences into a locker room for a pep talk, and when Main spoke, it was as though he transported audiences into a pub for a quick swig. They were the coach and the bartender.

After touting the Super Famicom's success in Japan, Main nonchalantly answered the question on everyone's mind by announcing that Nintendo would release a 16-bit system in America toward the end of the year. He then said that Nintendo had posted another record year in 1990; with 7.2 million Nintendo Entertainment Systems sold and millions of software titles, sales in the United States topped $3.4 billion. Yet despite the staggering figures, Main acknowledged that Nintendo had fallen short of analysts' expectations. "We weren't far off the mark," he explained. "What none of us could forecast back in June was the war in the Gulf, the economic volatility resulting from the current recession, and the combined impact of these two external forces." Kalinske took a moment to savor the irony in Main's blaming the conflict in the Middle East for disappointing sales figures, given that journalists had started referring to the Gulf War as the "first Nintendo war" for its videogame-like coverage.

Main nimbly navigated through the financial data and then spoke about Nintendo's bold plans to increase their already large pop-culture footprint. The company's after-school cartoon had been so gigantically successfully (with over forty million viewers per week) that Nintendo was already in development to make a big-budget Super Mario Bros. movie, slated for 1992. After Main finished speaking about Nintendo's unyielding expansion, he segued into answering questions from the audience. One of the first asked if Nintendo was trying to become the next Disney, and if they would soon have their own line of theme parks. Peter Main, who believed that the sky was the limit, didn't dismiss this possibility. "The value of characters like Mario is very strong," he said. "And as we go down the road, you're going to see many applications."

Kalinske, who couldn't help imagining his daughters begging him to take them to Nintendoland, had heard enough. He left the press conference and, after giving a pep talk to the Sega employees, spent the rest of the day wheeling and dealing at various booths across the show. When he had a free moment he'd watch visitors roam through Sega's home base. Each time someone snickered at their underwhelming roster of games, shrugged, or referred to Columns as a "retarded version of Tetris," Kalinske felt like he'd been punched in the gut. But by the end of the day he was still standing tall and ready to fight—just like Muhammad Ali.

That night, Nintendo relished the role of top gun and threw a glitzy party headlined by a performance from singer Kenny Loggins. After a lavish evening of dancing and drinking, Peter Main once again took the stage to address an eager audience. This time, though, in the comfort of being among colleagues, his words were as smooth, sleek, and casual as the black silk Nintendo jacket he donned for the occasion. Main was joined onstage by Howard Lincoln, who festively wore a spike collar and neon glow sticks around his neck. Celebratory glitter coated the heads of both men. They took turns speaking into the microphone to express gratitude, make jokes, and give away prizes (including a Chevrolet Geo to the winner of Nintendo's Campus Challenge). Afterward they called up to the stage their boss, Minoru Arakawa, whose fashionably teased hair and oversized fluorescent glasses signified that his typically reserved demeanor was on hold for the evening. As the DJ played the Hollywood Argyles' hit "Alley Oop," Main, Lincoln, and Arakawa launched into a song and dance. The roaring applause for these three giddy, gleeful, and uncoordinated middle-aged men was proof of what everyone in the room already believed: Nintendo was invincible.

Meanwhile, down the road, Kalinske took his team out for dinner and drinks at an Italian restaurant in a strip mall near their hotel. Sensing a bit of disappointment amongst the troops, he tried his best to play the role of Mr. Bright Side. He lifted a cheap glass of pinot noir and addressed the two dozen Sega employees squeezed around a pair of large, circular tables in the back of the restaurant. "Ladies and gentlemen of Sega," Kalinske said, "I want to personally thank everyone for all your hard work. And I don't just mean for this week, but for all the months of effort that have gone into molding Sega into what we are today."

"And what are we?" an employee heckled. "Nintendo's chew toy?"

The room filled with good-natured laughter, Kalinske's included. "Nah, too ambitious," he said. "Dogs actually acknowledge their chew toys." More laughter ensued. "Hey, I will be the first to admit that this wasn't Sega's finest hour, but let me tell you, our time is coming. And it's right around the corner. Six months from now, at summer CES, Nintendo won't know what hit them." A flood of applause swept the tables. "On that note, I wanted to inform you all that I managed to catch part of Peter Main's speech today, and he said two things that got me really excited. The first was that Nintendo had another record year of sales in 1990 but had fallen just short of expectations. Can you believe that? They had a record year, doing over $3 *billion* in sales, and that wasn't good enough!" Kalinske went on to explain that Nintendo was a victim of the worst enemy of all: high expectations. This was a burden Sega didn't have to carry. They were underdogs through and through, and this was their greatest advantage. "We have nothing to lose," he said. "And that's how we're going to win."

After a chorus of cheers, Nilsen asked about the other thing Main had said.

"Oh, right. The other thing is that Nintendo is planning a big-budget feature-length movie around the Mario Brothers. And everyone knows that when Hollywood gets involved in anything, things get needlessly complicated," Kalinske said with a smirk. "Anyway, my hand is getting tired of holding this wineglass, which means that I've been going on for way too long. So let me just say congratulations to all of us for surviving the Consumer Electronics Show. Cheers!"

Glasses of cheap red wine clinked and a sense of camaraderie permeated the room. As the first course was served, Kalinske regaled his employees with old stories of Mattel in a wistful, fairy-tale tone. Rioux, having been there, jumped in from time to time, while Toyoda wore a smile on his face and occasionally shook his head in disbelief.

At the other end of the table, Al Nilsen was sitting between Hugh Bowen and Ed Annunziata.

"Did you run that idea by Al?" Bowen, on Nilsen's right, said to Annunziata.

"No, no, he won't like it," Annunziata, on Nilsen's left, replied.

"Tell him, tell him," Bowen said, egging him on.

"Nah, man, I'm telling you, he won't like it," Annunziata replied.

Nilsen realized right away that this conversation had been rehearsed and that he was being set up. But the pasta primavera looked good, and it occurred to him that the only way he'd be able to enjoy his meal in peace was to let Tweedledum and Tweedledee finish their skit. "All right, Ed. Tell me all about whatever it is that I won't like."

"Okay, so it's like this . . ." Annunziata started, his eyes shining with excitement. Ed Annunziata was a self-taught programmer from New York whose laid-back demeanor and free-flowing vibe made him feel at home when he moved to California. He had been hired by Sega's head of product development Ken Balthaser in 1990 to become the first producer at Sega of America. For the most part, his job entailed "localizing" games from SOJ, meaning that he slightly altered titles like Ghouls 'N Ghosts and Phantasy Star 2 so they could be understood and enjoyed better by Western audiences. But he hadn't joined Sega just to be SOJ's errand boy and was finally starting to produce his own games. At the moment he was working on Spider-Man vs. the Kingpin and was looking forward to working on original projects as SOA gained more autonomy. "I got this idea for a game, but it's unlike anything you've ever seen before."

"He's not kidding," Bowen echoed. "It blew my mind."

"Forget about Mario and Sonic. Forget about saving the princess and racing through levels to stop the bad guy. This isn't about good or evil. This is about life. Not life as we know it, because we all play that game every day, but life beneath the surface of the ocean. Infinitely long and infinitely deep, where beauty meets danger, and the everyday isn't corrupted by the nonsense of words. This is the world's last great unknown, but not for much longer. Because we're going to let people pick up a controller and transform into a dolphin."

For the next forty-five minutes, Annunziata told the incredible tale of a dolphin who gets caught in a storm and loses contact with the other members of his pod. With nothing but the power of his sonar, he has only one chance of reconnecting with his beloved pod: by going on a quest across the ocean. He'll need to trek to the Arctic and find a revered whale, swim into a deep cavern and meet the oldest creature on earth, and eventually find his way to the lost city of Atlantis.

When he finished, Nilsen was at a rare loss for words. "Wow."

"I told you!" Bowen said.

"Just . . . wow," Nilsen said again. "How did you come up with that?"

"An artist never reveals his secrets," Annunziata said before chugging the rest of his glass of wine. "But I'm getting nice and drunk, so all such rules go out the window. It's just been, like, growing in my mind for months. It started when I read this great book called *The Founding*, which was from the POV of a humpback whale. Then I got into reading all this stuff from John Lilly about taking LSD and going into a sensory deprivation tank. The guy spent his entire life trying to communicate with dolphins. And then one day I just asked myself: how can I translate all that into a side-scrolling game?"

"Let's do it," Nilsen said, prompting Bowen and Annunziata to nearly collapse from shock. Nilsen knew this was exactly the kind of risk that Kalinske had been talking about, an opportunity that would redefine Sega. Though nothing like this had ever been done before, Nilsen didn't even have a doubt. He reasoned that if Annunziata could pull off even half of what he described, Sega would still have an incredible game, something special that could spawn a whole new genre while helping to differentiate Sega from the companies flooding the market with me-too games.

And as the evening wore on, any remaining seeds of doubt from the Consumer Electronics Show were washed away by a mighty river of wine. Kalinske watched with pride as his employees spent the long evening sharing stories that had been locked away in the attics of memory, hopes and dreams from the past, and hopes and dreams for the future. For the rest of the week, Sega's team of dreamers would endure every punch that the industry had to throw. But six months later, at the summer Consumer Electronics Show in Chicago, it would be a completely different type of fight.

17.

SHOWDOWN

It really was quite lovely, this Super Nintendo device. And durable too, thought Olaf Olafsson as his fingers danced atop the machine's pale gray exterior. Of the hundreds of people who had shown up for the great unveiling of the Super Nintendo at the 1991 summer CES at Chicago's McCormick Place, it was safe to say that Olafsson was the only one pedantically focused on aesthetics. The rest were there to see the games, the graphics, and the next five years of their lives begin to take shape. That's why, by eight in the morning, hundreds had already arrived and thousands more were on their way. The wait was finally over.

Having satisfied his curiosity, Olafsson moved through Nintendo's massive booth. No, "booth" wasn't the proper word. Their display was more like a towering black fortress that had erupted out of the gray-carpeted floor; at least five times the size and twice the height of any other videogame company's booth. The height not only enhanced Nintendo's prominence but also blocked out the sharp halogen lighting above. As a result, the shade, the black color scheme, and the dark floor all served to masterfully highlight the bright, bouncing colors of their videogames.

It struck Olafsson as both odd and impressive that Nintendo had so ably managed to balance the dark with the light. Not just today, but in general. To suit their needs, Nintendo expertly toggled between coming across as a fun-loving toy company and presenting themselves as a gravely serious tech firm. While it was hardly unusual for an organization to wear two faces, it was rare for one to go so long without having to choose one path or another. But then again, there was no competitor out there who could force Nintendo's hand.

Olafsson checked his watch and realized it was time to go. Nintendo's press conference would begin shortly, and it was important that he be there. He adjusted his tie and moved against a strong current of smiling faces funneling in. Normally he was immune to such infectious excitement, but the grin on his face revealed that today was a special occasion.

Less than twenty-four hours earlier, Olafsson had given a press conference of his own, in which he announced that Sony would be getting into the hardware business. The journalists were immediately intrigued by the prospect of Sony competing against Nintendo—two giant Japanese firms going head-to-head. He could already imagine the overly dramatic articles that would be published, filled with comparisons to battles between Godzilla and Mothra. But actually it was the opposite: Sony was going into business with Nintendo.

In late 1992, Sony would be releasing the Nintendo PlayStation, which would be a peripheral device that attached to the Super Nintendo and play games on CD. At this time, it was generally understood by both experts and laypeople that CDs would soon become the standard delivery mechanism for all entertainment: music, movies, and videogames. It just made too much sense. A CD could hold ten times as much information as a 16-bit game cartridge at one-tenth of the price. Perhaps there was a whimsical charm to game cartridges, but this was a matter of technological Darwinism. And Sony was thrilled to be evolving with Nintendo.

The alliance was fantastic in so many ways. Working with Nintendo gave Sony clout in the videogame space. The creative relationship would ensure that the PlayStation had top-level games (which was especially key because Sony's software publisher, Imagesoft, was still struggling on that front). And last but certainly not least, this could be a financial windfall. Not only would Sony make money on each PlayStation sold, but just like Nintendo (and Sega) they would get to play the role of toll collector and collect fees from software companies who wanted to create games for their CD-shaped roads.

Olafsson's announcement yesterday had created a stir, but not nearly the whirlwind that he anticipated after Nintendo did the same at their press conference. After all, a proclamation from the king carried more weight than one from the noble prince. And as Olafsson took a seat in the front row of the conference room at McCormick Place, he eagerly awaited being handed the keys to the kingdom.

As per usual, Nintendo's press conference was packed. This one, however, carried an extra flair of excitement and felt like the beginning of a grand new era. Olafsson was unsure who the speaker would be. Typically, that honor would go to the president, in this case Minoru Arakawa, but he didn't enjoy speaking in public. Instead, Nintendo's official proclamations were made by either Peter Main or Howard Lincoln. Olafsson was curious how they decided who would speak at which occasion, and he made a mental note to explore this in future meetings.

At 9:00 a.m. Lincoln stepped to the podium. He welcomed the audience, invited them to check out Nintendo's booth after the speech, and then spoke at length about the Super NES, which would hit stores on August 23, 1991. All systems would come with the groundbreaking new Super Mario World game, while four others would immediately be available for purchase: F-Zero, Pilotwings, Gradius III, and SimCity. Their library would quickly grow, with eighteen games available by Christmas.

Lincoln confirmed that, as with the Super Famicom, there would be no backward compatibility and the 16-bit Super NES couldn't play 8-bit NES games. Sensing a degree of dissatisfaction, he quickly assured everyone that Nintendo was still very much committed to supporting the 8-bit system. He expected at least forty new 8-bit games in the second half of 1991, though obviously the company's focus would shift to the more advanced Super Nintendo. Based on the perpetual sold-out situation in Japan, Nintendo had nothing but the highest of expectations. They estimated selling two million units by the end of the year, and already anticipated shortages during the Christmas season. Blah, blah, blah, Olafsson thought. Hand out a memo, smile for the camera, and let's move on to more important matters. Everyone already knows what you're about to say: Sony + Nintendo = CD-ROMance.

"Compact discs will play a key role in Nintendo's vision for the future," Lincoln finally announced, now ready to reveal the plans for Nintendo's new CD unit. Olafsson stirred in his seat as the crowning moment inched closer.

"And who better to partner with than the company that invented the audio compact disc: Philips Electronics."

Wait, what? A tremor of shock and confusion swept through the room as journalists raced to take note that Lincoln had said Philips and not Sony. After Lincoln said it again, confirming that his words were not a slip-up, all eyes turned to Olaf Olafsson, who tilted his head and furrowed his brow. Was he shocked, appalled, furious?

In truth, he was none of those things. He was merely plotting his next move.

"Did he just say Philips?" Nilsen whispered to Kalinske as both men stood in the back of the room. "As in *not* Sony?"

"It would appear that way," Kalinske whispered back, speaking a little louder than he intended due to the unexpected smile on his face.

Kalinske kept his eyes fixed on Olafsson. He didn't personally know Sony's president of electronic publishing, but he was impressed by the man's reaction, which showed class, tact, and diplomacy. In fact, by the time Lincoln concluded his press conference, Kalinske could have sworn he detected a hint of amusement in Olafsson. He wanted to go over and introduce himself, but realized that it wasn't the right time. Reporters were swarming around Olafsson, looking for an emotionally charged quote, and Kalinske and Nilsen should have returned to Sega's booth fifteen minutes ago.

What neither of them knew was that only a few weeks earlier, Arakawa and Lincoln had flown to the Philips world headquarters in Eindhoven, Netherlands, for a meeting with Gaston Bastiaens, head of the Compact Disc Interactive (CD-I) group. The Nintendo executives had traveled there at the behest of Yamauchi, who was growing concerned about an alliance with Sony. He had realized that the deal he signed in 1988 gave Sony the right to control software for a joint CD venture. This detail hadn't appeared to be a problem back then, when Sony was exclusively a consumer electronics company, makers of televisions, stereos, and music devices like the Walkman and MiniDisc player. But with their acquisition of CBS Records and Columbia/TriStar films and now the creation of an electronic publishing group, they were growing too ambitious for Yamauchi's tastes. Sony was already the supplier of the key audio chip in Nintendo's 16-bit console, and he didn't want to enter any alliance that would grant Sony any additional power.

This change of heart led Nintendo to go behind Sony's back and sign a deal with Philips. As per their arrangement, Philips would create a CD-ROM drive that hooked up to the Super Nintendo to play games on CD. Additionally, the CD titles that Nintendo created would be compatible with Philips's CD-I players. Naturally, Nintendo would control the licensing rights to all CD games regardless of which system they wound up being used on. Because Japanese contracts tended to be succinct, with a large reliance on good faith, Yamauchi felt that he could break the contract and ditch Sony without penalty. He also decided not to inform Sony of his side deal with Philips, to ensure that Sony's humiliation was made public.

Although nearly every journalist was now locked on Olafsson, a reporter from *Fortune* swam against the current and ambled over to Kalinske and Nilsen in the back of the room. "How about this for a headline?" the reporter asked by way of an introduction. "Sega Execs Crash Nintendo Press Conference: Leave Cowering in Fear."

"Well, well, well," Nilsen said. "If it isn't Nintendo's biggest fan?"

The reporter blushed. "Nonsense. I worship at the altar of journalistic integrity."

"Well, what can we help you with?" Kalinske asked.

The reporter looked at Nilsen with a glint of gloat in his eyes. "The Super Nintendo features 32,768 colors, 256 of which can appear on-screen at the same time, eight highly sophisticated sound channels, and a clock rate of 3.58 megahertz. How does Sega plan to compete with this?"

Kalinske raised an eyebrow. "Journalistic integrity, eh?"

"Hey," the reporter said with a smirk, "these are just the facts."

"Follow us," Nilsen said.

The reporter reluctantly followed Kalinske and Nilsen under a golden "Seal of Quality" and into Sega's booth. Like Nintendo, their color scheme was black on black, but that was all they had in common. Sega's setup, bathed in sunlight, exuded an elegantly zany joie de vivre. There were bright colors blinking everywhere, upbeat music throughout, and a giant blue hedgehog standing by the entrance to greet guests as they entered. Sega did a good job of positioning themselves as the offbeat alternative to Nintendo's autocratic reign, but that alone didn't make their booth spectacular. Any mutt trying to out-bark the top dog would have done the same. What Sega did that no other company would have dared to do was acknowledge that it was a dog-eat-dog world.

At the heart of their booth was a television displaying highlights from Super Mario World. Directly below it was a television showing off Sonic The Hedgehog. In an industry where Nintendo coated the ground in eggshells and cautioned all to walk slowly, Sega was going head-to-head at full speed. The differences between the two games were self-evident: Sonic ran laps around Mario. The Super Nintendo was still three months away from being released, and already it looked extinct.

"Nintendo may have 32,768 colors," Nilsen said to the now speechless reporter, "but I think it's safe to say that Mario literally pales in comparison."

After considering several creative ways to meet Kalinske's challenge to prove that Sega had the better game, Nilsen had hired a team of researchers and set up play tests around the country, where boys and girls were brought in to play Super Mario World and Sonic The Hedgehog and decide which was the better game. Since Sonic had been kept under wraps and the Super Nintendo had not yet been released, none of those involved in the experiment had had any experience with either game, though of course most already knew of Mario from previous games. And that's exactly what Nilsen wanted. He specified to the researchers that 90 percent of the subjects should be NES owners and at least 75 percent had to consider one of the Mario titles to be their favorite game. He wanted to stack the deck so that the results of these experiments didn't merely tell him what he wanted to hear but would prove to Kalinske that, beyond any shadow of a doubt, they had a winner. By the end, 80 percent chose Sonic.

"So what do you think?" Kalinske asked.

"It's great," the reporter conceded. "But devil's advocate: it's just one game."

"So young, so naive," Nilsen said, walking him through the rest of the booth. Sonic was certainly their best game, but it was only one of many. The various displays showed off other titles that would be released later that year, like Mario Lemieux Hockey, ToeJam and Earl, and Quackshot Starring Donald Duck. Sega's already bountiful library of games appeared even more impressive beside those from Electronic Arts, which had been churning out hit after hit exclusively for Sega. The unusual romance between EA and Sega turned out to be a godsend for both companies; amazingly, Joe Montana Football had not only wound up being a megahit, but was in some weeks even outselling the Madden game it had been derived from.

"But don't just take my word for it," Nilsen said, pulling out a brochure called *The Nintendo Guide to Choosing a Videogame System*. "Listen to my very intelligent competitor, who specifically suggests buying the console that has the most games. So I encourage everyone to take Nintendo's advice and buy the Sega Genesis."

Over the next few hours, word began to spread. As Nilsen read passages from Nintendo's brochure with near-religious conviction, visitors came from across the trade floor to witness Sega's coronation. They wanted to see the charming blue hedgehog, get a look at the 16-bit system that cost $50 less than the Super NES, and see what else was on the horizon from the company that had stolen at least some of Nintendo's thunder.

Like a baseball team whose pitcher was throwing a no-hitter, the Sega employees tried not to talk about what they were witnessing, to keep it all business as usual. But they had waited so long for this to happen, never being sure that it actually would, that many were guilty of a private smile, fist pump, or high-five from time to time.

It was Kalinske's job to act like this was exactly what he had always expected, but even he couldn't hide the hints of laughter on his face. Feeling good, he walked tall to the reception area at the center of Nintendo's booth, where he asked to speak with Arakawa. Not to gloat, but to show his opponent that he meant no disrespect. Unfortunately, he was deprived of a friendly all's-fair-in-love-and-war conversation due to Arakawa's busy schedule.

"Is he available sometime later today?" Kalinske asked hopefully.

"Um, no, he's booked solid," the receptionist said, knowing exactly who he was. "Sorry."

Kalinske tried for the next day, the day after, two months from now in Seattle. No, sorry, Mr. Arakawa was a very busy man. Kalinske nodded, realizing who and what he was actually up against. He thanked the receptionist and looked around Nintendo's colossal booth, scanning for Arakawa. He had to be there somewhere, lurking amongst the consumers who had excitedly come out to get their first look the perhaps-not-so-super Super Nintendo.

Eventually Kalinske gave up and walked away. As he moved past Nintendo's already obsolete monument to itself, he had the wonderful revelation that Sega actually, legitimately, inconceivably stood a fighting chance. Hide behind those consumers while you can, Kalinske thought, because sooner or later I'm going to steal them all.

18.

THE UNDERDOG DAYS OF SUMMER

"I'd love to sit here and promise you the world," Kalinske said, addressing a conference room full of employees readying for battle. "Because, in my opinion, that's what you each deserve for all the hard work you've put in. But the truth is that I can't even promise you that a year from now Sega is still going to be making consoles."

Now that Nintendo's 16-bit machine had a release date (August 23, 1991), Sega was preparing to make the most of every minute before then. During this critical period, which they called the "Sixteen Weeks of Summer," Kaliske, Rioux, and Toyoda would authorize a series of unorthodox hirings, promotions, and marketing strategies to blunt the impact of the SNES.

"Unfortunately," Kalinske continued, "there's not much I can promise in this room today. But I'll tell you one thing I know for sure: we have this summer to give ourselves a chance at actually competing against Nintendo, and I can't even begin to imagine what this group would be capable of accomplishing with such an opportunity." Kalinske paused and in one swift look around the room seemed to make individual eye contact with everyone. It was time for the summer games to begin.

Week 1: Radio Killed the Video(game) Star

Nintendo's Peter Main and Bill White ushered in the warm weather by announcing a three-month $25 million marketing blitz to promote the upcoming SNES. Because Sega's advertising budget for the entire year was less than what Nintendo had allocated for a single quarter, Sega's strikes had to be more strategic, and their missiles more heat-seeking. In this case, the heat that Sega sought was the older, wiser, and more wiseass-ey crowd of teens, college kids, and rebellious adults. As Sega continued to define its image, this demographic was no longer simply an economic necessity but had become an audience that helped sell a narrative vision for Sega: a technologically superior company whose advanced, offbeat products could be appreciated only by those mature enough to handle its power. Sega aspired to be not just the name on a product but the secretly whispered password of coconspirators involved in a revolution.

To reach this audience, and to do so with a fraction of Nintendo's budget, Sega kicked off the summer with a pair of ambitious marketing campaigns in June. The first was "Graduate to Genesis," which aimed to hammer home the concept of Sega representing the next phase of videogame evolution while also catering to a time of year filled with graduation ceremonies. Though "Graduate to Genesis" presented another opportunity to define Sega, the real goal was to further cut into Nintendo's hold over third-party software developers. The promo offered a free third-party title (produced by one of nine companies including EA and Namco) upon the purchase of a Genesis. By doing this, Sega was able to reward the third parties who had taken a chance on them and demonstrate to other game makers that perhaps it was time to consider defecting from Nintendo.

With a campaign in place that aimed to chisel away at Nintendo's strength, Kalinske wanted another that exploited their weaknesses. One place where Nintendo had little presence was on the radio, so that's where Sega struck next. Nilsen put together a list of radio stations that best exemplified Sega's desired cool, in-your-face identity. He zeroed in on Los Angeles's 102.7 KIIS-FM and partnered with them to do an LA-wide "Sixteen Weeks of Summer" campaign with round-the-clock promos, giveaways, and updates from the Sega-verse. In addition, the station would help expand Sega's visibility by setting up playable Genesis displays during the station's on-location events at beaches, concerts, and hot spots around the city. Similarly named and equally invasive assaults were launched in Chicago and New York. In addition to dominating a medium that Nintendo had ignored, there was an unexpected benefit to the radio maneuver: Cheryl Quiroz, the senior account executive at KIIS-FM, was so smitten with Sega's products that she reached out to Blockbuster Video about bringing them into the mix. At the time, Blockbuster was in the midst of a lawsuit with Nintendo, who had taken a harsh stance against videogame rentals, and so they were eager to help out. Blockbuster would gladly set up in-store displays, tout Sega's latest games, and hold gaming contests all summer long (though they wanted to dub the promotion something that honored 102.7 KIIS-FM). Sega was gladly willing to forgo the semantics, birthing "102 Days of Summer."

Whether it was sixteen weeks or 102 days, the summer would undoubtedly be long and stressful. Luckily, around this time, Kalinske felt incredibly rejuvenated. After nearly a year apart, his wife and three daughters were finally moving up to live with him in the Bay Area. Things appeared to be finally coming together.

Week 2: Mr. Extremely Dangerous

While Kalinske and his family searched for a dream home in the Bay Area, Sega began aggressively putting together a dream team of new employees. Like the brand they were cultivating and the hedgehog they had helped create, Kalinske, Rioux, and Toyoda searched for people with a distinct Sega-ness: sharp, scrappy go-getters who craved long odds and last-second victories.

As Kalinske continued to look for ways to embody the tone depicted in that Reebok "Extremely Dangerous" bungee-jump ad, he went straight to the source: Reebok's marketing manager, Steve Race. "We're building a team to go to war against Nintendo," Kalinske said. "We're going to take over the videogame market. Will you join us?"

Race was a clever, foul-mouthed prankster who used his boisterous personality to hide the fact that he was a brilliant marketing strategist. An impossible challenge? Shit, of course he was in. But Race wasn't yet ready to jump feet-first into the pool, and for the time being he wanted to run Sega's marketing department as a consultant. If that was what it took, then Kalinske was okay with it.

"Just one question," Race said. "What the hell is a videogame?"

Kalinske's eyes went wide. Even he had known more than this going in.

"Kidding," Race said with a cheeky smile. Not only did Race know what a videogame was, but upon joing Sega he'd immediately become the company's most senior expert. Back in the early eighties, he had been the vice president of marketing and communications for Atari's International Division (selling in territories where, notably, no crash had been caused) before cofounding Worlds of Wonder, the toy company that first nationally distributed the NES. "Don't worry about me," Race explained. "I was selling videogames back when you were pimping out plastic dolls."

Week 3: The Electronics Expert

Sega was adamantly focused on positioning their products as being more than just a simple toy, unlike Nintendo's; they were consumer electronics and should be marketed and sold as such. To make inroads into that realm of retail, Sega's executives targeted Richard Burns, a VP of sales at Sony who moved and spoke with the quiet but powerful demeanor of an assassin.

"We don't just want you, Richard," Kalinske said. "We need you."

Burns scratched his forehead, intrigued by the idea of moving from bleak New England to sunny California, but concerned about entering a business that overlapped with the pesky terrain of the toy world. "Why me?" he asked. "What I know about videogames could sit on the end of your finger."

Kalinske nodded. "Good. You'll be perfect."

Burns agreed to take the leap and run the sales department, thus beginning the mission to position the Genesis (and videogames as a whole) as consumer electronics, no different from the stereos, VCRs, or camcorders he'd sold during his days at Sony. Burns had been around long enough to feel confident in his ice-to-Eskimos sales skills, but he quickly discovered that getting retailers to take Sega seriously was only half of the challenge. The bigger problem was that his predecessor had either had terrible organizational skills or was an outright anarchist. A lack of any sort of filing system was merely a nuisance, but not having a formalized sales structure was just plain leaving money on the table. There were limited records of which systems and games retailers had ordered from Sega, and almost no information about what had sold through, what had been marked down, and what had been returned. In an environment where it took about eight months to develop a game and two months to produce and ship it, knowing these kinds of things wasn't just helpful but mandatory.

Week 4: Tiny Billboards

The sheer speed of Sonic The Hedgehog was bound to make it a hit, but that wasn't the only goal here—not even close. Kalinske wanted Sonic to become an instantly recognizable cultural icon who could define the decade and eventually grow into a multibillion-dollar intellectual property that would continue to pump money into Sega for decades even after he'd left the company. This was why Sega of America had been so protective of Sonic. They didn't want him to join that long list of videogame characters whose innovative gameplay had made them celebrities but whose lack of dimension had caused them to fade away. They had to make sure that Sonic would find a better fate than one-hit wonders like Dig-Dug, Frogger, or even Mr. & Mrs. Pac-Man, all of which had aged with the ungraceful gawkiness of a former child star.

Grand aspirations were certainly admirable, but without proper execution they were nothing more than delusions of grandeur. Transforming a 16-bit critter into the next Mickey Mouse, however, presented the same problem as marketing against the Super Nintendo: money. Without a war chest full of financial resources, Sega relied on the kindness of strangers. Or, more specifically, writers from the most popular gaming magazines of the era: *GamePro*, *VideoGames & Computer Entertainment* (*VG&CE*), and *Electronic Gaming Monthly* (*EGM*), which had been created to fill the growing appetite for videogame previews, reviews, and rumors. Though they differed in subtle ways (*GamePro* slanted younger, *VG&CE* skewed older, and *EGM* swung for mainstream), the editors at each all had one thing in common: a distaste for Nintendo. To Nintendo's credit, *Nintendo Power* editor in chief Gail Tilden consistently churned out a colorfully brilliant issue each month. But they kept the best content for themselves, and so it was exceptionally difficult for the other magazines to adequately cover the industry when the already close-lipped company who dominated over 90 percent of the market felt little incentive to share information.

Up until now, Kalinske's formula for success had always been to rely on his charm, wit, and facility with public speaking, but he was quickly discovering that none of those talents could compare to the power of harnessing the hatred for his competitor. In the same way that Blockbuster had been eager to rain on Nintendo's parade, Sega hoped that the scorned magazines could help advance their agenda. Kalinske knew that the gaming magazines typically appealed to only the most hard-core fans, but it wasn't just them he wanted to reach. The magazines' greatest asset wasn't the readership but rather the physical space they occupied. With wide circulations, these magazines populated newsstands, kiosks, and drugstores around the country, which made each cover almost like a tiny billboard. So maybe John Doe didn't care at all about videogames, but when buying his paper each morning, he would inadvertently notice the bright game magazine covers and a small imprint would be created. It would be only seconds each day, but they would add up.

To make the math work, Kalinske relied on Nilsen. Ever since joining Sega in 1989, Nilsen had always made it a priority to build strong relationships with the press. He made it a personal policy to return every call, from anyone at any publication, and when doing so he would always have a memorable quote ready. He was all about going the extra mile, whether that entailed flying out to Los Angeles to have lunch with writers from *VG&CE* or trekking out to Lombard, Illinois, to meet new members of the *EGM* team. Nilsen took great pleasure in seducing the tastemakers, but what really made his tactics work was that they were not tactics at all. As he saw it, these people were devoting their lives to writing about what he did for a living; they made his life easier, and he wanted to return the favor. It was less about sneakily seeking competitive advantages and more about demonstrating good manners. And if his sentiment contrasted with that of Nintendo, then that was just the cherry on top.

This all put Nilsen in a great position to ask for assistance when it came time to make Sonic a star. He coordinated a three-pronged attack of magazine covers, hitting *EGM* in May 1991 and then *VG&CE* and *GamePro* one month later. In addition, Sega released a sixteen-page promotional Sonic comic, which not only grabbed more eyeballs on newsstands but was

Trojan-horsed inside other publications like *Disney Adventures* and an issue of *Superman*.

As the warm days grew longer and Sega of Japan neared completion of Sonic The Hedgehog, Kalinske didn't quite know what to expect. But he was feeling optimistic, and pleased that Nilsen had acquired so much tiny but omnipresent real estate.

Week 5: Superstars Wanted

In addition to attracting eyeballs, Kalinske also craved hands. More specifically, he wanted to get Sega products into the hands of people who personified verve and coolness. In a perfect world, Sega would have hired young celebrities to star in commercials. In a less perfect world, they would have at least run ads during shows that featured those young celebrities. In reality, however, money was tight, airtime was expensive, and networks weren't in the business of granting discounts to unknown companies. Still, Kalinske knew that if Sega could only afford an ad buy during one primetime show, it was still worth the risk. They'd just better be damn sure they found the right show. ABC's *Full House*? NBC's *Saved by the Bell*? CBS's *Major Dad*?

But why pick only one when you can have them all? In the Sega spirit of killing a flock of birds with a single stone, Kalinske and his team developed a crafty solution. Instead of running an ad during a hit show, they would create a hit show of their own. Nilsen spearheaded a deal with producer Richard Rovsek to create a syndicated prime-time special, shot at Universal Studios, in which young sitcom stars would compete against one another in a series of zany athletic events for their favorite charities. Sega financed the operation, and between naming the special and including challenges like hedgehog races and Game Gear duels, they benefited from a parade of product placement. It would have been easy to dismiss this as shameless promotion, but with such a bright constellation of trendsetting teenage stars, no one complained.

They would all be there, the princes and princesses of the TV universe. Hunks and babes (like *Saved by the Bell's* Mark-Paul Gosselaar and Tiffani Amber Thiessen), loving on-screen sisters (*Full House*'s Candace Cameron and Jodie Sweetin), squabbling on-screen brothers (*Home Improvement*'s

Jonathan Taylor Thomas and Zachary Ty Bryan), and even a few second fiddles from *Blossom*, *Growing Pains*, and *Who's the Boss*, each hoping to seize the spotlight. All of them had eagerly signed up to star in the *Sega Star Kid Challenge*, breaking free of laugh tracks and very special episodes to compete in raft races, obstacle courses, and tug-of-wars set above a pool of whipped cream.

Though the show would not air until June, filming had taken place on April 18–19 at Universal Studios Hollywood. Kalinske and Nilsen had flown down to oversee the proceedings, along with an excited Toyoda, who brought his daughter to marvel at Sega's intersection with glitz and glamour. As they had hoped, it turned out to be a wonderfully raucous event, tied together by host Scott Baio and his happy daze.

Filming had gone off without a hitch, and now, two months later, they tuned in to see the fruits of their labor. The high ratings, coupled with a costumed Sonic The Hedgehog persistently lurking in the background of most scenes, indicated to Sega of America that things were looking rather peachy.

The only foreseeable pit was running out of money. Even on a shoestring budget, their plans to launch Sonic (and, in effect, relaunch Genesis, the Sega name, and their own careers) were adding up very quickly to some serious cash. Still, Kalinske knew that now was not the time to back down. What's another hundred thousand dollars today, he thought, to build a hundred-million-dollar property tomorrow? Nakayama, however, didn't see things as clearly when Kalinske called seeking additional funds.

"We have budgets for a reason," Nakayama reminded him. "Correct?"

"Believe me, I know that you don't love getting calls like these," Kalinske said. "But it's my job to evaluate opportunities and then ask."

"As with Wal-Mart?" Nakayama asked with a tiny groan.

Kalinske was rendered momentarily speechless, as things had still not progressed on that front as well as he hoped. While Sega had been running the Genesis store in Bentonville and taken over the city's billboards, for the time being Wal-Mart continued to stonewall Sega, though Kalinske was convinced they were about to crack. "Hey," Kalinske said, "it takes money to—"

"To make money, yes."

"I was going to say that it takes money to take people's money, but I admit my wording was primarily a function of attempting to spice things up."

Nakayama chuckled, though his skepticism was still evident. Neither man said anything for a while. Finally Nakayama broke the silence. "Okay, Tom, keep doing what you are doing. Sega of Japan will help out."

Nakayama hung up, though Kalinske could still hear Nakayama's groan ringing in his head.

Week 6: Altering the Beast

"Make a fucking decision already!" Steve Race shouted and then slammed down the phone. Race's words were loud enough to draw Kalinske to his office, but rational enough not to cause him concern.

"What's wrong?" Kalinsked asked.

"I don't know," Race said, trying to shrink complex frustrations into a string of intelligible words. "Just having my patience tested by our friends in Japan. First they won't show me game footage, then they won't even share a synopsis, then all of a sudden they might send me some screenshots, but a synopsis is still off the table. How the hell do they expect me to sell a mystery box with goddamn question mark inside?"

"Don't worry about that nonsense right now," Kalinske suggested. "There's a whole other beast I could use your help with."

With the marketing plans coming together, Sega was now faced with the logistical conundrum of getting Sonic from production lines in Japan into living rooms in America. Because the game would be bundled with the Genesis, this wasn't simply a matter of sorting, shipping, and selling. The situation was made more complicated by the fact that at this very moment Sega's warehouses were overflowing with about 150,000 unsold Genesis systems containing Altered Beast, with about 100,000 more collecting dust on shelves at retailers all across the United States. Financially Sega couldn't afford to just write off a quarter of a million systems, but commercially they couldn't in good conscience sell soon-to-be obsolete systems to customers who would feel foolish or deceived that they didn't get Sonic. Sega could have waited until the rest of the Altered Beast systems sold out, but they weren't selling all that fast to begin with. Besides, the whole point of unleashing Sonic now was to cuckold the Super Nintendo.

The whole situation had a vaguely solve-for-X middle-school-algebra feel to it and ideas were tossed around for days. Finally, they found a way to kill two birds with one stone, and set a schedule to make it happen:

June 15: Lower price of Genesis + Altered Beast to $149.95
June 30: Third-party "Graduate to Genesis" promotion ends
July 1: Begin promotion entitling customers to receive a free Sonic game by mail
Mid-July: Begin shipping Genesis units + Sonic The Hedgehog to retailers
Mid-July to mid-August: Ship remaining Genesis units + Altered Beast to select retailers
September 15: Exclusively sell Genesis + Sonic The Hedgehog at $149.95

This final strategy would offer the best of all worlds. Customers would be happy to get two games for the price of one, and retailers would be happy to unload the old hardware systems. Meanwhile, while the retailers were busy unloading the 100,000 units of old inventory, SOA and SOJ were busy with their own 150,000 units, though the product didn't physically travel far. The employees on both continents would open boxes of the old systems, remove the console, and then repackage it into a brand-new box containing Sonic on the cover and the new game inside.

Week 7: Sonic Boom

Unlike movies, books, and music albums, in 1991 there was no official release date for videogames. When a game hit stores was a matter of logistics, not premeditation. There were just too many variables and too many unaffiliated retailers; besides, mostly the product came in from Japan in dribs and drabs. As a result, there was no game-changing D-day for Sonic The Hedgehog but rather a period of several weeks in late June and early July when the blue blur started showing up in stores. Nevertheless, as soon as Sonic sped into the homes and hearts of a few players around the country, word spread exponentially—in schoolyards, on college campuses, and around watercoolers. And because Kalinske had been granted approval from Japan to pack Sonic with the system, it wasn't just $50 games that were flying off the shelves, but Genesis consoles that cost three times that much. And when people bought a Genesis, not only would they end up buying more games later but, most important, they likely would not purchase a Super Nintendo. A line had been drawn in the sand, and the only way to hang out with Sonic was by stepping onto Sega's side.

As sales of the Genesis doubled, tripled, and then quadrupled, Kalinske couldn't help but stare at the figures in his office and secretly want to see the faces of Sega's board of directors in Japan. He knew how stupid they'd thought he was to give it away for free; he remembered how condescending their smiles had been when they shouted at him in Japan. What were they feeling at this very moment, Kalinske wondered, and what would they be feeling a month from now when Sega of America's success continued to grow? Kalinske allowed himself just a moment to gloat.

Then Kalinske reminded himself that Sega was one company, and together SOJ and SOA were inciting a pop-cultural revolution. And yet, even as he had this thought, a small part of him couldn't help but root for Sega of America to beat the living daylights out of Sega of Japan and make those directors choke on their condescending smiles. It was only a small part of him, but it was a part of him nonetheless.

Week 8: The Happiest Place on Earth

Financial reports, sales figures, and market breakdowns can capably tell a story, but the power of numbers will never compare to that of anecdotal evidence. And in the weeks following Sonic's release, everyone at Sega had their own story. A friend called to say that his son kept curling up in a ball and trying to zoom around the house. Some kids at the mall were tapping their shoes like Sonic. The guys at the comic store were arguing about who would win a race between Sonic and the Flash. The realization amongst Sega's employees that what they did in this small office made real-life ripples filled their lives with an anything-is-possible excitement that most of them had lost at some point during their childhood.

Kalinske collected similarly inspiring anecdotes of his own, though his favorites were the secondhand stories that his daughters shared about their

Sonic-loving friends at summer camp. To celebrate Sonic-mania, he took his family to Disneyland. Kalinske and Karen locked arms and led the way, the giddy girls scampering along by their side. Together, they strolled through the crisscrossing little streets of Kalinske's favorite section of the park, Fantasyland. In addition to the teacups, it had the Matterhorn, Mr. Toad's Wild Ride, and It's a Small World. He knew that it was en vogue to mock It's a Small World, to call those animatronic dolls creepy or brush off the music as maniacal, but he loved how it was one of the few rides that actually tried to impart a message: peace, love, unity, community. It strived for more and maybe it failed, but there was something respectable about how it tried.

Kalinske was humming the ride's hypnotic music when Karen gently nudged him. "Look," she said.

Kalinske assumed that it must be another Sonic The Hedgehog devotee, doing something hedgehoggy. But it wasn't. He followed the path of his wife's gaze to a father and daughter moving in the opposite direction. The father was sweaty and tired but trying his best to remain enthusiastic as he pushed a pale little girl in a wheelchair. It was Bruce Kaspar and his daughter Anique, Kalinske's former neighbors from Los Angeles.

Karen flagged them down, and they all talked and laughed and remembered when. A couple of times Tom and Karen kindly tried to probe about what was wrong with Anique, but Bruce brushed off the question and explained that she was just sick. Following this encounter, the Kalinskes would learn that Anique had pediatric AIDS, but even before they were aware of the diagnosis, they could tell that the situation was bad. Despite the gravity of her illness, however, Anique wore the biggest smile of them all. Happy and upbeat, she shone with a joy so real that it was contagious.

The two families spoke for a while, vowed to keep in better contact, and then went their separate ways, each enjoying a day in the happiest place on earth.

Week 9: Humans Against Genesis

"Play it again," Kalinske instructed during a meeting between Sega's marketing team and account executives from Bozell, the advertising agency that Michael Katz had worked with to create Sega's previous "Nintendon't" campaign. The executives were presenting the first national Sonic The Hedgehog commercial. In the spot, a fashionably bespectacled woman dressed like a librarian sits at a desk and speaks to the camera in a saintly, nun-like tone of voice. As the president of a fictitious organization called HAG (Humans Against Genesis), she denounces Sonic for his blazing speed and smarty-pants attitude, and ends by asking why he can't be more like "that nice boy Mario." After running the ad for a second time, the executive stopped the tape. "So?" he asked expectantly.

As all eyes redirected toward Kalinske, he remained silent for a moment. "So?" Kalinske finally said emotionlessly, matching the executive's cadence. "That about sums it up. So what? So this woman is telling us to ignore Sonic and we're supposed to care? What's the point of this?"

Kalinske looked around. "I mean it. What's the point of this commercial? What's the message? What do we want people to feel when they watch this?" The room was stunned into silence. They had never seen Kalinske like this before, at least not in the office. He continued on, questioning the reason for the commercial's existence—not in a particularly cruel way, but not in a particularly kind way either.

"It's supposed to be funny," an executive finally said in defense.

"Yes," Kalinske said. "It's supposed to be funny. But this is just derivative. It's an obvious rip-off of *Saturday Night Live*'s Church Lady. But at least she's less likable. She's prissy without any glimpse of warmth. She's angry on the outside, not the inside, and that makes the HAG joke fall flat."

"Kalinske's spot-on," Race interjected. "If we're serious about going to war with Nintendo, then it's time for us to start launching grenades."

"There you go," Kalinske said, nodding to Race. "Someone who gets it."

"But you approved the HAG concept!" the executive reorted.

"Yes," Kalinske said with a slap in his voice. "And I'd approve it again and again. It's a clever idea if done right. It's supposed to make Genesis owners feel proud to be outsiders, but this is just confusing. It doesn't work." Kalinske shook his head. He had a lot more to say but saw no point in saying it. It was too late—this was the spot they had produced. "I don't like it," he said, and left the room.

Kalinske and Race went to his office and began scribbling ideas on a pad. None great, but some decent. Better than the Church Lady ripoff, at least. He kept filling up the page until he felt certain of the answer. He didn't yet know the what, but he knew the how, and he needed to speak with Shinobu Toyoda. "Hang tight," Kalinske said to Race. "I need to speak with our friend Shinobu."

"Be careful what you say around him," Race cautioned.

Kalinske paused before opening the door. "Why's that?"

"Oh come on," Race said, as if it were obvious. "The guy's a banana. Yellow on the outside, white on the inside; who knows where his allegiance truly lies."

Kalinske rolled his eyes, left the room, and made his way to Toyoda's office. He knew that some at Sega still questioned Toyoda's devotions, but there wasn't a doubt in Kalinske's mind which side he was on. That's why he was completely unafraid to approach Toyoda with what he was about to ask. "I need your help."

"Of course," Toyoda said, waving him in. "Tell me what is required."

"That ad is all wrong. I know it's tongue-in-cheek, but we're going after ourselves when we ought to be going after Nintendo. Steve is right, we need to go negative, and I need you to get Japan on board."

Toyoda gave a noncommittal nod. "There is a chance this can be done," he said. "But hard because they did not even like us to say 'Genesis does what Nintendon't.'"

"Good," Kalinske said, "because that'll be complimentary compared to what we need to do. I'm talking head-to-head, in-their-face, no-turning-back stuff. Like what we did at CES, but on a national scale."

Toyoda adjusted a shirt cuff as he thought this over. "I know how we can do this," Toyoda said at last with a proud smile. "But we can only get away with something like this once."

Week 10: Attention Nintendo Console Owners

Toyoda's strategy for slipping a new commercial past the Japanese gatekeepers was quite simple: he wouldn't tell them until it was too late. He and Rioux would move money around to pay for the new ad, which could be done with relative ease due to the skyrocketing sales of the Genesis. Meanwhile, Sega's marketing team would work with Bozell to create the new ad, which would be ready in time for the release of the Super Nintendo. Three days prior to the start of the campaign, Toyoda would inform Japan about the spot. He would act as if he had been caught off guard, and would offer to do everything in his power to pull the ad. In reality, however, it would be too late to stop the process, and the commercial would air, at least for a few days. If SOJ was angry, at least SOA would get a few days out of the ad. And if SOJ was okay with the ad, so much the better. Either way, the commercial would air.

Kalinske was in his office working with Nilsen on a concept to deliver an uppercut to Nintendo on national television. Or, rather, that's what they were supposed to be doing. Instead they'd gotten sidetracked as Nilsen paged through a newspaper in search of an article that might be worth reading aloud in his best radio-announcer voice. For most of July, newspaper-fishing had become a favorite pastime as a string of small but unfavorable developments at Nintendo translated into lucky breaks over at Sega.

It had started on July 5 when a federal judge ruled in favor of Galoob Toys, makers of the Game Genie, in a copyright infringement that would likely cost Nintendo of America $15 million. Then on July 19, Mike Tyson, the face of Nintendo's popular boxing game, was arrested and charged with the rape of Miss Black America contestant Desiree Washington. Two days later, on July 21, for the first time in three years, a Nintendo product was not the number-one-selling toy in the country. They had been displaced by a gigantic water gun called the Super Soaker. Kalinske made it a point to use every piece of good news, no mater how small, to motivate the troops. He would also clip these articles and send them to Wal-Mart each week along with a barrage of sales reports to show them how quickly the Genesis was flying up the sales charts.

"Anything good in there today?" Kalinske asked.

"Let us see, let us see," Nilsen said, flipping through the newspaper. "Here's something," he said, though didn't seem particularly pleased. He laid the paper on the desk for Kalinske to see for himself. It was a notice that occupied more than a quarter of the page and began with the headline "Attention Nintendo Console Owners." In the line below that, it read:

> Did you buy a Nintendo Entertainment System game console between June 1, 1988, and December 31, 1990?
>
> If so you are entitled to a $5 coupon.

The notice went on to explain how the attorneys general from all fifty states had brought a price-fixing case against Nintendo, which had been settled when Nintendo agreed to pay back $25 million to their customers—in coupons.

"Those brilliant bastards," Kalinske said, more impressed than annoyed. For years Nintendo had been fighting federal pressure regarding their business tactics at the retail level. Some called their tactics monopolistic, some believed Nintendo was merely being admirably aggressive, and others believed the government's case was nothing more than a witch-hunt, another case of Americans trying to stymie the influence of Japan (an argument partially aided by the fact that the initial charges were brought against Nintendo on December 7, 1989, which also happened to be the anniversary of the 1941 attack on Pearl Harbor). Whether the charges were valid, the government's threat was undoubtedly real, and for years Nintendo had been operating with the sword of Damocles dangling over its head. Now, however, it seems they had managed to steal away the sword and invert the entire situation into an advantage. Instead of facing stiff, crippling penalties like those inflicted upon AT&T and General Electric in years past, Nintendo's punishment was to offer $5 off to customers, who'd have to spend at least $50 to use the coupon. That wasn't even a slap on the wrist—it was more like a government-issued printing press allowing Nintendo to keep minting money.

"You worried?" Nilsen asked.

"Worried?" Kalinske said. He believed that the 8-bit NES was nearing the end of its life cycle, so the coupon itself didn't particularly bother him. What nagged him, however, was Nintendo's political savvy. It was as impressive as it was ominous. "Worried? Hardly!" Kalinske belted. "The stronger they are, the sweeter it'll be when we finally take them down."

"My thoughts exactly," Nilsen said, sporting a feisty smile. "And for what it's worth, no amount of bureaucratic mumbo-jumbo can change the fact that when a kid walks into a store and sees a Genesis and a Super Nintendo sitting right next to each other, they'll know exactly what to do. You know what I mean?"

Kalinske knew exactly what he meant but failed to reply because his mind was busy replaying the scenario Nilsen had just mentioned. A kid walks into a store . . .

Week 11: Spy vs. Spy

A kid walks into a store.

No, Kalinske thought, that's not right. We're going after teens and adults, the kids are just a bonus. Okay, so someone walks into a store. Someone? Really? That's so vague. Who, then? A teen wearing a leather jacket? A jock wearing a sweaty uniform? A curvy college coed wearing . . . not that much? Don't want to alienate any segment of the market, so maybe all of them should walk into the store? Nah, too crowded, too rehearsed, too diversity-for-the-sake-of-television. Kalinske felt like he had the commercial on the tip of his tongue, but every time he opened his mouth to try to let it out, the idea slid down his throat and hid. Okay, let's try this again: Someone walks into a store. Wait, what kind of store is this anyway?

A tap on the frame of his already open door dragged Kalinske out of his head. It was Toyoda, standing in the doorway with what was meant to be a blank expression. Over the past year, however, Kalinske had grown familiar with the nuances of Toyoda's seemingly empty expressions and had grown adept at filling in the blanks. This one, for example, said: good news. Not great news, but better than bad news, no?

"What is it?" Kalinske asked with a hint of concern in his voice. Though Kalinske believed he had decoded Toyoda's expressions, he made sure to keep this to himself. Perhaps it was due to his Japanese heritage, or perhaps it was meant to serve as an ode to the furtive nature of an idealized businessman, but Toyoda seemed to place a premium on ambiguity, and Kalinske was happy to play along. "Is everything okay?"

Toyoda stepped forward, and broke into an unambiguous smile. "Nintendo has made the price official. It shall be $199."

Kalinske matched his smile. "Just as we had expected."

"Just as we had hoped," Toyoda said. Kalinske was unsure if Toyoda's words were meant to echo enthusiasm or remove any insinuation of overconfidence, but they managed to accomplish both goals.

Kalinske nodded humbly. "You are absolutely correct. This is great news." Kalinske thought once again of his teenager, jock, or curvy coed walking into a store. They would see a Genesis and a Super Nintendo and have to make a choice. The Genesis would be cheaper and faster, and it would feature a much bigger library of games. No, Kalinske thought, with a subtle shake of the head. The library didn't matter, at least not at the moment; all that mattered was the single game that came with the console. "Hey, did Nintendo announce which game would be bundled with the system?"

Toyoda lightly shook his head. "Not yet."

"Thanks for the update," Kalinske said. "Why don't you go fill Paul in?"

As Toyoda moved toward Rioux's office, Kalinske dialed up Nilsen. "Good news, pal. As expected, Nintendo's pricing their machine at $199, though there's been no decision on which game yet."

In his own office, Nilsen nodded as he listened to Kalinske. A price of $199 was perfecto! Dear Nintendo, thank you for digging your own grave. But as quickly as Nilsen's mind swelled with excitement, it just as quickly filled with a strange flavor of disappointment. The high price point was good news, but it wasn't anything that he had earned. It was Nintendo's decision, pure and simple, and would have been just the same in a Nilsen-less universe. This prompted Nilsen to decide that he would take action. What kind of action he didn't yet know, but he was committed to giving his colleagues something to be more excited about than just the expected. He thanked Kalinske for the intel, hung up the phone, and walked out of his office like a man on a mission.

Not knowing where to go, Nilsen followed his feet as they led the way. Eventually, after wandering through the building in a nebulous quest for the unexpected, he found himself inside the office of Richard Burns.

"Um, Al?" Burns asked. "Do you need something?"

Nilsen's mind whirred as he looked around the office, playing a light-speed game of I Spy (stapler, family photo, Wite-Out), which evolved to thoughts of Spy vs. Spy. Like an actor onstage who remembers his line just in time, Nilsen asked Burns if he knew of any retailers who were especially loyal to Nintendo, "guys who you think would pass along to the other side any information you give them."

Burns thought a moment, then said he did.

Nilsen's eyes lit up. "Here's what I want you to do. Call them up and tell them that we are desperately worried that they will pack the new Mario game with the Super Nintendo. And then make it sound like you just slipped up and told them something you shouldn't have." Burns chuckled and agreed to put on his best dramatic performance.

A few weeks later, Nintendo announced that Super Mario World would be bundled with the $199 SNES. Nilsen realized it was unlikely that his ruse had been the reason for this, but at least there was a chance it was because of him. A war was coming, and he didn't just want to be on the winning side—he wanted to be the reason for victory.

And now that Nilsen knew for sure that everything would come down to Sonic vs. Mario, he had another idea: something big, something memorable, something unexpected. Across the office, Kalinske was having his own why-didn't-I-think-of-that-earlier moment. Many years ago, during his Barbie days, a toy company named Topper began making dolls based on a character named Dawn. Like Barbie, Dawn was pretty and friendly-looking and had many outfits, but unlike Barbie, she sold for less. By 1972, the Dawn dolls started to sell pretty well—until Kalinske intervened. He created a promotion where, for only two bucks, customers could swap their cheap Dawn dolls for a luxurious Barbie. One year later, Topper went out of business. Now Kalinske had an idea to do something similar to Nintendo, but he needed Steve Race to help him pull it off.

Week 12: The Haves and the Have-Nots

There are only two occasions when it is perfectly acceptable for a grown man to yelp like a little boy: on New Year's Eve and when watching sports. Tom Kalinske and Steve Race took advantage of the latter when a harshly hit line drive drove in two runs and gave the San Francisco Giants a late-inning lead. They were not alone in momentarily unleashing their inner children, as everyone in the forgettable bar they'd found after work seemed to hoot, holler, and high-five like they were ten years old and had just learned how to make s'mores. Ravenous excitement filled the room until a strikeout by the next batter led to a commercial break and, in unison, all the buzzed businessmen snapped back to their real age.

"Did you ever play?" Kalinske asked Race as they sat back down at a small table with a pair of nearly empty beer bottles between them.

"Damn straight. I was very nearly going to play center field for the New York Yankees," Race said wistfully. "Except of course they had no idea about this."

Kalinske laughed. "Reminds me of the time I almost dated Kathy Ireland."

"Well put," Race said, and then leaned back in his chair with a smile that seemed to say, This is what life's all about: beer, broads, and baseball. "So, what's on your mind? You're looking rather exhausted these days. Exuberant, but exhausted."

"You think so?" Kalinske asked, trying to shield any hint of vulnerability.

"You hide it well, but it's there to see at certain angles."

Kalinske shrugged. Maybe he was more exhausted than he realized, but he had a job to do. "I have a wife and three daughters. I'm fairly certain that my days of not looking exhausted are long gone."

"Fair enough," Race said. "So have you brought me out tonight to play marriage therapist? I'm okay with that role, but if that's the case, then expect a hefty invoice."

"Ha," Kalinske said. "Not quite. But I did want to discuss your role. I obviously appreciate having you here as a consultant, but I think it's time for you to officially come in-house and take control of marketing. We're on the cusp of doing something extraordinary, and I want you leading the charge." It was Kalinske's job to always believe that Sega was on the cusp and to make others believe it too, but for the first time the facts were starting to look like they warranted his confidence.

Ever since Sega of America had dropped the price of the Genesis and put Sonic in the box, units had been flying off the shelves. Sales of the console skyrocketed throughout July: 20,000 sold one week, 25,000 the next, 30,000 the week after that. Sega of America was on pace to sell more consoles in the summer of 1991 (500,000 units) than they had sold in all of 1990 (400,000 units). And the best part was that with each Genesis sold, the consumer would typically buy more than three games per year. No, strike that—the best part was that each sale of the Genesis likely meant one less sale of the Super Nintendo. You were either a Sega person or a Nintendo person; you couldn't choose both. Videogames were quickly becoming a religion, and luckily for Sega, the company was offering a 16-bit console that could be worshipped today, while Nintendo's system wouldn't be accepting prayers until early September. "This year is going to be good," Kalinske said, "next year is going to be great, and I don't know what adjective will describe the year after that, but I'm honestly thrilled to find out."

"I think you're absolutely right," Race said. "And I'd like to be at Sega for a while, but every time I start to think this is the right place for me, I run into some bullshit with the Japanese that makes me want to quit on the spot."

"Don't let that stuff get to you, Steve. I can help with SOJ."

"Believe me, I already know how much shit you shield us from. I don't know how you put up with Nakayama and his cohorts, but I tip my cap to you."

"Come on, I can't really believe that your dislike of the Japanese—"

"It's not like that. I'm not racist or anything," Race broke in. "It's the whole culture. It's the passive way they do things. It's the fact that when I come into work in the morning, it's not uncommon to discover that they've made a decision in the middle of the night that cancels out everything I did yesterday."

Kalinske shook his head, disappointed but unable to fully disagree. Nakayama had certainly given him the leeway to do things his way, but many of his decisions still led to unnecessary battles, even if they did work out in Sega of America's favor. "I admit that it can be tough at times, but when push comes to shove we've always been allowed to do things our way. Just look at the Genesis. They let us drop the price, they let us put Sonic in the box, and you see how well that's going."

"I do see that and I think it's great," Race said, unconvinced. "But did you ever wonder what they think when they see Sega of America doing so well? In my experience, stories about the haves and the have-nots don't end too well."

The situation between Sega of America and Sega of Japan was hardly as dire as Race insinuated, but there was certainly noticeable difference in recent sales results. Sega of Japan had chosen to give Yuji Naka, Sonic The Hedgehog's chief designer, a few additional weeks to fix any kinks and then launched the game in Japan in late July 1991, one month after Sega of America. In addition to the delay, the other notable difference was that in Japan Sonic did not come included for free with the hardware. The game was sold separately for 6,000 yen, and though it quickly became Sega of Japan's best-selling software title, it paled in comparison to Sonic-mania in America. The first week Sonic The Hedgehog was released in Japan, it sold 7,178 copies. The next week it held steady at 7,062 copies until the following week sales dipped down to 6,086 units. By the end of the year, Sonic The Hedgehog would eventually become a hit in Japan, but not the mega-hit that it was in America. And because the game didn't come with the hardware, it didn't dynamically increase the installed base. Though it was rare to hear anyone at SOJ acknowledge the sales disparity, when they did they attributed the weaker results to Nintendo's iron grip over retailers, a lack of third-party support, and the subtle ways that SOA had altered Sonic to work in America and not in Japan.

"It's not a matter of haves and have-nots," Kalinske said. "We're one company, and the better we do, the better they do. Besides, Nakayama is thrilled by how we're doing, and his is the voice that counts. Trust me, Steve, this is going to work. And when the roads get rocky, we can lean on Shinobu for help."

"Shinobu? Their spy? Don't get me started on him."

"Fine, fine," Kalinske said. "Why don't we table the issue for a little while and focus on less abstract matters."

"I thought you'd never ask," Race said. "What's next?"

"Do you remember a doll named Dawn?" Kalinske began, though he could immediately feel Race's interest slipping. "Anyway, I think we should run a promotion where customers can turn in their old NES and get a brand-new Genesis."

"I love it."

"Yeah?"

"Absolutely. But I'm fairly certain that's illegal these days, pal. And I'd go to war for you Tom, I really mean it, but not jail."

"Damn," Kalinske replied, and then tried not to appear too deflated. "Then let's talk about this Sega World Tour," he quickly suggested with an optimistic smile. This was Nilsen's big idea.

To convince anybody still waiting for the SNES that it was time to stop that nonsense and buy a Genesis, he'd dreamed up a nationwide thirty-city mall tour. Much like at CES, the goal was to set up a side-by-side comparison between Sonic and Mario and let players decide which game was better.

"The Sega World Tour?" Race echoed, rolling his eyes. "Maybe we should go back to discussing the many merits of working for the Japanese."

"What? You don't like Al's idea?" Kalinske asked.

"The idea is excellent, but the execution is a nightmare," Race said. He was right. The dare-to-be-bold idea looked great on paper, but it was positively hellish to actually put together. Different cities, different mall managers, and different setups posed challenges, which were compounded by the fact that to truly steal Nintendo's thunder the events of the tour should take place at multiple locations on the same day, which meant different crews and multiple equipment rentals. Nilsen had an undeniable talent for thinking big, but what Kalinske needed right now was for Race to take Nilsen's oversized thought bubbles and transform them into something real.

"Execution is certainly the issue," Kalinske said. "So can you think of anyone that we can hire to be our, um, executioner?"

"Let me think about that," Race said, before suddenly becoming distracted by the baseball game on television. The Giants were still winning by a run, but now the Dodgers had a runner in scoring position and a hot hitter at the plate. Kalinske followed Race's eyes to the television and eagerly watched the at-bat. Strike one, strike two, and then Race threw Kalinske for a loop when he said, "Let's go, Dodgers. Come on now, a single here brings in a run."

"What? I thought you were a Giants fan?"

"Nah," Race said. "I root for whoever is losing."

"So if the Dodgers take the lead, then you'll root for the Giants again?"

"What can I say? I like comebacks and exciting endings," Race said.

Strike three. Like many in the bar, Kalinske emitted a tiny yelp. Race, however, just shrugged, his mind already spinning on something else entirely. "For the Sega World Tour, I have an idea about the job: EBVB."

"EBVB?" Kalinske asked. "Is that a person, place, or thing?"

Week 13: EBVB

Few people know that Lake Tahoe's Squaw Valley was the site of the 1960 Winter Olympics. Even fewer know that it has a base elevation of 6,200 feet, a vertical rise of nearly half that, and an annual snowfall of 450 inches. Ellen Beth Van Buskirk, however, knew all of this and much more. She knew that there were twenty-six chairlifts and a state-of-the-art gondola, and that not only had it hosted the 1960 Winter Olympics, but also that those were the first Olympic events to be televised. She even knew that the broadcasters had freaked out because there was no snowfall until the week before the games began. When it came to Squaw Valley, EBVB knew it all.

It wasn't that she was a devout skier (her sports were basketball and distance running) or some kind of skiing-trivia savant. But as marketing services manager of the brand-new Resort at Squaw Creek, she felt compelled to memorize everything there was to know. A lesser manager, or a less neurotic individual, might have been content to know that many of these facts were safely tucked away inside a brochure somewhere, but Van Buskirk felt more secure when information was locked up in the vault inside her head. It was more challenging this way, more like a game, and that helped reduce the reality that her life had come down to filling 405 rooms at a ski lodge.

She had grown up in an era where it was considered more adorable than ambitious for a woman to work. She always found that notion ridiculous and took pride in never just wanting *a* job, *a* place at the table, *a* spot on the team, but wanting *the* top job, *the* best seat at the table, *the* starring role on the team. True, she rarely got the *the* and often had to settle for an *a*, but the no-no-no of the status quo didn't extinguish her fire. She developed a precocious talent for spinning negatives into positives, and interpreted her 5-foot 11-inch frame to be the Lord's not-so-subtle way of telling her to keep thinking big. No matter how many times the world tried to teach her that she was lucky just to participate, she never stopped trying to teach the world that it was flat-out wrong.

Until now, perhaps. Van Buskirk hadn't quite given up, but there were signs: she heard her own voice quieting, saw her worldview blurring. To her, moving to Squaw Valley meant raising the white flag. Throughout 1990, she had been able to do her job from an office at the Rincon Center in downtown San Francisco, where they had a small, street-level retail storefront decorated by photos, sketches, and a massive architectural model of the resort (so large, in fact, that it had to be placed in the office before installing the windows because the doorway was too narrow). In 1991, however, the resort had opened to the public, and they wanted Van Buskirk to move out there permanently. She thought the mountain was beautiful but dreaded the move because she believed San Francisco to be the center of action. Unfortunately, however, there didn't appear to be any other option besides occasionally staring at the phone and willing for it to ring.

Eventually it did, and it was Steve Race. "What are you up to?" he asked.

"Oh you know," Van Buskirk said, "everything, nothing, and the occasional something." She had known Race for nearly five years now, and he still had a knack for both impressing and bewildering her with his brashness. They had met back in 1986 at Worlds of Wonder, when she was learning the ropes of the public relations game. He used to say that big things were in her future; she didn't know if this was just something he said to everyone, but the way he said it made her believe it to be true. "I'm not up to much. Just home, watching TV, thinking about going for a run."

"I meant more generally," he said. "Job-wise."

"Ah, okay. I'm managing a ski resort in Lake Tahoe."

"Oh," he said, seemingly as surprised as she was that this was her job. "What about the rah-rah ladies' PR agency?" Race was referring to Van Buskirk, Morris, Webster & Smith, a firm that she had formed in 1988 with three other female rising stars in the PR industry. It had begun with the best of intentions but had come apart a couple of years later amidst mistrust and finger-pointing. Van Buskirk wasn't too pleased about how the agency had fallen apart, which explains why Race didn't know that it had.

"What happened with that?"

"Long story," she said with a familiar sigh. "But the short story version is that women are crazy. Myself included."

"I could have told you that!" Race bellowed.

"I could have told myself that," she replied. "Oh, well. Lesson learned."

"You sound borderline convincing, EB," he said. "Anyway, this is good. Well, not for you, but for me. I need your help." Race went on to provide a brief overview of Nilsen's mall tour and then sold her hard on Sega, using many of the facts that Kalinske had been using on him. "I realize that the job is below your pay grade, but we're about to give Nintendo a run for their money, and who knows what happens next?"

"What's the name of the company again?" Van Buskirk asked. She couldn't believe it. Amazingly, an opportunity had come along at the eleventh hour, but why oh why did it have to be with a company she had never heard of before? The only thing she feared more than moving to Squaw Valley was taking a job with a company that would be extinct only a few months later. And even if this thing did work out, she was quite sure that "mall tour organizer" didn't exactly pop off a résumé. "Can I think about it?"

"Think about it? What are we, Greek philosophers?" Race retorted. "I prefer you talk about it, Lady Socrates. I'm setting you up on a lunch date with this Nilsen guy."

"Okay, cool," she said, relieved. She thought the world of Race, loved how he could motivate people to walk through walls for him, but knew that he tended to charge in and out of companies, either because he'd grown bored with the lack of challenge or had worn out his welcome. As a result, she realized that even if she decided to join this Sega place, there was a chance that Race might already be gone, so she was happy to meet with someone who likely went through life and jobs in less of a hurry than Race did.

Nilsen met Van Buskirk for lunch at a Chinese restaurant in San Francisco's South City. Before going, he was given explicit instructions by Race to "reel her in." Nilsen felt confident in his metaphorical fisherman skills but wasn't sure that he'd need to use them. He urgently wanted someone to help bring his mall tour to life but was uncertain that he and Race valued the same qualities in a potential employee. Race was aggressive, impulsive, and unapologetic, leading Nilsen to half expect that he'd be meeting someone who looked and sounded just like Race, but with a ponytail and painted fingernails.

She was, happily, nothing like that. Van Buskirk was elegant, insightful, and self-aware in the best way possible—not self-conscious, but gracefully conscious of herself. If Race ran through life as if it were a sprint, Van Buskirk moved with a well-measured marathon approach. After an exchange of names, biographies, and small talk, she handed Nilsen her portfolio, which was mostly filled with stuff about Squaw Valley. He opened it, scanned through it in a matter of seconds, and then handed it back to her.

"That bad, huh?" she asked.

He opened his mouth to respond but then decided not to, curious if this would ruffle her feathers. Nilsen liked testing people in small and strange ways, believing that most of life was small and strange and that things like this revealed a lot about a person. She didn't seem bothered whatsoever, already happily discussing the etymology of the term "duck sauce." "I thought it was such a strange name," she said. "Soy sauce is made from soy, hot mustard is both hot and made of mustard, but duck sauce . . ."

Nilsen inspected the tiny bowl of orange sauce. "I venture to say that no ducks were harmed in the making of this sauce."

"My thoughts exactly!" she said. "So I went to the library and looked this up. Turns out that in Hong Kong and southern China, whenever you order roast duck it comes with this sauce. They give it out to mask the gamy duck flavor and also hide the occasional taste of fat. So even though it's actually made with pickled plums, sugar, vinegar, and occasionally pickled pits, they called it duck sauce, and the name just stuck."

Nilsen was impressed. Not only did Van Buskirk crave answers to the most wonderfully inane questions, but she actually went the extra mile to get answers. From there, he bombarded her with a barrage of questions. The more he liked what she said, the harder the questions got. He didn't want good, he wanted awesome.

"Where's the best place to start the mall tour?"

"You want me to say New York or Los Angeles," she said, "but I'd opt for somewhere close to Nintendo. Seattle, maybe?"

"There's a mall in Bellevue, Washington," he said with a smile. "Five miles from Nintendo's headquarters." Since initially conceiving the idea, Nilsen had added some flourishes to his master plan. For one, he wanted to open the Sega World Tour in his competitor's backyard. Another change was that he didn't just want mall-goers to see Sonic vs. Mario and get the point; he wanted that point to be reportable and irrefutable. To accomplish this, visitors would not only play both games but also be asked to vote on which was better. "Players Enjoy Sega 16-Bit System at Mall" would have made for a nice story, but "80 percent Choose Sega over Nintendo" would be a headline for the ages. With this alteration to the plan, whoever ran the mall tour would not only have to serve as executioner but also arrange for a judge and jury.

After lunch, Nilsen met with Race to talk it over. They agreed that she was overqualified for the job, but both thought that she was wise enough to see the potential of Sega. After a pleasantly awkward moment where both men realized they were in a rare state of complete agreement, Race gave Nilsen the green light to hire Van Buskirk.

"What about approval from Tom and Paul?" Nilsen asked.

"No problem," Race said. "I already spoke with them."

"When?" Nilsen asked, not quite sure when that could have happened, but happy to have the answer he wanted.

"Don't worry about it," Race said. "We're good."

Nilsen set up another meeting with EBVB and started off by handing her a small box. "What's this?" she asked. Again he ignored her, and again she passed the test she didn't know she was taking by opening the box with curiosity as opposed to caution. Inside was a big fishing lure. She turned to Nilsen, not knowing what to make of this.

Nilsen sported an isn't-it-obvious smile. "I've been told to reel you in."

Van Buskirk laughed, then smiled, and then did both at the same time. There was no logical reason to join Sega, but then again, there was the fishing lure in her hand, and the fun, puns, and blissful insanity that it no doubt symbolized. How could she resist?

Week 14: Rolling Thunder

From his office across the hall, Kalinske couldn't make out many of the words between Al and Paul, but he could hear the thunder of their conversation. Al wanted money for something, and Paul was responding in his normal frigid, frugal manner. Normally, Kalinske didn't mind hearing Rioux's bad-cop routine, but today he didn't want to deal with any distractions. With the Super Nintendo launch just around the corner, time was feeling less like an abstract concept and more like a noose. He knew that they needed to have a new commercial ready soon if they wanted to avoid being hung. He had been working with the ad guys at Bozell but continued to be less than pleased with their work. They were good guys with good ideas, but the days of "good enough" had burned away with the summer. Kalinske thought about switching agencies, as he had been (reluctantly) green-lit to do by Nakayama and the board of directors in Japan. But a proper agency review takes months and a good chunk of money, two resources that Sega couldn't afford at the moment.

What they could afford, just barely, was a head-to-head against Nintendo. There was still a decent chance that Japan would pull the commercial after Toyoda revealed what Sega of America had planned, and there was also a chance that Nintendo would sue them for showing Nintendo products in their ad, but Kalinske was intent on moving forward with an ad they referred to as "The Salesman." In the spot, a pushy salesman ferociously tries to hawk the Super Nintendo to a customer who keeps getting distracted by the many benefits of the Genesis: cheaper, faster, et cetera. Though he felt that Bozell's execution was a bit hokey, he believed that it did the trick of throwing down the gauntlet at Nintendo's feet. With the what, where, and when coming together, his remaining concern was the who. Whom did Sega want to identify as their ideal customer? The teen wearing the leather jacket? The jock wearing the sweaty uniform? The curvy college coed?

Kalinske was contemplating this and looking over the storyboard when he heard footsteps approaching his office. It was probably Rioux, headed his way with someone's head on a platter. Or maybe it was Nilsen, bursting in with another idea about how to get Nintendo's goat. Or maybe it was Race, pissed off by another roadblock in his way. Unless it was Burns, with new Sega of America sales data to send down to Wal-Mart. Or Toyoda, with new Sega of Japan sales data and a grenade of disappointment they'd have to fall on. Or maybe it was . . .

Anyone. That was the answer Kalinske had been seeking. Nintendo wanted kids, but Sega wanted anyone, and that's exactly what the commercial should show. If the ad was shot from the perspective of the customer, then the customer would be anyone—the teen, the jock, the college coed, and millions of others. The ad that Bozell had readied accomplished this perfectly. Though the salesman in the ad looked like he'd graduated from the University of Used Car Salesmen, by having the customer be no one instead of someone, Sega's ideal customer would be anyone and everyone.

As the commercial went through postproduction, Sega of America worked on a plan to deploy their head-to-head missile. For budgetary reasons and out of fear that Japan would pull the plug, they developed a strategy called "Rolling Thunder." Instead of evenly broadcasting the commercial over several weeks or spacing it out for a slow build of momentum, they would air it as much as possible at first, and then sporadically after that. Start with a bolt of lightning, make sure everyone notices it, and then reinforce the message with smaller thunderclaps in the following weeks.

Well, assuming that Shinobu Toyoda could successfully hold off Sega of Japan.

Week 15: Coke vs. Pepsi

The moment of truth was upon him. For months, Toyoda had been watering down information and finagling reports to Nakayama and Sega of Japan so that he and his colleagues could proceed as they saw fit. This was true not only in regard to the new commercial but also when it came to Sega of America's unabashedly aggressive attitude in general (CES, the mall tour, and so on). Now, however, with the Super Nintendo about to be released and

Sega of America's plans too far along to be stopped, the time to come clean had finally arrived.

It was just after midnight, and as usual, Toyoda was the last one left at the office. He searched for a comfortable position in his desk chair and prepared for his nightly update call to Nakayama, who would just be waking up. Toyoda was not one to assume the worst, but given Nakayama's tendency to be temperamental, he knew there was a chance this could be his last day at Sega. He looked around the office, filled with Sonic mementos and photos of his family, and felt good about all that he had done. If he wasn't fired outright, there was also a chance he'd be recalled back to Japan. If so, he'd likely be forced to endure the cultural practice of *murahachibu*, in which a dissenting employee is overtly shunned by his colleagues until it is decided that he's paid the price for his transgression or has appropriately demonstrated his loyalty.

Whatever the outcome, Toyoda had decided that he would not let this deter his American dream. No matter what, he would find a way to stay in America, provide for his family, and elevate himself through effort, excellence, and entrepreneurial spirit.

Toyoda called up Nakayama and spoke clearly and confidently, revealing everything as if he had just now learned of these details. Once it was done, he instinctively adjusted the collar of his shirt, bracing for the worst. It was all out in the open now, and even as the ensuing silence chipped away at Toyoda's confidence, he regretted nothing. This was America, the land of opportunity.

"Very interesting," Nakayama said in a perky tone. "Those are good ideas!"

"You think?" Toyoda asked, unsure if Nakayama was being sincere. After all, Sega's head honcho had a flair for the dramatic, and nothing made for better drama than building someone up before watching him fall.

"Most certainly," Nakayama explained. "I have just finished reading this great book by Mr. John Scully, who is now the president of Apple, and before this he was the president of PepsiCo. He talks at length about making bold moves to go head-to-head against market leader Coca-Cola. Our Sonic against Mario, it is very much like the taste-test challenge that he describes."

"Oh, yes," Toyoda said. "I can see how they are similar." He was so relieved that he was amazed he could even speak. Nakayama was not always the easiest man to deal with, but at this moment Toyoda felt overwhelmingly fortunate to be working for him. He was a rare breed, one of the few Japanese business leaders who not only had an admiration for Western business tactics but was willing to admit it. With Nakayama in Japan and Kalinske in America, Toyoda saw no reason that Sega would not dominate for decades to come.

Week 16: After the Summer Comes the Fall

Shortly after Toyoda's surprising conversation with Nakayama, Kalinske enjoyed an unexpected one of his own. He was in his office reading the September 2 issue of *Forbes*. Well, not reading so much as rereading the same sentence over and over. "We don't regard [Sega] as a competitor in the U.S.," Nintendo's Yamauchi had said in a recent interview. Kalinske chuckled at the words and then began to read them again when he was interrupted by a call from Wal-Mart's electronics merchant. And, as was always the case, the man spoke with a level of stress-induced grouchiness that Kalinske had nearly come to appreciate.

"My nemesis," Kalinske said, sitting upright. "To what do I owe the honor of receiving a call from the beloved 501 area code?"

In addition to sending Wal-Mart a packet of sales reports and news clippings each week, Kalinske had been calling the merchant to share marketing plans and product development information, attempting to walk the fine line between persistence and pestering. This was no easy thing, particularly in light of the fact that he never acknowledged that Sega had been trying to take over Bentonville, Arkansas, for almost a year now. The endless struggle sometimes got to Kalinske, and on his worst days he viewed himself as Don Quixote pathetically tilting at windmills with a sword made of rubber. Still, despite the occasional doubt, something deep inside him welcomed the challenge. Kalinske had never tried so hard to get a retailer on board, and as a result, he had never wanted it so badly. "Did you get the latest figures I faxed—"

"Look," the electronics merchant said, interrupting without remorse. "We give up. We'll carry Sega. Just close that damn store already and stop all the advertising in Bentonville. My boss and his boss are driving me crazy with questions about why we're not carrying the Genesis, and I can't take it anymore. You win."

And just like that, Sega was in Wal-Mart. Not all stores at first, just initial test regions, but where the product was sold mattered less than the fact that it was being sold at all. If mighty Wal-Mart was carrying Sega products, then other retailers had no excuse not to do the same. The curse had been broken.

It had been a long summer, but things were really starting to come together. Electronic Arts enjoying their most profitable quarter to date, and Nakayama had recently persuaded Acclaim to risk Nintendo's wrath and start publishing games for both systems; it was only a matter of time before all the third parties followed suit. And any retailers, third-party publishers, or consumers who remained unconvinced of the future that Sega promised would be swayed soon enough—if not with games or advertisements, then by EBVB's execution of Nilsen's fantasy.

"You did it," Nilsen said, smiling like a proud papa bear whose cub has just swiped his first fish out of the water. He was standing toward the front of a rapidly growing crowd as Sega's side-by-side world tour was getting started at the Alderwood Mall in Seattle. In the center of the hubbub, below a giant Genesis banner, was a black-on-black stage set up with several stations for people to discover the difference between Sonic and Mario. In just a few moments, kids, teens, and adults of all ages would be invited to step right up and see for themselves which 16-bit system was worth buying. "It's like you found a secret portal into my mind, took photos of my imagination, and then brought what you saw to life in a way that actually made sense."

Van Buskirk put an arm on his shoulder and nodded with pride. "Busted."

After Seattle, the mall tour would go on to hit twenty-four malls in sixteen cities, converting the masses from Peabody, Massachusetts, to Torrance, California. To get even more mileage out of Sega's avant-garde traveling circus, Nilsen sewed up a partnership with Nickelodeon to "borrow their cool" and start encroaching on Nintendo's grip on six-to-twelve-year-olds. Nickelodeon, the number one kids television channel, would be running "Where Is the Sega Tour Now?" spots throughout their programming, and also run a promotion called the "Slime Time Sweepstakes" where viewers could win Sega products if they answered questions correctly live on air when randomly called by a television host.

This was what Nilsen loved most: taking a big idea and stretching it bigger and bigger until it enveloped the world. After he brought on Nickelodeon, EBVB recognized the immense opportunity and was granted permission to hire a new PR firm to help rebrand Sega. She selected Manning, Selvage & Lee (MS&L), whose Sega efforts would be led by Brenda Lynch, a peppy publicist with the compulsive habit of getting others to see the world in whatever shades of gray that she thought were appropriate. Lynch took it upon herself to make sure that everything Sega did, made, or thought would be seen by the world in the context of Sega vs. Nintendo: hip vs. lame, new vs. old.

Kalinske hoped that with so much momentum behind it, the mall tour would prove to nongamers what videogame players were quickly discovering: Sega was a force to be reckoned with. The new price point of the Genesis seemed to have hit a consumer sweet spot, and the incredible launch of Sonic The Hedgehog had given Sega the Mario-killer they had so badly craved for years. Well, that was true in America, at least. In Japan, Sega's 16-bit sales were improving, but they were still unable to reach more than about 10 percent of the market. Kalinske was a bit troubled by the disparity but figured that if push came to shove, then Sega of Japan could always just follow Sega of America's blueprint for success: aggressive pricing, aggressive marketing, and unrelentingly aggressive employees who refused to rest until Nintendo had been toppled.

"What a second," Nilsen said, pulling Van Buskirk away from the hoopla in progress. Ten people had already tested out both consoles and eight of them chose Sega. "Is that who I think it is?"

Van Buskirk squinted and couldn't help but cackle. "I realize this kind of makes me sound like a witch," she admitted, "but if ever there were a time, right?"

Lurking behind the crowd and moving with the slow stalk of a shadow was a familiar face. It was Nintendo's Peter Main, and he did not appear to be particularly amused.

19.

THE ENEMY OF MY ENEMY

In October 1991 Minoru Arakawa was asked under oath if he considered the inventor of Sonic to be a "genius," on par with the likes of Shigeru Miyamoto, who had created Mario, Zelda, and other classic games for Nintendo. Arakawa looked around the San Francisco courtroom and carefully considered the question, though there was a lot on his mind. This trial, for one thing, had become a nuisance.

The litigation between Nintendo and Tengen had been dragging on since 1988. After the infamous reverse-engineering incident, Nintendo had sent letters to retailers such as Toys "R" Us and Bradlees threatening legal action against anyone who continued to sell unauthorized NES-compatible game cartridges. Tengen responded by suing Nintendo for unfair competition and violations of section 2 of the Sherman Antitrust Act for the alleged monopolization of the videogame console market. Nintendo countered by bringing a suit against Tengen for patent infringement, breach of contract, and RICO violations. Beyond compensation for damages, Nintendo sought an injunction that required Tengen to remove its unlicensed games from stores. Though it had been three years and there was still no resolution, the litigation had effectively crippled Tengen.

The legal issues weren't Arakawa's only problem, however. They were annoying, yes, especially the corrosive manner in which they dragged on and on, but the more pressing matter was the Super Nintendo. It had been released on August 23, 1991, and, in only two months, Nintendo of America had already sold 500,000 systems. Though the numbers appeared to be strong, they were falling slightly below the initial forecasts. This angered Yamauchi, who had grown accustomed to expecting nothing but the best out of his son-in-law—especially following the frenzy that the Super Famicom had stirred in Japan.

There were two reasons the American launch had failed to replicate the excitement found in Japan. The first was an issue of public perception. Nintendo of America had known that the lack of backward compatibility would be a problem but hadn't anticipated the scope of the backlash. Around the country, parents reacted as if they had just discovered that Nintendo was operating some kind of electronic Ponzi scheme. This outrage prompted a carousel of newspaper headlines like "Parents Say Nintendo Isn't Playing Fair" (*Kansas City Star*), "Nintendo Game Plan Infuriates Parents" (*Atlanta Journal*), and "Parents Vow to Resist Onslaught of New, More Costly, Nintendo" (*Patriot-News*). Parents weren't just upset by the compatibility issue but also irate that Nintendo didn't sell a converter that could resolve the compatibility problem—especially when such a device was being sold by Sega, who presented the other major obstacle to Nintendo's American launch.

For $35, Sega sold something called the Power Base Converter, which allowed games designed for their 8-bit Master System to work on their 16-bit Genesis. Even though very few people owned the Master System and the Power Base Converter was hard to find in stores, Sega flaunted the fact that such a device existed as proof that they cared more about their customers than Nintendo did. It quickly became Sega's most valuable, if worst-selling, product.

The converter, however, was only the tip of the iceberg when it came to Sega, the much bigger problem was Sonic The Hedgehog. The character had been only a mild success in Japan for Sega's Mega Drive but was an instantly adored phenomenon in America. It was as if after a decade of conservative politics under Presidents Ronald Reagan and George H. W. Bush, Sonic's combination of speed, attitude, and energy seemed to embody the promise of the 1990s.

Although Minoru Arakawa would never admit it to Kalinske (he wouldn't even meet with Sega's president) nor ever voluntarily proclaim it in public, on the witness stand he had to tell the truth. "Yes," he said. "They looked at Super Mario. They wanted to come up with something similar." After answering, all he could do was sit there and wait for the next question. There would always be more, it seemed, because this battle with Tengen would never come to an end.

Later that month, Kalinske flew to New York City with Nilsen, Toyoda, and Burns. With the holiday season approaching, they all had a lot on their plates, but for the next few hours they would have only one thing on their minds: Tengen.

After touching down at JFK, the four of them hustled through the airport with their carry-ons, sneering at the suckers who had checked bags, and flagged down a heavily dented yellow cab. As Kalinske got into the passenger seat, his three employees took deep breaths and thought thin thoughts in order to fit into the backseat. "We need to go to the Coliseum in Columbus Circle," Kalinske told the driver, before taking a mental photo of Nilsen, Toyoda, and Burns all crammed together. "And at the risk of sounding like a movie cliché or, worse, a tourist, feel free to step on it."

"How are we doing on time?" Toyoda asked, out of breath.

"Bordering on screwed," Burns replied.

"Nah, don't worry about it," Kalinske said in his breezy manner. "We'll be fine. Besides, they won't start without us."

Nilsen looked out the window, his eyes taking in the towering buildings. "We've done all we can. Our fate is now in the hands of Trafficles," Nilsen said. When the others didn't seem to get it, he elaborated: "You know, the Greek god of gridlock."

Kalinske, Toyoda, and Burns chuckled, which eased somewhat the anxiety they'd each been feeling. Nilsen had a real talent for detecting tension and knowing how (and when) to defuse it a little. It was a skill that had proved to be quite valuable at Sega over the past few months, when things had quickly gone from calmly good to chaotically great. By July, word of Sega had started spreading throughout the country, among consumers and business executives alike. By August, customers who had been holding out for the SNES started giving in to their 16-bit desires and purchasing a Genesis. And by September, retailers were experiencing periodic stockouts. To keep up with the increased demand, the company had to grow quickly—not just in terms of facilities and number of employees but also creatively and in the employees' mind-set. More people, more press, more games—it would have been a nightmare, if it all weren't so wonderful.

Proof of success could be found in the little things. The speed at which calls and faxes were returned. The respect, unspoken but visible on people's faces, when you told someone that you worked at Sega. And the unexpected moments when you were in line at the supermarket or video store and some random kid was explaining to his parents how important it was that he get a Genesis for Christmas. All these things were more than just a pat on the back—they served as a push on the shoulder to keep Sega moving faster, faster, faster. Sega now had its foot in the door, but Nintendo still owned the house.

Kalinske was willing to try anything as long as it didn't go against the Sega narrative that he was trying to sell to the world: fun, scrappy, edgy company revolts against the tyrannical status quo and brings entertainment to the next level. Naturally, the idea of jumping into bed with an enemy of Nintendo fit right into his mission. So he was not at all surprised when Toyoda came to him expressing a desire to work with Tengen.

At the behest of Nakayama, Toyoda had been testing the waters with Tengen's CEO, Dan Van Elderen. On the surface, a relationship between Sega and Tengen made a lot of sense. Sega needed more games, and Tengen could provide great titles. Below the surface, a relationship between both companies made even more sense. What the public didn't know, and what Kalinske had just found out, was that Nakayama had been helping out Tengen behind the scenes for years. In an effort to weaken Nintendo, he had offered financial resources to help foot their mounting legal bills. Kalinske wanted to know how long this had been going on. He also suspected that Sega might have even been the reason that Tengen had reverse-engineered the NES in the first place, but he knew he'd never find out for sure. Getting straight answers out of Nakayama was like catching a shadow and pulling its teeth with a needle from a haystack. Kalinske respected his boss greatly but hated the fact that he'd been at Sega for a year now and still felt like there were secrets hidden behind every door. Still, there was no use fixating on something he couldn't control. And even with the occasional corporate surprise, Nakayama had been true to his word, granting Kalinske the autonomy to run Sega of America as he saw fit. Besides, in this case Nakayama's penchant for chicanery had worked out to his advantage, as Sega had signed a very favorable deal with Tengen.

As New York whizzed by outside the windows of their cab, Kalinske, Nilsen, Toyoda, and Burns scrambled through files and folders to review talking points for today's joint press conference with Tengen.

"Which sounds bigger?" Burns posed. "'Tengen will be making forty games for Sega over the next two years' or 'Tengen will be making twenty games per year for the next two years'? Forget it, that sounds redundant."

"Territory question," Kalinske put in, flipping through a folder. "Nothing in the language with regard to Asia, right?"

"That is correct," Toyoda said, not needing to consult any notes. He knew this deal through and through. "It applies only to America and Europe."

"If you feel the need to mention a title off the cuff," Nilsen said, scanning a partial list of Tengen's games. "I would go with the R.B.I. Baseball series. The Pac-Man stuff is great, but I think it'll come off as old hat with this crowd."

Kalinske nodded, now reviewing his speech. Even though it would be short and sweet, the text was sharp, witty, and well put together. Kalinske had gotten help from Ellen Beth Van Buskirk, who'd been lending a hand to PR endeavors and would be staying with Sega after the mall tour to run communications inside the company. The mall tour was almost done, and the results showed that bringing her into the fold was a no-brainer. The Sega World Tour had already reached over 100,000 people (with 63 percent of that audience being children and teenagers). Of that total, 88 percent chose the Genesis over Super Nintendo. In addition to the staggering results of Sega's "taste test," articles about the showdown in major national and regional newspapers were on pace to reach ten million people. Van Buskirk was most definitely a keeper. With a small smile on his face, Kalinske reread the opening to the speech she had helped write and nodded to himself. "She's real good."

"Who?" Toyoda asked, always alert.

"Ellen Beth," Nilsen said, knowing exactly whom Kalinske meant. "EBVB." Nilsen had immediately recognized that she was a star from how cleverly she had brought his vision for the Sega World Tour to life to the way she could speak at length to anyone, from any walk of life, and never sound like she was talking up or talking down to them. "She was incredible at the malls," Nilsen said. "She just gets it. She really does."

"She's making a dent," Kalinske said. "Next time we should bring her along."

"Definitely," Burns echoed. "That is, you know, if she isn't busy hobnobbing with the world's greatest athletes."

The others laughed. EB had done such a bang-up job in her first couple of months at Sega that SOJ had entrusted her with an important assignment, or at least something they considered to be a top priority. With SOA making a name for itself, some of the folks at SOJ decided to play a game of anything-you-can-do-we-can-do-better. So they stepped into SOA's sandbox and searched for a splashy marketing opportunity. As lovers of golf, they looked into sponsoring an event on the PGA tour. But that turned out to be pricy and difficult to procure. Instead, they decided to sponsor an LPGA event in Atlanta. When Kalinske found out, he thought it was a prank. Beyond the LPGA's lack of popularity among the general public, Sega wasn't even selling a golf game, let alone a women's golf game. But before he or anyone else at SOA could intervene, it was already done. And EB, a former athlete who had demonstrated an ability to pull off unusual events, was put in charge of this one. The first annual Sega Women's Championship would tee off on April 19, 1992.

"Jeez," Kalinske said. "I sure hope a good excuse to skip that LPGA event crops up between now and then."

After a nod of camaraderie, they returned to their last-minute homework for the press conference. Kalinske would be the only one of them speaking, but he wanted everyone to do as many press interviews as possible. So they all pored through the information they had on hand and made sure they had a few quotes ready.

They made it to the Coliseum on time, and it was a good thing they did. The lavishly decorated room was filled with a large crowd of journalists, financial analysts, and potential investors (Tengen was rumored to be seeking a round of additional capital). The event was kicked off by the two men that Toyoda had been negotiating with: Dan Van Elderen, Tengen's tall CEO, and Ted Hoff, their bespectacled executive vice president of sales and marketing.

After they proudly announced a strategic alliance with Sega to produce forty games, Kalinske came to the podium and uncorked a bottle of his steely enthusiasm. The content of the speeches that day was secondary to the visual of Nintendo's rivals shaking hands, though Kalinske made waves with his bold claim that Sega was already outselling Nintendo. Whether this was true or not was a matter of debate, but it foreshadowed a contentious game of numbers that Sega and Nintendo would play over the next few years.

Following a series of photo ops and press interviews, Kalinske and company went out for a round of celebratory drinks with Van Elderen and Hoff. The Tengen guys were a lot of fun, and it was nice to clink glasses and speak in the buzzed, conspiracy-laden hush of new friends with big, bold plans. Nintendo was in the crosshairs, they decided, and would finally get what was coming to them.

"How's the never-ending trial going?" Kalinske asked, sipping a beer.

"I'd be more than happy to bitch about it," Van Elderen said with a grumble. "But unfortunately we're not allowed to discuss the proceedings. I will say, however, that it was funny to see Arakawa have to share his thoughts about Sonic under oath."

After a gurgle of laughter, Kalinske checked his watch and announced that he had to step out for a few hours.

"You have a date?" Hoff asked.

Kalinske nodded to his guys, who knew where he was headed. "A big one," he told Hoff, and then said his goodbyes. Before leaving New York, Kalinske had another one of Nintendo's enemies to meet with, someone with a little more muscle than Tengen: Sony.

A few months earlier, Kalinske had received a call from Olaf Olafsson, who was at Sony's Imagesoft offices in Santa Monica. Olafsson had wanted to meet Kalinske, and with the feeling being mutual, they agreed to get together for lunch. They bonded over their outsider status in the videogame world, as well as a mutual love for honesty, innovation, and unconventional ideas. Thanks to their similar personalities, it was quickly evident that they were destined to be friends. The only question that remained was whether or not they would also be business partners. Olafsson offered Kalinske a standing invitation to meet with him and his boss, Mickey Schulhof, the next time he was in New York—which turned out to be for the Tengen press conference.

The three of them decided to meet for dinner at the private club atop the Sony offices at 55th and Madison. Kalinske entered the white-facaded building and took the elevator all the way up to the thirty-seventh floor, where Olafsson and Schulhof were waiting for him. "Right on time," Olafsson said, flashing a half smile. "Tom, meet Mickey Schulhof; and Mickey, say hi and be kind to my friend, the incomparable Mr. Kalinske."

"A pleasure," Schulhof said, extending a hand. He was a handsome man with perfectly parted hair, a tan, and a bright white smile. Though his soft handshake might have suggested that he was another of the dime-a-dozen pampered executives, he glowed with a gritty, omnipotent confidence. "I've heard way too many great things about you from Olaf, so I already know they can't all be true."

"I like it when the bar is set high," Kalinske replied. "Otherwise, what's the point?"

Kalinske, Olafsson, and Schulhof were seated by a window overlooking the city, where they enjoyed an elegant seafood meal with a fine chardonnay. They started off the evening by sharing rehearsed chapters from their individual life stories, and eventually progressed to sharing impromptu anecdotes from their travels around the world. Somewhere between five and ten bites into their main course, though, the conversation shifted to business matters. "So," Schulhof began, "Olaf tells me that you plan to turn the videogame industry on its head."

"That's the plan," Kalinske said with a smirk.

"Excellent," Schulhof said. "That's what I like to hear."

"The fact is, we strongly believe in the business," Olafsson clarified. "The games themselves, I could take or leave. But the industry as a whole—there will be a gold rush."

"In many ways," Kalinske said, "it has already begun."

"True, true," Olafsson said, nodding. "Perhaps my vantage point is a bit affected by Sony's angle on all of this."

Kalinske leaned back in his chair to get a better look at both of his hosts. "So tell me, then, what is the ideal scenario for Sony?"

Schulhof gladly fielded the question. "When I first began at Sony in the late seventies, we made our bones selling televisions and stereos. We did quite well. Well enough that it would have been easy to stick our heads in the sand and keep cashing paychecks for the foreseeable future. If it ain't broke, don't fix it, right?"

The three used the rhetorical pause to take a long sip from their wineglasses. After a satisfied sigh, Schulhof continued, "But then in 1978 an audio engineer named Nobutoshi Kihara invented a small, portable stereo that allowed people to listen to music anywhere they went. Many questioned this logic and wondered why anybody in their right mind would possibly want to enjoy music outside the living room. Sony, however, didn't bat an eyelash, and moved full steam ahead with this device called the Walkman. Without any need to take the risk, we pushed all of our chips to the middle and changed the way that music is enjoyed around the world."

"I remember getting my first Walkman," Kalinske said nostalgically.

"Of course you do," Schulhof said. "Because if Sony does something, we do it memorably. The whole nine yards, or nothing at all."

Olafsson smiled. "I believe the popular parlance is 'Go big or go home.'"

"It's true," Schulhof said. "Columbia Pictures. Compact discs. CBS Records."

"You don't need to paint me a résumé," Kalinske said. "I'm already thoroughly impressed. But I'd like to know where Sega fits into all of this. Is Sony looking to make an acquisition?"

"Not at all," Schulhof said, shaking his head. "I truly believe in synergy. Not the bullshit buzzword version of the term, but the real-life implications of finding situations of mutual benefit."

Olafsson elaborated. "Sega has the experience and is gaining the credibility. Sony has the tech and financial resources. But what we share is the intelligence to realize that multimedia is the inevitable future of entertainment. On that note, is it safe for me to assume that Sega has some kind of CD-based gaming system in the works?"

Kalinske considered a vague response but decided to reveal his hand. If there was any chance of building something with Sony, he couldn't start with smoke and mirrors. "Yes, that's accurate," he said. "Sega is planning a CD attachment to Genesis."

"How far along is it?" Olafsson asked.

"Nearing the end," Kalinske said. "We want to go to market in late '92."

"Sega will beat the drum about this at CES in Vegas?"

"That's a fair assumption."

Olafsson pursed his lips, calculating the ramifications. "Good, that's exactly what Sega ought to be doing," he said. "And now that we know hardware needs are off the table, then any convergence between ourselves would be focused on software."

"I hear that's where the money is anyway," Schulhof chimed in.

Olafsson subtly nodded to himself. "Is it also safe for me to assume that with the Genesis surging and all hands on deck to support the system, there is an increased need for outside parties to provide Sega with CD-based software?"

"Definitely," Kalinske said. "I won't even consider bringing a CD add-on to America until we have the right software in place."

Olafsson looked at Schulhof, hinting that there was potentially something here. They paused again for a bite or two of their delicious seafood dinner before Olafsson spoke up once more. "We're obviously not at a place to jump into anything of substance here," he said. "But we have no reservations saying that we like you, and I believe that feeling is mutual."

Kalinske nodded; it certainly was.

"So we're two teens in love," Olafsson mused, "ready to offer the world to each other, but any consummation would require parental approval."

"And Japanese parents have a reputation for being strict," Kalinske added.

"Tell me about it," Schulhof murmured, shaking his head.

"Nevertheless," Olafsson concluded, "with some massaging, I think that there very well could be a way for Sony to help supply Sega's software needs."

"I'll work on things from my end," Kalinske commented.

"And we'll do the same," Schulhof replied.

"Fantastic, then," Olafsson said. "But I would like to note, in the spirit of complete and utter honesty, that a working relationship would likely not preclude the possibility of Sony one day looking to get into the console business."

Kalinske shrugged. "I wouldn't expect it any other way. But for now, I think we both have a lot to gain by holding hands. Down the road? Maybe that's something we explore together, or maybe we go our separate ways. But I'm okay with that 'maybe.'"

"Us too," Schulhof said.

"Good," Kalinske said, and raised his glass, prompting Olafsson and Schulhof to do the same. "To 'maybe,' and all the wonderful possibilities that the word may create."

20.

WORTH WAITING FOR

Thick trees with raggedy brown leaves. A graying road. And another parking lot. Although the view from Kalinske's new office could hardly compare to the princely panorama from atop the building where he'd dined with the Sony executives a few months earlier, it felt a good deal more satisfying. This was a view that he and his employees had earned, the reward for their furious innovation, imagination, and experimentation. Their efforts had precipitated a blizzard of success, requiring Sega to move into a new building in late 1991.

It was a long, gray, two-story building in Redwood City, with the downstairs dedicated to operations, sales, market research, and human resources, and the upstairs devoted to marketing, legal, and the executive staff. The Redwood City office was a bit more formal than the last one but still had a humble work-in-progress feel that precluded any sort of Jeffersons-like moving-on-up moment. Besides, with things just beginning to heat up, everyone was too busy to spend time patting each other on the back.

Sega went into Christmas 1991 by continuing to curve upward at a near exponential rate and, by now, had captured about 25 percent of the market. Stores were selling out of the Genesis with such frequency that it began to become a problem. It was great to feel like the popular, unattainable guy at the party, but not at the risk of suitors settling for a more available, second-rate friend. At first Sega managed the problem by air-freighting in systems from Japan—a costly proposition, but worth the additional overhead to avoid losing customers. But, like a Band-Aid that's lost its sticking power, this was no longer a viable option as the holidays approached and demand skyrocketed. To maintain the momentum, Nilsen spearheaded a marketing program intended to prevent parents from settling for Nintendo at any cost. The initial concept was for a preorder reservation kit that would ensure customers a Genesis by a certain date and reward them for waiting with a free, collectible T-shirt. But with the crazy Christmas season around the bend, it was impossible to get retailers set up in time. So Nilsen took the idea and stuck it in his pocket for a rainy day, at the time not realizing how well it would serve Sega in the future.

Instead, with the same *c'est la vie* spirit that Kalinske and Nilsen had taken one year earlier with the Buster Douglas game, Sega decided to embrace the dilemma. Starting on December 9, Sega launched their "Worth Waiting For" campaign. They aired ads on national television and radio (print would have taken too long) to speak directly to customers and let them know that even though the Genesis was red hot and might be hard to find, they shouldn't give in and buy an SNES. Sega revealed that they were airfreighting in new systems every day, and suggested that customers keep checking with their local stores, turning a frustrating shortage into something of a rousing scavenger hunt. The ads also publicized a toll-free number that allowed callers to voice their grievances and, at the same time, receive a special offer to buy one of four titles directly from Sega with guaranteed arrival before Christmas. As a result, Sega managed to dominate the holiday season and edge out their hated rival.

Or did they? Kalinske stared out the window of his new office, positively convinced that Sega had stolen Christmas from Nintendo. Eight hundred miles north, however, it was fair to assume that Peter Main was drinking a cup of coffee and reveling in Nintendo's destruction of Sega.

Nintendo claimed to have its goal of 2.1 million Super Nintendos sold by the end of 1991. The problem was, however, that even though Nintendo had sold all these systems to retailers, those retailers had only sold about 70 percent of the product. Meaning that on the day after Christmas, stores still had 30 percent of their Nintendo inventory on the shelves, which they could return, mark down, or continue to sell. Sega, on the other hand, sold through a staggering 95 percent of their systems, totaling 1.6 million by the end of the year. Though numbers can get slippery, Sega's 1.6 million was higher than Nintendo's 1.4 million (70 percent of 2.1 million), making them the clear victor. Then again, Nintendo had only been selling the SNES since September, just a third of the entire year, further devolving the issue into a heated cup of number soup. "True or false?" Steve Race asked, sidling into Kalinske's office. "The real reason we moved from South San Francisco to Redwood City was because Nakayama believed the wind here would be kinder to his combover."

"I can neither confirm nor deny that," Kalinske said as Race took a seat in front of his desk. "But if I had to lean one way or the other, it would not be toward deny."

"Perfect," Race said. "I've always believed that there's an illustrious career waiting for me in the conspiracy-theory racket."

"Oh, yeah?" Kalinske asked. "Well, there's an illustrious career waiting for you here at Sega if you're ready to make the move to full-time."

"Jumping right into the ball pit, are you?" Race mused. "How dare you deprive me of that patented Kalinske small talk? For example, how was your Christmas?"

"You want small talk?" Kalinske playfully replied. "All right, here you go. During the break, Karen comes up to me with this sneaky look on her face. It's a classic of hers, one of my favorites, and she tells me that she has news. Which do I want first, the good news or the bad? Of course I want the good, so that's what I say, and back comes that sneaky look and she says: 'it's a boy.'"

"Congrats!" Race exclaimed. "That's a hell of a gift."

"Sure is," Kalinske replied. "And what about you? How was your Christmas?"

"There may have been some seasonal debauchery," Race murmured, "but nothing too crazy. In fact, for the most part, it was a very Sega-filled holiday season. I started really digging into the handheld market."

"Is your disciple not working out?"

"Rather the opposite. Thus far Diane has been a godsend," Race said, referring to Diane Adair. Race had hired her in late November to replace Bob Botch, Sega's director of marketing responsible for the Game Gear, who left to become president of a software company called US Gold. Though Race was admittedly not brokenhearted to see Botch go, as he had grown tired of breaking up shouting matches between Botch and Nilsen, the timing couldn't have been worse. With Sega growing at Sonic speed, Botch's departure threatened to throw a wrench into things. Luckily, Adair fit right in and hit the ground running. She was a doe-eyed woman with a happy-go-lucky voice, and she wasn't afraid to dive headfirst into problems or get her hands dirty. With Nintendo now in the 16-bit console business, Sega's executive staff wanted to extend the lines of battle and make their opponent fight a war on two fronts, hence the emphasis on the handheld Game Gear.

At this point, the Game Gear had been in stores for over six months, and sales were strong but not spectacular. Part of the reason was continued problems with battery life, but a larger part was the lack of software, visibility, and brand identity. Kalinske hoped that Adair could pull a rabbit from her hat and do for the Game Gear what Nilsen was doing for the Genesis. Sega had been able to come up with creative solutions to some of the challenges (for example, an accessory could be purchased to extend battery life) and found imaginative ways to take advantage of the Game Gear's un-Nintendo-like ability to display colors (a tuner could be purchased that allowed one to watch TV on the handheld). During her second week on the job, Adair became fast friends with Race's other key hire, EBVB. Both women had been sent to Los Angeles for media training sessions with their PR guru, Brenda Lynch, at MS&L. There, Adair was wowed by EBVB's on-camera presence and her talent for compressing complicated topics into easily digestible media bites. EBVB, equally smitten, perceived Adair as a model of efficiency, always planning ten steps ahead with another five steps in reserve as a backup in case things went awry. Mutual admiration quickly led to daily morning jogs, which ultimately led to long-lasting friendship.

"She's a playmaker," Race said, who had known Adair since the late eighties, when she had worked for him at a failed start-up called Homestar that sold housewares by phone. "She's just as good in the worst of times as she is in the best of times. That's rare, trust me. She'll be a big part of what you're putting together here."

"I don't doubt it," Kalinske said. "But what about you?"

"Right now, the sixty-four-thousand-dollar question," Race said with a grin.

"There's a lot more money in it for you than sixty-four grand."

"It's not about the money."

"It's always a little bit about the money."

Race bobbed his head. "True. But obviously that's not the issue here."

"So what's the issue?" Kalinske asked. "The budgets are too low?"

"No," Race said dismissively. "You know I like doing more for less."

"You don't like the games?"

"I don't like any games," Race conceded, "but ours seem better than most."

"The marketing team? I've let you stack the deck with your own aces. All key personnel, except Al," Kalinske said, considering this. "Is he the problem?"

"Nilsen?" Race said with a chortle. "He's a strange dude, that guy. You would think he has a hidden laboratory where he pulls the wings off butterflies."

"Hey," Kalinske said, defending Nilsen, "if he keeps churning out these great wild ideas, then I'll buy him a case of exotic butterflies."

"Save the call to your butterfly broker," Race said. "Nilsen is an odd duck, but a good guy nonetheless. And he's drunk the company Kool-Aid, which never hurts."

"Most people have," Kalinske noted. "Myself included. And it tastes great."

"I'd love nothing more than to do the same," Race said. "But if you look on the label, you'll notice that it says 'Made in Japan' right there beside the ingredients."

"Seriously?" Kalinske asked. "The Japan thing again?"

"Trust me," Race explained, "I don't give a shit about the color of someone's skin or the slant of their eyes. All I care about is doing business, and our friends in Japan seem intent on fucking us at every turn."

Kalinske nodded, because he shared a lot of the same frustrations. He also wasn't completely surprised to hear this. He knew that Race had been locking horns with some of the folks at SOJ, who had taken to calling him "racist" in Japanese behind his back. Kalinske just hadn't realized this had mounted to the point that it was preventing Race from accepting the opportunity of a lifetime. "I understand where you're coming from," Kalinske said. "Believe me, I really do. Sometimes it drives me crazy, but at the end of the day Nakayama has always been fair to me and let me run this how I want to."

"LPGA Tour?" Race asked.

"Fine, almost always," Kalinske said with a lighthearted roll of the eyes. "But don't let that stuff drag you down. I can protect you."

"Can you, though?" Race asked. "And even if so, for how long? Do you really think that success is going to keep that at bay?"

"I do, I really do," Kalinske said. "But I realize that I'm not in a position to offer you a lifelong guarantee. What I can offer you, though, is the opportunity to lead a marketing assault unlike anything the world has ever seen before."

"Go on," Race said, raising an eyebrow.

"A little over a year ago, when I went to Japan, Nakayama gave me the green light to hire a new ad agency and go to town. I've been waiting for the right moment, and I'm thinking this would be a great way to ring in the New Year. I want to get going on this right away, but before getting started I want to make sure that you'll be here through the whole agency review process. We need a leader here. Can I count on you?"

Race was flattered but didn't answer right away. "I don't know," Race said, playing out various scenarios in his mind. "But I think we've both been around the block enough times to realize that anything short of 'yes, sir, absolutely' is probably going to come back and haunt one of us down the road."

"You're right," Kalinske said with a sigh. "Thank you for being the one to say it. Would you like to continue on in some sort of part-time consulting role?"

"I think a clean break would be best," Race said. "Though, if possible, I'd like to hang around for another few weeks. Train the new guy, show him around CES, and just generally plant seeds of dissension where I see fit."

"Of course," Kalinske said. "And I want you to know that even though this didn't quite work out as I had hoped, you were worth the wait."

"Thanks, Tom," Race said. "I know you have a very high opinion of me, probably inaccurately so, but I want you to know that I think the same of you."

Kalinske nodded, touched.

With the sappy stuff out of the way, Race leaned back in his chair, making himself at home. "Now on to more important matters. You excited for CES?"

21.

THE HEART AND THE BRAIN

Big Bang Beat flooded the Alexis Park Hotel's echo-friendly Parthenon Ballroom with an energetic smorgasbord of rock, soul, disco, and swing. The twelve-piece party band played hits from the sixties, seventies, and eighties with such shimmering gusto that it felt like a bar mitzvah, graduation party, and wedding all rolled into one. This was Sega's opening-night party at the 1992 winter Consumer Electronics Show, and in many ways it was all of those things. It represented an unofficial coming-of-age ceremony, a graduation to the major leagues, and an irreversible marriage into the videogame industry's royal family. This was everything Kalinske had been waiting for.

In an elegant sport coat and playful Sonic The Hedgehog tie, Kalinske danced under the high ceilings of the gold-paneled ballroom. He normally didn't like to get up and dance, but tonight the auditory ecstasy was contagious. It was as if the music exploded out of the band's oversized instruments, zigzagged around the crowded room, and latched on to the wrists, hips, and ankles of the guests, pulling them to the dance floor. The music captured Kalinske, his employees, and even the jet-lagged representatives of SOJ. It infected members of the press, third-party developers, and Sega's corporate partners. The music even managed to seduce the retailers, an old-school breed of businessmen who rarely unleashed their embarrassing array of dance moves.

Van Buskirk was dancing beside Kalinske and Adair. "I take it you're enjoying your first CES?" Kalinske asked her.

"Sorry," Van Buskirk said, looking around the room to see which of the dancers looked most ridiculous. "Mental photography session in progress. I wish I had a real camera with me—I can't believe how uncoordinated some of these guys are!"

"And you?" Kalinske asked Adair.

"Considering that a couple of hours ago I thought I was fired," Adair said, "I would say that I'm having a grand old time."

"Fired? What? Why?"

To the sounds of a high-energy rendition of "I Will Survive," Adair relayed the details of her afternoon. She had been in the middle of taking buyers through the upcoming titles for Game Gear at Sega's booth when someone came by with an ominous-sounding message slip for her: Nakayama was in town for the occasion and wanted her to come up to his suite immediately. After instantly going pale, she excused herself and went to Nakayama's room as instructed, believing that she was about to be canned six weeks into her stint at Sega.

When she got to his suite, Nakayama quizzed her on a multitude of topics. Why had Nintendo been successful? Why did she want to join Sega? What were her parents like? After the lightning round ended, Nakayama nodded and congratulated her on a job well done. Before he could dismiss her, though, faint tremors from a small earthquake shook the room. Adair brushed off the jolt, but Nakayama immediately dove headfirst underneath a nearby conference table. Adair bundled together all the strength she'd gathered from her adult life, thanked Nakayama for his time, and ran back to Sega's booth, cackling all the way. Even now, two hours later, she couldn't help giggling at the memory.

"No way," Kalinske said. "Just no way."

"Yes way!" Adair said. "And if you're imaging what it looked like, the answer is yes, his combover did get all ruffled in the process."

"Now it's *really* too bad you didn't have a camera with you."

"I know!"

"Where is our mutual friend?" Kalinske said, scanning the ballroom. "I haven't seen him this evening."

"Oh, God," Adair said. "You can't tell him that I told you. Please. I know I should have prefaced the story with that, but please don't tell him."

"Of course not. We're a team," Kalinske said. She smiled with relief, and they went back to dancing alongside Rioux, Toyoda, Schroeder, and Nilsen, whose giant public smile kept being interrupted by a sly private one whenever he thought about an incident that had occurred earlier in the day. He had been walking through Caesars Palace with Ken Balthaser and Clyde Grossman, normal as a day at a Vegas casino can be, until someone noticed the three of them were wearing Sega jackets. After it was discovered that they worked at Sega, they were instantly treated liked celebrities; someone even offered to comp their gambling if they were willing to reveal Sonic game tips while playing cards.

It really was a night to remember, with nobody able to resist the silky allure of the music.

Except for Nakayama. It appeared that no amount of corporate success could coerce his feet into dance steps. Yet it was clear from his shining eyes, judicious nods, and executioner's grin that he was pleased by the world around him. He celebrated the occasion at a table in the corner of the ballroom, speaking passionately with a thick-jawed man in a sharp suit from a decade long gone. This was Nakayama's friend and business partner, and Sega's original founder, David Rosen.

Out of the corner of his eye, Kalinske saw Nakayama and Rosen seated together. He waved in their direction, but they did not wave back; they probably didn't notice, he told himself, but a small part of him suspected that they had snubbed him as some kind of maneuver in their series of bizarre head games. Of all the business relationships that Kalinske had seen in his time, the one between Nakayama and Rosen easily qualified as one of the strangest. There was clearly a good deal of respect between the two men, but their opposite personalities, the geographic distance between them, and decades of drama made for an unusual dynamic between the two most important figures in Sega's history: the man who had created the company and the man who had shaped it in his image.

Like most tangled relationships, the one between Nakayama and Rosen began years before the two ever even met. In 1949, David Rosen, then an eighteen-year-old ambitious pragmatist, enlisted in the United States Air Force. His unit was sent to Asia and stationed at various bases throughout the continent before, during, and after the Korean War. When his service ended in 1954, the Brooklyn-born entrepreneur remained in Japan and formed Rosen Enterprises, whose greatest success came in the photography business. At the time, the Japanese had a strong need for ID photos (to be used for anything from school applications to rice ration cards), but photographs were expensive and required a substantial amount of time, with the typical photo studio charging 250 yen and taking about three days. Rosen's solution was to import coin-operated photo booths from America and set these up throughout Japan. By charging less and delivering photos instantly, the fully automated booths became an instant success. From there, Rosen Enterprises began importing all kinds of coin-operated products like jukeboxes and pinball machines.

In 1965, Rosen Enterprises merged with Nihon Goraku Bussan, a prominent Japanese jukebox company, which, most important, had a large manufacturing facility in Tokyo that was idle. Rosen was made chairman of the new venture, which was named after the English translation of Nihon Goraku Bussan, "Service Games"—hence SEGA Enterprises. A year later, Sega created its first original product, an electromechanical submarine game. Without the luxury of digital code, computer chips, and colorful screens, electromechanical games ran on a tightly rigged system of switches, relays, motors, and lights. In the evolutionary track of interactive entertainment, they were nothing but monkey business compared to the arcade videogames that would soon follow. But in 1966 Sega's electromechanical contraption was leaps and bounds ahead of the pinball machines and skee-ball games of its era. Their original creation was a bulky red and gray machine that stretched to ten feet long and six feet wide, highlighted by an actual periscope in front. The game consisted of players peering into the viewfinder and swiveling a torpedo-shaped light to aim at enemy ships. Successful assaults were rewarded with flashing red lights and a gurgled explosion, while failures received nothing more than a dismal whoosh. Though it cost twice as much as any other game, Sega's aptly titled Periscope was a major hit in Japan. In addition to providing cash flow, it turned out to be an undefined variable that allowed Rosen to flip the international business equation. Instead of importing games from America and selling them in Japan, Sega would now make their own games in Japan and export them to America.

Periscope quickly became a major hit in the United States as well, generating so much demand that it became the first electronic game to dare to charge 25 cents per play. Sega followed up Periscope with a slew of similarly sophisticated games with equally uninspired titles, such as Basketball, Drivemobile, and Helicopter. With a reputation for quality, innovative technology, and a cross-continental distribution network, Sega became attractive to multinational corporations with an appetite for potential. In 1969, Rosen agreed to sell Sega to Gulf and Western on the condition that they kept him on as the company's CEO. Under his command, Sega continued to find steady revenues with a succession of electromechanical games like Soccer, Sea Devil, and Lunar Rescue, but jumped to new heights of influence with the creation, popularization, and global sensation of arcade videogames. Following the success of Atari's Pong in 1973, Sega manufactured their own unabashedly similar arcade game, Pong Tron. The game was a smash hit, prompting Sega to shift its focus to videogames and, very nearly, do more than just copy Atari. In 1976 Rosen negotiated with Nolan Bushnell for Sega to acquire Atari, but on the day that they were to draft the contract, Bushnell backed out because he learned that his company had successfully developed a new console that could play more than one game (through the innovation of cartridges). That became the Atari 2600 and led to untold riches, but Sega continued thrive in its own right as well. Rosen's Sega spent the decade mastering the art of videogames, progressing from a me-too mentality (with games like Pong Tron 2 and Fonz) to an artistic devotion to the craft (with games like Blockade and Monaco GP). As Sega's success grew, so did the constant problem of bootlegging, which led Rosen to cross paths with Nakayama.

In Japan's highly stratified society, Hayao Nakayama aspired to rise to the top. But for someone with humble origins, an unremarkable family name, and a memorable rough-and-tumble personality, social elevation posed a major challenge. No matter where he went or what he did, it seemed that he'd always be walking around with the concrete blocks of his past tied around his ankles. But that didn't stop him from trying. If the past would not afford him access to certain banks, restaurants, or other places of status, then Nakayama's passport to the world would have to be his cunning resilience.

He attended college to become a doctor, but wound up dropping out midway through the process. To find the next rung in his ladder to the extraordinary, Nakayama skimmed the newspaper's classified ads and pounced on the opportunity to become a salesman for a jukebox leasing company. As opposed to the fat cats upstairs who did little more than plan cushy vacations for themselves, Nakayama learned the amusement industry from inside the belly of the beast. With a preternatural talent for detecting failures, inefficiencies, and misgivings in others, he quickly discovered many ways that his company should change, and he particularly touted the need to get into the arcade business. His insights fell upon deaf ears, so he left to start his own arcade distribution company, called Esco Trading.

Nakayama lived vicariously through Esco Trading, which serviced and repaired arcade cabinets for companies like Sega. And with arcade fever sweeping the globe, life was good. But Nakayama wanted more than just good. He expanded Esco's operations, first to buying and selling used equipment and eventually to bootlegging American arcade games. Consequently, Esco became a threat to Sega, which prompted Rosen to ask for a meeting with the bandit who was cutting into his business. Rosen was not predisposed to like what he found: a street-smart, business-savvy guy with a gluttonous appetite for risk. But Rosen also believed that Nakayama had a marvelous sense of what machines the market would accept, which convinced him to make Nakayama an offer instead of threatening him. In 1979, Sega Enterprises purchased Esco Trading and made Nakayama the head of its Japanese operations.

By joining forces, they could focus on keeping Sega ahead of the curve while also thwarting an array of copycats who had recently cropped up, looking to make a quick buck in the arcade business with a horde of derivative games, companies like Irem, Nichibutsu, and Nintendo. By the early 1980s, Sega was generating revenue of over $200 million per year with hit games like Astro Blaster, Head-On, and Zaxxon. To outsiders impressed with Sega's success, it was easy to compartmentalize the credit and declare Nakayama the heart and Rosen the brains—the doer and the thinker, the yin and the yang. But the truth was that despite their very different personalities, they were kindred spirits, each with an unflinching hand on the

Ouija board of success. Which made it all the more startling when Rosen retired in 1982, voicing concerns that the arcade business was headed for trouble. He turned over all control to Nakayama, then moved to California with his wife for their happily-ever-after.

Nakayama continued to expertly steer the ship in Japan, but the videogame crash of 1983 devastated the American business. With Sega reeling, Gulf and Western looked to unload the company. There were few suitors looking to get into the videogame business while everyone else was getting out, but Rosen still believed in what he had started, and so he formed a buyers group with Nakayama and Isao Okawa, whose company, CSK, put up most of the $38 million they paid to buy back Sega. Under the new arrangement, Nakayama would run the majority of Sega's operations from Tokyo, Rosen would run the company's small American subsidiary from Beverly Hills.

While Nakayama and Rosen were busy putting the pieces back together at Sega, their progress was dwarfed by Nintendo's success, which forever reconfigured the videogame landscape. In a few short years, Nintendo found viability in the once-barren handheld game market with its Game & Watch, conquered the arcade industry with Donkey Kong, and resurrected the home console business with the NES. What Nintendo managed to do in only a few short years made Sega's decades of success look like a child's artwork on the refrigerator of life: kind of pretty, but also kind of pitiful. Not only had Nintendo single-handedly revived videogames in the United States, but their utter domination transformed it from a niche industry into big business.

There was a janitor's ring's worth of keys to Nintendo's success, but beyond the lucky alignment of people, places, and things, their invasion of America appeared to be the critical element. Success abroad changed the perception of Nintendo from a trendy gizmo company into a cross-cultural powerhouse. In addition to worldwide clout, Nintendo's American pipeline tripled revenues, quadrupled the customer base, and immeasurably multiplied the reach of their soon-to-be iconic intellectual properties. Nakayama and Rosen couldn't turn a blind eye to such scorching success, forcing them to play a game of follow-the-new-leader.

In 1985, two years after Nintendo's Famicom launched in Japan, Sega released an 8-bit console of their own. In 1986, one year after Nintendo's

U.S. invasion, Sega set their sights on also infiltrating the American market. In an attempt to replicate their competitor's staggering success, Rosen poached Nintendo's VP of sales, Bruce Lowry, to become Sega of America's first president. Lowry was handed the keys to the car, with Rosen supervising from the passenger seat as Sega of America's chairman and CEO. They zipped off to a fast start in June 1986, selling 125,000 units of Sega's Master System in the first four months. Even so, Sega could hardly compete with Nintendo, which had sold two million NES systems over the same period. In the coming months, Sega fell even further behind, and by 1987 Nintendo commanded 85 percent of the market. Unable to continue pouring money into a losing proposition, Sega threw in the red-white-and-blue towel and chose to focus on coin-operated arcade games. They granted all of the marketing and distribution rights for their Master System to the toy manufacturer Tonka, who had no experience in the videogame industry and over the next two years did little to change that reputation. Part of this was due to the fool's errand it had become to compete against Nintendo, and the other part was due to Nakayama's aggressively marking up Sega's products to the point at which Tonka was left with almost no profit margin.

One year after conceding the market to Nintendo, Sega returned to the party in a snazzy new outfit. Under the direction of Hideki Sato, the head of R&D, a team of engineers were successfully able to condense the technology inside Sega's arcade boards to fit into a sleek black box. Amazingly, Sega had created the world's first 16-bit console, but Nakayama and Rosen were torn on how to proceed. Should they jump back in the ring and try to knock out Nintendo, or play it safe and license the technology to another company? After their previous console debacle, they opted for the latter, leading Rosen to approach several prominent companies, but only one seemed foolish enough to think they could actually beat Nintendo: Atari. He met with the head of their videogame division, Michael Katz, who quickly became convinced that Sega's 16-bit system could be the key piece in Atari's resurgence to the top. Rosen and Katz excitedly prepared a presentation for Atari's president, Jack Tramiel, who took one look at the thing and passed. Though Atari wouldn't be launching the 16-bit console, Michael Katz would: Rosen hired Katz to become Sega of America's second president and release this new console that Rosen had named Genesis, in the hopes that it truly would represent a new beginning for the company.

As before, Rosen was content to supervise from the sidelines, but Nakayama wasn't particularly impressed with this hire. Katz may have been clever and experienced, but he didn't know how to manage people. He was too lax, too indecisive, and too predictable. He also shied away from office politics in the worst way possible: he nobly thought that he was above it. Worst of all, Katz was a videogame journeyman at a time when Sega needed fresh blood. As head of the parent company, Nakayama could have nixed Katz at any time, but he didn't want to undermine Rosen—at least, not directly. Instead of throwing Katz off the cliff, he decided to clip his wings and then see if he could still fly. If Katz wanted to internally develop a certain kind of game, then maybe he'd be told it wasn't Sega's forte. If he found an outside developer who could do the game he wanted, then maybe Japan would scoff at the cost, or chastise him for seeking outside help when Sega was perfectly capable of doing it themselves. Further adding to Katz's hornet nest was the reverse-engineering situation with Electronic Arts. The night before the 1990 Summer CES, Trip Hawkins had met with Rosen and Nakayama to let them know that EA had reverse-engineered the Genesis. Not only that, but EA had supposedly set aside ten million dollars to fight any potential litigation. Nakayama was furious, as was Rosen, but as the cooler-headed of the two he was able to turn a potential shouting match into a more cordial negotiation that would last several months. This would eventually lead to a mutually beneficial arrangement, but for the time being it was just another headache for Michael Katz. Rosen would try to support him, but even that became a double-edged sword. Because Katz was his guy, Rosen was less willing to allow him to take risks—like when Katz wanted to hire a new ad agency and Rosen nixed the idea because their current firm, Bozell, was located nearby in Los Angeles and, even if they weren't the best, their work was reliable. Through no fault of his own, Katz became embroiled in something of a philosophical chess game between Nakayama and Rosen.

To keep tabs on Sega of America's new president, Nakayama relied on a tenacious young manager he had hired one month earlier: Shinobu Toyoda, who up until this point had served as chief lieutenant to SOA's interim president, Dai Sakarai. Coming from the Aerospace Division of the prestigious Japanese conglomerate Mitsubishi, Toyoda knew nothing about videogames but learned quickly from Sega's American president. Katz taught him what made for a good game and, more important, the value to Sega in offering a matrix of diverse titles (i.e., sports, puzzles, role-playing games). Toyoda was thankful for the tutelage, but gratitude didn't prevent him from agreeing with Nakayama. Six months into the job, Toyoda strongly urged Nakayama to find a replacement for Katz, preferably someone who could more smoothly manage internal conflicts and external relations.

Without notifying Rosen, Nakayama traveled to Hawaii and tracked down Kalinske. Throughout the courting process, Nakayama consistently asked Kalinske not to interact with Rosen. This seemed odd to Kalinske until he better understood their bizarre relationship: puppeteers who fought for control of their marionettes. Or at least that's how it appeared. Nevertheless, Kalinske took the job, Rosen appeared thrilled to have him on board, and now less than two years later, Sega began cutting into Nintendo's lead. Nearly forty years of drama, comedy, and innovation had led to this moment, with all of them in a ballroom in Las Vegas, celebrating past, present, and future.

Once again, Kalinske waved at Sega's masters of destiny, who this time took notice and gestured for him to join them. After a sequence of smiles, handshakes, and pats on the back all around, Kalinske took a seat beside Rosen and Nakayama. He allowed himself a moment to imagine Nakayama diving under a table in slow motion, then forced himself to push the image out of his mind.

"Pretty amazing, isn't it?" Kalinske asked.

"I am certainly amazed," Rosen said, "but certainly not surprised."

"And this is only the beginning!" Nakayama proclaimed.

"Well, now, let's not get ahead of ourselves," Rosen said. "I feel good about where we're headed, but for now we ought to just soak it all in."

"Yes, of course," Nakayama replied. "All I'm saying is just that this is tiny compared to where we are going."

Rosen rolled his eyes, which prompted Nakayama to do the same. Kalinske wondered how many times this kind of exchange had occurred before. For a moment, none of them said anything, and they simply looked out onto the dance floor at what they had built together.

"All right now, that's enough," Rosen finally said to Kalinske, using a paternal tone that he rarely invoked. "Don't waste another moment with us fossils. This is your moment. Go have some fun."

"Stop it," Kalinske said. "I could sit here for hours."

"Well, you'd be sitting alone, then," Rosen said, standing. "Hayao, you can stay if you want, but I'm going to retire to my room and see what's playing on the television."

"This is a good idea," Nakayama said, sporting a grin fit for a Bond villain. "I've got some other business to attend to."

They said their goodbyes and left Kalinske alone at the table, where he remained for some time. With a peaceful sigh, he gazed back out onto the dance floor to take in the wonderful sight of his employees letting loose. It was rare for him to see them and them not to see him; usually it was the reverse. Kalinske was used to being in the spotlight and liked it that way, but tonight was a monument to Sega's success, which couldn't have been accomplished without them. No matter how big or small their contributions, every single one of them was a necessary part of the equation.

"Pretty amazing, isn't it?" someone asked, snapping Kalinske out of his trance. It was Emil Heidkamp, senior vice president at Konami of America. Heidkamp was a kindhearted, thoughtful salesman, who had been with Konami since 1986, from the very first day they started making home videogames. He'd been responsible for the launch and success of some of Nintendo's best-selling titles, like Castlevania, Contra, and Teenage Mutant Ninja Turtles. Under Heidkamp's leadership, Konami had made Nintendo wealthy, and vice versa. As a result, he'd grown close to Nintendo's top brass, fostering a relationship that he'd recently put on the line by agreeing to make games for Sega.

"I am certainly amazed," Kalinske said, echoing Rosen's words from earlier, "but certainly not surprised."

"Nor should you be," Heidkamp said, gently patting Kalinske on the shoulder. "You and your team do good work. I have nothing but respect for the guys in Redmond, but I'm glad that Sega managed to enter the picture while I was still here."

"Still here? Why? Do you have plans to go somewhere?"

"Not at the moment. But I think that my time is coming to an end."

"Emil, is everything okay?" Kalinske asked with sudden and genuine concern.

"Oh, no, nothing dire," Heidkamp said, realizing how ominous he had sounded. "Did I never tell you about the deal I made with Mr. Kozuki?" he asked, referring to Takuya Kozuki, the president of Konami.

Kalinske shook his head.

"It was shortly after I joined Konami," Heidkamp said, squinting a little as he reached back for the details. "I found the Lord and became a born-again Christian. I said to Kozuki-san that there had to be a certain kind of purity to our games. I didn't want us going down the road of lowest common denominator with blood, nudity, and debauchery. After all, we're in the business of selling entertainment to kids. We have a certain responsibility, don't we?"

Kalinske nodded, captivated. Heidkamp's story put words to a collection of tiny doubts that had been building up in the back of his mind. He hadn't been able to put his finger on the nature of these barnacles of dread until now. "What was the agreement?"

"I made a deal with Mr. Kozuki that I would stick around and continue to take Konami to the top as long as we never did anything worse that cartoon violence. He didn't hesitate for one second and agreed to the deal. Even more important, he backed up his words with actions. Right around that time we had a game out of Japan called Dracula Satanic Castle, and he let me rename it Castlevania and make other slight modifications. I consider Mr. Kozuki a great friend and I have no doubt that there is eternal truth to his words, but as I look around this industry that we're all creating, I can't help but realize it's only a matter of time."

"It doesn't have to be," Kalinske said, shaking his head. He wasn't sure if he disagreed with Heidkamp's outlook or if he just didn't want it to be true.

"No, it doesn't have to be, but likely it will," Heidkamp said. "Have you seen what's coming into the arcades these days? The most popular game is Street Fighter, where the entire purpose is to clobber your opponent. I'll admit that beyond the premise the game itself is tame. But how long do you expect that to last? The world is full of slippery slopes, and once you start going down . . . well, there's only one way to go from there."

"Right," Kalinske said vaguely, looking like he'd seen a ghost.

"Oh, Tom, I'm sorry. I didn't mean to burden you with my personal woes," Heidkamp said, shaking his head. "All I meant to do was come over here and congratulate you on a job well done. Don't give another thought to this, okay?"

Kalinske nodded, trying but failing to put it out of his mind.

"Come on," Heidkamp said, standing. "Let's get back on the dance floor."

"Sure thing," Kalinske said. "I'll meet you out there in a couple minutes."

Heidkamp nodded and joined the swelling crowd on the dance floor. Kalinske watched from the table, trying to reconcile what he was seeing with what he had just heard. Had he been looking at this night all wrong? Was it not a celebration of passion, creativity, and hard work, but rather a villainous triumph of pulling the wool over the world's eyes? At least when he'd sold toys and faced similar claims of impropriety, there was always the fallback consolation that in addition to entertainment he was helping to activate a child's imagination. But that wasn't quite the case with videogames, whose immediate feedback and preprogrammed outcomes could be seen as the opposite of imagination. Part of Kalinske wanted to laugh and discard such ridiculous thoughts, while another part wanted to dive under the table like Nakayama and hide from the world while he gave the nature of this business greater consideration. Back and forth the thoughts volleyed, playing out the struggle between heart and brain.

Eventually Kalinske shook off the dueling thoughts and let the music woo him back to happiness. Big Bang Beat was playing a fast-paced version of "Celebrate," which reminded Kalinske that tonight he could not be a man divided. Nor could he be one tomorrow, or any day after that. He owed it to himself, to make the most of this wonderful opportunity. He owed it to Nakayama, who had picked him to be his guy. And he owed it to all the people here tonight, people who meant a great deal to him, who worked harder every day, and who deserved everything they'd ever dreamed for themselves.

Kalinske refused to let a petty case of conscience jeopardize everything that lay ahead. First they would need to pull even with Nintendo, and next they would take the lead. He didn't know what would happen after that, but he vowed that it would be beautiful and extraordinary. Kalinske got up from the table and joined the bouncy, happy people on the dance floor. Tonight Sega had arrived, but as Nakayama had just said, it really, truly was only just the beginning.

PART THREE

THE NEXT LEVEL

22.

TARGET PRACTICE

"It's not arrogance," Peter Main explained to a room full of skeptical faces. When he delivered these words, he truly believed them with all his heart. The owners of those skeptical faces, however, felt fairly certain that Main's heart was made of stone. It was December 1991, three weeks before Winter CES, and he and Randy Peretzman were in Minneapolis for an emergency meeting. One week earlier, they had met with Target's buying group to discuss Nintendo's new returns policy and management was subsequently so incredulous that they were "invited" back for an encore performance—this time with the senior VP, EVP, and president. "Arrogance? Not even close, fellas. It's just good business sense, that's what it is."

"It's the truth," Peretzman added with a friendly nod. "If you would like, we're more than happy to send you a copy of the numbers."

It took a moment for Target's managers to fully absorb the news, and when they did, they were not particularly moved. "This is not about numbers," the senior VP explained. "It's about relationships. Our relationship with our guests, and our relationship with Nintendo."

"We understand, and we respect your point of view," Main conceded, torching his words with a light touch. "But we need you to understand where we are coming from as well. If this keeps happening, then it's only a matter of time before we go bust."

"Oh, come on!" someone scoffed, outraged by the notion that Nintendo's empire could crumble so easily. "It's a minor problem at best."

In the silence that followed, Main and Peretzman locked eyes in a moment of shared commiseration. They had not expected this to be an easy task, but anticipating the push-back didn't make enduring it any more enjoyable. "With all due respect, fellas, this is a major problem, and our new policy is the only conceivable way to fix this thing."

Whether minor or major, the problem itself was relatively easy to understand. Shortly after the Super Nintendo was released, customers began returning 8-bit NES consoles to their nearest retailer and then using this refund to buy the 16-bit system. From Nintendo's perspective, this was understandable, but only up to a certain point. If someone had bought an NES in August and then a month later, when the SNES was released, they wanted to return it for the newer model, that was fine. Nintendo didn't ever want their customers to feel cheated. But by the same token they didn't want to feel cheated themselves. And that's exactly how they felt when customers began returning NES systems they had played for years, some originally purchased as far back as 1985. In effect, they were flipping the old consoles into the new ones for nothing more than a small fee to cover the price difference. The reason they could to do this was because of the extremely liberal return policies that most retailers had in place.

In 1991, someone could walk into a store with no receipt and no original packaging—they didn't even have to have all the parts ("Um, I don't remember this NES coming with a controller")—and receive a full refund by claiming there was a defect. No proof had to be provided, nor did the retailers even really care. In their eyes, the customer was always right, and financially it didn't hurt them much because they would just send the supposedly defective product back to the manufacturer for their own refund. As a result of retailers empowering their customers with the gift of infinite rightness, companies like Nintendo would suffer the costs of that moral wrongness.

Perhaps Nintendo would have felt more sympathetic about the situation if their products had been considered notoriously defective. But that was not the case, not even close. Because of Yamauchi's fixation on controlling every tiny detail, and Arakawa's unyielding dedication to the user experience, Nintendo enjoyed a famously low defect rate (less than 1 percent). In addition, NOA was one of the first companies to devote an entire department to studying, analyzing, and managing the returns process. Given this unrelenting commitment to excellence, was it so much to ask that the retailers getting rich off Nintendo stop turning a blind eye to the abuses of that process?

Peter Main shook his head. "Just ask yourselves where this will end. Let's say that we continue the way we've been going. Then what happens five years from now when we release a 32-bit system? And then five years after that, when we come out with a new 64-bit machine? And so on, and so forth? Are you honestly going to let these people keep turning in something they bought in 1987 for the rest of their lives?"

"Well," someone said after a collective shrug, "what's the alternative?"

"I just told you the alternative!" Main declared. "Our ninety-day return policy. I think that you will agree that three months is more than enough time to figure out if something is defective. Is that really too much to ask of your customers?"

"*Guests,*" Target's senior VP said, forcefully pronouncing the word.

"Huh?" Main said, mentally replaying his previous sentence. "Which guests?"

"You said *customers*. But at Target we don't think of our visitors that way. We consider them *guests* and we treat them as such."

"All right," Main said, making every effort not to roll his eyes. "Your guests will be given ninety days, which certainly seems like a fair amount of time."

"Come on, guys, that's three whole months," Peretzman said. "Wars have been won and lost in much shorter amounts of time."

"Regardless," someone else said, "how would we even enforce such a thing?"

Finally, thought Main, a good question. This was another aspect where Nintendo demonstrated that they were logical innovators, not unreasonable dictators. To create a clearer picture of a product's life cycle, Nintendo had conceived, developed, and patented a system called POS electronic registration. This system allowed retailers to track a product from manufacture all the way until purchase, and also offered other features like real-time sales reports and trend analysis. For Nintendo, this methodology had been invaluable for managing inventory, and they hoped that retailers would find it similarly useful; ideally they would even use it for non-Nintendo consumer products as well. "The system," Main said after describing it, "is all about efficiency. But, as with all of our advances, it's about more than just Nintendo."

"Gentlemen," Target's senior VP stated, "I think you are missing the point."

Once again, Main and Peretzman briefly locked eyes, this time with mutual curiosity about how this could possibly be the case. "With all due respect," Peretzman said, "what is it that you think we're missing here?"

The senior VP squinted for a moment, evidently looking for the most delicate way to express himself. "The returns issue may be a big problem, but forgive me for saying that it merely feels like today's big problem. First it was the allocations. Believe me, I get it. 'There's only so much product to go around! There's a chip shortage in Asia! There's no way we can give you everything you want because that's what Atari did and look at what happened there!' Next it was getting rid of December 10 billing, which, once again, I get. You sell us something, you expect us to pay for it right away; that makes sense. So does North Bend, to a certain extent."

"North Bend" referred to Nintendo's newly opened distribution center in North Bend, Washington, which was the latest disruptive innovation from Minoru Arakawa. In a pattern that could be traced back to his engineering roots, Arakawa was forever seeking ways to improve efficiency. As physical proof of this obsession, he carried around a thick tan Ultrasuede portfolio that was always overflowing with cost sheets, inventory reports, and all sorts of additional data that would put most company presidents to sleep. But Arakawa relished the details; he loved knowing exactly how many cents it cost to purchase the little screws that held the game cartridges together, and he loved even more looking for places to buy them more cheaply, finding a route that shipped them faster, or any other way to improve productivity. Most of the time, Arakawa's attention to detail manifested itself in small ways, like a better deal on packaging, but occasionally it would result in an industry-wide shakeup, as it had with the new $60 million distribution center in North Bend.

From the moment game cartridges came into this 360,000-square-foot distribution center until the moment they were loaded out for delivery, the merchandise was touched only twice by human hands. Instead, the majority of the work was done by highly sophisticated robots, unmanned forklifts, and an enormous mainframe computer, which were collectively responsible for getting the products out to retailers as soon as (in)humanly possible. North Bend operated with a futuristic level of efficiency, but a part of that efficiency required that retailers had to use Nintendo's distribution center instead of their own. So instead of the retailer receiving their order at a central warehouse and then parceling it out to their stores as they saw fit, Nintendo now wanted to be the one doing the parceling. Nintendo would, of course, follow the retailer's instructions about which products go where, and they would do so with incredible efficiency, but from the retailer's perspective there was more at play here. Efficiency was a good thing, but what was most efficient for Nintendo was not necessarily what was most efficient for the retailer. Maybe the retailer wanted to hold off on selling a certain game. Maybe they didn't want to provide such great transparency into their operations. Or maybe they just didn't want to pay Nintendo for a service that they already provided, even if perhaps they didn't provide it as well. Whatever the case, retailers didn't have much of a choice; as with most of Nintendo's policies, it was Nintendo's way or no way at all. And the fact that they didn't have a choice often bothered the retailers more than the actual situation itself. The retailers believed this was just another example of Nintendo's arrogance. Meanwhile, Peter Main and Randy Peretzman believed this was just another example of Nintendo's innovation being misconstrued as arrogance. And though neither side would ever admit it, in a way they were both absolutely correct.

"The point is," Target's senior VP continued, "I can appreciate the intentions behind everything that Nintendo does. And, as a result, we've always found a way to comply with, shall we say, your pioneer spirit? Sometimes doing so has cost us money, sometimes it cost us manpower, and other times it cost us the flexibility to run our company in the manner we desire. But these are all costs that we are willing to accept. What we are not willing to accept, however, is now putting the cost on our valued guests. And this latest, it just goes too far."

"But your cust—" Main began, this time catching himself. "Your *guests* are abusing the relationship. You have to factor this in to some degree."

"What's to factor in? This is just the way things are done. Look at Nordstrom. They will take back shoes for up to ten years!"

"Yeah?" Main shot back. "Well, we're called Nintendo, not Nordstrom."

The senior VP let out a sigh. "Peter, Randy," he said, "we are not trying to be difficult. But please just understand that this policy really violates the whole tenet of how we view our guests. It's all about trust. And no matter what, we cannot violate that trust and interfere with the right of our guests to return anything they've ever bought from us."

Main nodded, scouring his mind for any last shred of common ground. "I hear you," he said. "And it is not our intention to interfere with what you've got going on. If you want to keep allowing them to return anything, go ahead. Heck, let them turn in their black-and-white televisions if you want. All I'm saying is that starting in 1992, you're not going to be returning any of that stuff to us anymore."

Although it had become one of those meetings where hands inadvertently curl into fists, there was enough respect between the two companies for things to end on a more positive note. "Give us the holiday break to think about it, okay?" the VP said.

"Sure thing," Main said.

"And call us with any questions," Peretzman added.

The guys from Nintendo bid adieu to their hosts and then caught a plane headed east, girding themselves for the same uncomfortable conversation with the next retailer on their list.

"Well," Peretzman said as he raised an airplane-sized bottle of vodka, "I don't think the retail community will ever vote Nintendo as most popular."

"No sir," Main said, raising his own tiny bottle of vodka. "But let me tell you, it sure is a hell of a lot better to be the guy with a target on his back, then one of the guys who has already been shot down."

They tapped their drinks together and toasted to the virtues of persistence, resilience, and misconstrued arrogance.

Shortly after the holiday, Target requested another meeting and also requested that Arakawa be there. With the beginning of the year chaotic for both companies, they decided to meet in Las Vegas on the eve of the Consumer Electronics Show.

The discussion was set to begin around 7:30 p.m. in Peter Main's hotel suite at Caesar's Palace, but because of a snowstorm in Minnesota the executives from Target were delayed. At some point, while Main, Lincoln, and

Arakawa passed the time, Arakawa decided to curl up on the couch for a nap. Finally, around 10:00 p.m., the Target guys arrived. After an exchange of apologies and pleasantries, they were ready to get down to business—except Arakawa was still asleep.

Main gently nudged his sleeping boss. "Arakawa? Come on, pal."

Nothing.

Main tried again, this time with a little more zest. "Arakawa," he said. "The guys from Target are here. You need to get up."

Without opening his eyes, Arakawa asked if they had changed their minds.

With the Target executives sitting only a few feet away, Main replied, "They're right over there."

"Yes," Arakawa replied. "But have they changed their minds?"

Main looked to the senior VP from Target for a response. "Well, no," the VP said. "But that's what we're here to talk about."

"Okay," Arakawa said. "Then I'm afraid we have nothing to talk about."

The following day, Target made an announcement stating that they would no longer be doing business with Nintendo. When pressed for a reason, the only answer they provided was a "difference of corporate philosophy."

For Nintendo, this news was not ideal, but they considered it nothing more than a temporary nuisance. Innovation always trumps stagnation, and Target would eventually realize the error of their ways. Until then, there was nothing for Nintendo to do but keep their eyes on the prize and keep doing what they thought was right. And if these were the crosses they had to bear, so be it.

Setbacks like these irked Peter Main, but one of his talents was the ability to put on blinders when he needed to. So when he delivered his address at Nintendo's 1992 Winter CES event, he was nothing but smiles, sunshine, and sophisticated sarcasm—with one exception. For the first time, Main directly acknowledged his competitor. He openly disagreed with Kalinske's sales figures, debunked his rival's rosy forecast for the upcoming year, and, upon introducing one of Nintendo's new peripheral devices, added one more amenity to its list of features.

"The Super Scope can do it all," Main proclaimed. "With a comfortable shoulder mount and an infrared transmitter, Nintendo's newest light gun has pinpoint accuracy from long distances." He then casually pointed the gun toward his competitor's booth. "It's also rather perfect for hedgehog hunting."

23.

SEQUELS, SKIRMISHES, AND BASEBALL IN SEATTLE

Kalinske sat at his desk, discussing early plans for Sonic 2 with Toyoda.

"Do you really believe it will be possible?" Toyoda asked. "For us to have the same magic touch as the first time?"

"Absolutely," Kalinske said. "I have no doubt whatsoever. As long as the game is half as good as the first, then we'll make it the most successful launch ever. Trust me."

It was a blustery day in late January, but, as always, the weather didn't matter. It could have been sunny, snowy, or apocalyptic outside, and that wouldn't have slowed down the scrapping, scheming, and strategizing that went on inside.

"You realize that without you, none of this would be happening, right?" Kalinske asked. "You absolutely saved the day."

"I just did what was best for the company," Toyoda humbly replied, unwilling to let Kalinske's sentiment rattle his aw-shucks demeanor.

"Yes, you did. But in the process you also saved the day. And I think the only fair thing to do now is to buy you a cape."

"Like Superman?"

"Exactly," Kalinske said. "Of course, yours doesn't need to be red like his. The color choice is completely up to you. But whatever you pick, it should match your boots."

"Boots too?" Toyoda asked.

"Look, I know that I've already thanked you a million times," Kalinske said, speaking more like a friend than a boss, "but you deserve every one."

"Thank you, Tom," Toyoda said, sounding more like a friend than an employee.

"Seriously, I don't even want to think about the mess we'd be in without Naka," Kalinske said, shaking his head at how such an alternate future would play out. He was referring to Yuji Naka, the creator of Sonic The Hedgehog, who was known for his genius programming skills, legendary perfectionism, and notorious temper. In late 1991, following Sonic's staggeringly successful release in the United States and its less overwhelming but still impressive release in Japan, Naka was discontent. He was fed up with Sega of Japan for a variety of reasons, including his financial compensation (around $30,000, though he did later receive bonuses), his treatment by management (which was giving him a hard time for taking fourteen months to complete the game, as opposed to the standard ten) and for needing a team of four people (as opposed to the usual three). Naka was also bothered by the lack of recognition.

This game and the whole Sega vs. Nintendo debate never would have existed without him, yet he was hardly ever referenced anywhere for his hard work. It's not like he was asking for his name to be in the title, but Sega of Japan wouldn't even allow his name to appear at the end of the game. It was company policy not to credit anyone from the development teams, partially to instill an all-for-one attitude, but mostly to prevent other companies from knowing who did what, which would allow them the chance to poach programmers. Naka understood the policy, but that didn't make it right. Artists sign paintings, writers get bylines, and a movie director's work is hailed as "A film by So-and-So." His desire for credit was less about ego and more about peace of mind. Naka felt so strongly that he found a secret way to credit himself and his team. At the end of Sonic The Hedgehog there is a seemingly blank screen. In reality, however, it's a black screen with black type on it, revealing the names of those that had built the game. Because it was black on black, no player would be able to read this, nor would Sega even be aware that it existed, but Naka and his team would always know.

Sega of Japan knew that Naka was displeased with certain aspects of the job, but they were shocked when he unexpectedly quit, exiting Sega just as they were hoping to begin development of a sequel. When Toyoda found out about this, he traveled to Japan and met with the now-unemployed Naka. With Kalinske's blessing, Toyoda did everything in his power to convince Naka to remain with the company. Naka appreciated the effort, but he'd heard it all before . . . until Toyoda suggested that Naka come to America and make the sequel there. They would pay him more, guarantee him better recognition, and allow him to bring any ten employees of his choosing. Naka and his team would conduct their work at the Sega Technical Institute (STI), the game studio run by Mark Cerny in Palo Alto. Cerny was a game prodigy famous for creating Marble Madness when he was sixteen years old; no doubt Cerny would be a kindred spirit who could help smooth out any wrinkles in the process. Naka accepted Toyoda's proposal and came to America with the so-called Sonic Team to create a sequel for fall 1992.

"Is he happy here?" Kalinske asked now. "Naka, I mean."

Toyoda closed his eyes and silently laughed to himself. "No. But that's Naka. He'll never be happy. This is part of his process."

"All right," Kalinske said. "Well, don't hesitate to let me know if there's anything I can do to make him less unhappy."

"Of course."

"I'll check in with marketing to see if they've made any progress on the sequel," Kalinske said. "What about the rest of the development slate? How are we looking?"

"We're very strong in original product," Toyoda said. "But we can use some more intellectual property. I'd like to spend more time in Los Angeles to see what we can get."

"Perfect. Do it," Kalinske said. "Are there any doors you'd like for me to open? It would be great if we could get something going with Spielberg. Or even Lucas."

Toyoda's eyes lit up at the mention of the famous directors. "I will let you know."

"Great," Kalinske chirped, always pleased to show off his Rolodex. "Then let's talk pricing for a second. Nintendo responded quicker than we expected, and we need to react." On January 10, mere months after releasing the SNES, Nintendo dropped the price of their console from $199.95

to $179.95. Peter Main could claim that Nintendo was crushing Sega all he wanted, Kalinske thought, but a 10 percent drop like that said otherwise. There were also rumblings from retailers that Nintendo would slash the price even further in the coming months. The price war had begun, and Kalinske looked forward to the many skirmishes ahead. "I like where we're at price-wise, for the moment, but we need to think about staying aggressive."

"How aggressive?"

Kalinske did some quick calculations in his head. "How about $99.95?"

Toyoda nearly laughed. "Come on, Tom."

"Not right away," Kalinske said defensively, but Toyoda remained unconvinced. "Fine, let's discuss with Paul, and get sales and marketing to weigh in. Either way, we should look to have something ready for Boca." Kalinske was referring to Boca Raton, the site of the first annual Sega Summit. Though Sega had found success at CES and could now be considered a big fish in the big consumer electronics pond, Kalinske and company thought it would be even better to have a pond all to themselves. Out of this desire came the Sega Summit, where hundreds of retailers would be invited to Florida for a week of fun, sun, and golf, with intermittent presentations for products, sales strategies, and marketing campaigns. The summit's retailer-friendly approach presented another chance for Sega to prove they were the anti-Nintendo, and that they were here to stay. Sega was evolving the industry, and there was now room for two.

"That is not much time," Toyoda said, now doing some quick calculations of his own. The Sega Summit would begin on May 11. "But I think we can make it work."

"Good," Kalinske said. "Speaking of price, has Japan realized yet that they need to be more aggressive themselves?"

Toyoda sighed, but before he could reply, a graceful giraffe scampered into the office. No, actually, it was EBVB, moving at a speed that betrayed her usual elegance.

"What's going on?" Kalinske asked.

She ignored him for a moment to pat down his desk in search of something.

"Hello, Ellen Beth," Toyoda said politely.

But he too was ignored as her mysterious quest continued.

"What do you need?" Kalinske asked.

She stopped mid-frisk. "I'm looking for your remote control."

"Sure. What for?" Kalinske said, reaching into his desk drawer.

"I would like to perform a magic trick for the both of you," she explained. "In the blink of an eye, I will magically dispel the notion that all press is good press."

"Okay," Kalinske said, already smiling as he handed her the remote.

"Thank you, sir," she said, turning on the television. "Now, as you can both see, this is an ordinary television. Made in Japan, but plays garbage made in America." She rapidly flipped through the channels before stopping on a news station. "And without further ado, bippity boppity boo."

Van Buskirk tried to return the remote to Kalinske, but he was too focused on the breaking story to take it back from her. He could hardly believe his eyes and ears.

Howard Lincoln gallantly took the podium at the Madison Hotel ballroom and announced that executives from several of the area's leading companies would be presenting an offer to buy the Seattle Mariners. "All of the shareholders of the new company will be Seattle-area residents," Lincoln stated, "and all the decisions will be made here." He then explained that this newly formed ownership group, the Seattle Baseball Club, would be presenting an offer of $125 million to purchase the team. Unsurprisingly, his remarks set the ballroom filled with businessmen, journalists, and policy makers abuzz. The uproar was partially based on local fervor for the baseball team, but more so because the Seattle Baseball Club's lead investor was Hiroshi Yamauchi, Nintendo's Japanese-born president. And if his bid was accepted, then Yamauchi would become the first non-American to own a piece of America's pastime.

In truth, Yamauchi cared little for baseball. His game of choice was Go, the ancient Chinese board game renowned for its difficulty. But despite his lack of interest in balls and strikes, Yamauchi had agreed to bid when Arakawa revealed the opportunity one month earlier. Arakawa, like his father-in-law, also lacked a love for the game, but he believed that buying the Mariners and keeping them in Seattle would support a community that had already lost a baseball team once before.

It had happened back in 1970, just one year after Seattle had gotten their first professional baseball team, the Pilots. Following the Pilots' disappointing inaugural season, the team's owner, Dewey Soriano, realized that perhaps he had bitten off more than he could chew. The team was losing money, primarily because they played their home games at a decrepit minor league stadium that seated only 19,500 and had a reputation for losing water pressure by the seventh-inning stretch. It was a bad situation, but help was on the way. Just one year earlier, citizens of Washington's King County had approved a bond for $40 million to build a domed stadium, which played a heavy role in Major League Baseball's decision to grant Seattle a team in the first place. The stadium was expected to be complete by 1972, at which point the franchise would likely recoup its losses and begin to flourish. Soriano, however, couldn't afford to wait. The ownership group had already run out of money by the end of the season, and with a recent petition by stadium opponents delaying construction, Soriano wanted out. After the season, he engaged in a series of secret meetings with former Milwaukee Braves minority owner Bud Selig, who was leading an effort to bring baseball back to Wisconsin.

Several weeks later, during Game 1 of the World Series, Soriano agreed to sell the Pilots to Selig for $10.8 million. Despite having a deal in place, baseball's owners voted down the sale as a result of vigorous pressure from Slade Gorton, Washington's attorney general. Gorton believed that Seattle was being swindled, and he refused to allow the city to lose what it had fought so many years to get, especially when they had already earmarked $40 million in public funds to build a multipurpose stadium. Gorton urged a wealthy community member to buy the team, and he got his wish when Fred Danz, a local movie theater chain owner, stepped up to the plate. Danz strongly believed that having a major league team would make Seattle a major league city, and he also felt that the domed stadium would bring a lot of revenue and jobs into the community. In November 1969 Danz signed a contract to buy the team for $10 million and was quickly approved by the American League, but one month later the deal fell through when he revealed that he didn't actually have the money. The drama continued into 1970, and in early March the Pilots reported for spring training still unsure where they would play. With the season scheduled to begin in weeks, the owners granted approval to the original deal they had nixed, which would move the team to Milwaukee.

Once again Gorton protested. He filed an injunction on March 16 to stop the deal, but his legal action was trumped by another: Soriano's ownership group filed for bankruptcy and announced that they would be unable to pay coaches, players, and office staff. Up until this point, the team's equipment had been sitting in Provo, Utah, where drivers awaited instructions on whether they should drive to Seattle or Milwaukee. They got their answer on April 2 when the Pilots were officially declared bankrupt, five days before the season was set to begin. In the next five days, Bud Selig, the franchise's proud new owner, changed the team name from the Pilots to the Brewers, in honor of the Milwaukee minor league team that he had cheered on as a boy. Though he was able to change the name, there was not enough time to order new uniforms with the navy and red colors from those Brewers teams of yesteryear. Instead, the newly minted Milwaukee Brewers were forced to adopt the blue and gold of the Seattle Pilots, a color scheme that the team still wears to this day, and on April 7 this new-old team squared off against the California Angels. They lost 12–0.

Though the former Pilots lost the game that day and Seattle was devastated to have lost its chance at becoming a major league city, Slade Gorton refused to throw in the towel. He obtained the services of famed trial attorney Bill Dwyer and, on behalf of the city, they filed a suit against the American League for fraud, breach of contract, and violations of the Sherman Antitrust Act. They claimed that as a result of promises made by Major League Baseball, the city of Seattle had made several large expenditures, notably $1,115,000 to purchase Sicks' Stadium and $1,800,000 more to make improvements demanded by the league. The legal drama persisted for years until the American League voted on January 14, 1976, to give Seattle an expansion team in 1977. Baseball in the city was officially reborn on April 6, 1977, when the Seattle Mariners, rather fittingly, lost to the California Angels, though by a score of only 7–0 this time.

Attorney General Slade Gorton, now a local hero, became Senator Slade Gorton in 1981. And while it's impossible to say that baseball would never have returned to Seattle without his persistence, it's fair to assume that when he watches the Mariners take the field, he feels the pride of a parent watching his son play ball. For this reason, he was particularly troubled to learn on December 6, 1991, that current owner Jeff Smulyan had given up on Seattle and would be moving the team to Florida unless a local investor bought the team by March 6, 1992. His ultimatum tugged at the city's heartstrings, but despite communal love for the Mariners, nobody seemed willing to make a bid for the team—until Arakawa, who understood the emotions at stake, brought the situation to Yamauchi, who agreed to buy the team so that it could remain in the city where it belonged. To Gorton, who received the good news on Christmas Eve, this was utterly incredible, but the exclamation point came with an obvious question mark: would Major League Baseball approve the sale to a Japanese investor?

In order to increase the likelihood of acceptance, he met once again with the same local businessmen who had declined to buy the team earlier and asked these true-blue, apple-pie-loving Americans if they would consider purchasing a minority stake. Through these efforts, the Seattle Baseball Club was formed, with Arakawa serving as chairman, Lincoln as its spokesman, and a group of investors that included Hiroshi Yamauchi, Chris Larson (Microsoft's senior program manager), John McCaw (executive vice president of McCaw Cellular), Frank Shrontz (chairman of Boeing), and John Ellis (chairman of Puget Power). Of these seven individuals, only one was not at the press conference: Yamauchi, whose absence further ignited skepticism about Japanese investment in American baseball. After all, the only thing more suspicious than finding an elephant in the room is being invited to an elephant's party and finding no elephant at all.

Lincoln was aware of this fact and worked to downplay Japanese involvement with the club's operations. He explained that the financiers were "passive investors who intend to leave the operation of the club to professional baseball people." Though concern about foreign ownership would linger through the press conference and likely into the months to come, it was clear that Howard Lincoln was thrilled by this opportunity and the good that Nintendo was able to do for its home state. Unlike Yamauchi and Arakawa, Lincoln loved baseball, so perhaps it was fitting that he was the one up there on the stage. Or perhaps it was more fitting that game-changing news for a team that was born out of a lawsuit would be announced by a litigator whose legal mind was instrumental in building an 8-bit empire that had just shocked a three-dimensional world.

24.

FYRIRGEFNING SYNDANNA

Olafsson's thin fingers danced across the keyboard, the thrill of this ritual evident on his face. From his apartment on New York City's Upper West Side, he spent the morning's pre-dawn hours tapping away with the dizzying grace of a concert pianist. There was a certain satisfaction in seeing one's progress scroll across the screen, as well as the ability to delete unwanted thoughts. Word processors, in a sense, had given man the power to reinterpret the entire human experience.

In addition to being a physicist, Olafsson had taken up a side career as a writer. While rising up the ranks at Sony, he continued to pursue his passion for storytelling, and in 1991 the Icelandic publisher Vaka-Helgafell released his first novel, *Fyrirgefning Syndanna* (The Forgiveness of the Sins). The 286-page thriller tells the story of Peter Peterson, an expatriate businessman living in Manhattan, who suddenly passes away. In addition to bequeathing his two children a vast fortune and a Park Avenue apartment, Peterson has also left them a grave secret. They discover a sheaf of pages, written weeks before his death, that reveal a crime of passion committed in the throes of unrequited love, a crime that had burdened Peterson for his entire life. The story they unravel spans their father's boyhood in Iceland, the Nazi occupation of Denmark, and his business career in modern-day Manhattan. Olafsson's debut novel was well received, drawing comparisons to the work of Ibsen and Dostoevsky, and in December 1991, *Fyrirgefning Syndanna* was nominated for the Icelandic Literary Prize.

Writing and the art of translation were both difficult undertakings, but at least with those endeavors the artist remains in control of the outcome. The same, however, could not be said about the world of business. Commerce was a kitchen filled with so many cooks, each of varying competence and inclination, that rarely did a dish fully satisfy the diner's hunger. Nowhere had this been more evident than with regard to Sony's future in videogames. In June 1991, Nintendo had publicly humiliated Sony. Now, six months later, many of Sony's senior executives were pushing to work out a new deal with Nintendo, one that would allegedly offer a "nongaming role." In contrast to the title of his book, Olafsson believed that forgiving Nintendo would put Sony's software aspirations in purgatory and effectively kill any future hardware plans. Though the majority of executives disagreed with Olafsson, he had some strong allies on his side, the most important of which was Sony president Norio Ohga, and it was becoming clear that two distinct forces were growing within Sony: the old guard, who wanted to work with Nintendo or not be a part of this industry at all, and a newer generation who believed that videogames were the future of consumer electronics. The old guard held much of the clout, but the newer generation had Ohga's support. Eventually these opposing forces would meet head-on and alter Sony's trajectory.

Olafsson did not know when this showdown would occur and where those ripples would leave Sony. But as he typed through the morning, his mind would occasionally drift to business and a plan that he had been developing. It was a strategy that he believed would both appeal to his forward-thinking brethren at Sony and also appease the dinosaurs who wanted to work with a company of Nintendo's caliber. It was a risky plan, with lots of moving parts, and the key to making it work hinged on Sega.

25.

BACK 2 WORK

Madeline Schroeder pushed a loose strand of hair away from her eyes, squinted in deep thought, and then slowly raised an eyebrow. "The Seattle Koopa Troopas," she said with a cheerful one-upmanship in her voice. "Or even the Seattle *Super* Koopa Troopas." She sat across from Nilsen in his office, their usual cave for swapping ideas. For the past few hours they had been working on an initial marketing presentation for Sonic 2, but had swerved into discussing Nintendo's recent bid for the Seattle Mariners, whether the new ownership group would rename the team, and if so, what that name might be.

Nilsen judiciously considered her suggestions, tapped a finger against his chin, and nodded lightly. Not bad, not bad at all. "What about the Seattle Princess Chasers?"

"Or how about the Tetrises?" she said.

"Well done," he said. "But I think the word you're looking for is 'Tetri.'"

"Exactly," she said, pointing at him. "The Seattle Tetri."

"Or maybe they keep it simple: the Super Nintendos," he suggested. "*But* they change the team name every time they come out with a new console."

"I like that," she said. "But I also like the Nintendo R.O.B.'s. Gives a little shout-out to their forgotten robot from yesteryear."

"Why dive so far into yesteryear? How about the Nintendo Power Gloves?"

"Or the Nintendo Power Pads!"

Nilsen was about to make a Zelda reference, perhaps something about gold cartridges perhaps, but suddenly the look in his eyes changed. Schroeder picked up on it instantly. How could she not? She'd practically been in a bunker with Nilsen for the past year, and as a result, their brains had seemingly fused together. Not only did she notice things about him, but sometimes she even noticed them before he did. "You got it, don't you? That perfectly clever, conversation-ending idea."

He did, and for this reason spoke in a slow, overly dramatic, I'd-like-to-solve-the-puzzle tone. "If Major League Baseball approves the sale, the Seattle Mariners will henceforth be known as . . . the Seattle Mariners."

"The Mariners?" she asked, feeling like she'd missed the joke. "That's boring."

"Exactly!" Nilsen said. "When was the last time Nintendo did something other than maintain the status quo? Two weeks before never?"

"Spot-on. Shame on us for ever thinking otherwise," she said with a laugh. "I think the better question is, what would the San Francisco Giants be called if SOJ bought them?"

Nilsen considered this. "It certainly is a better question, I'll give you that," he said. "But I refuse to answer on moral grounds."

"Oh, yeah?" she asked. "And why is that?"

"Because as cool as it would be for Will Clark to bat third for the San Francisco Sonic Speedsters, we should never resort to following in Nintendo's footsteps. Originality or bust!" he proclaimed. "But a football team, on the other hand . . ."

When the banter finally came to an end, they resumed their conversation about marketing plans for the Sonic sequel. The game was scheduled to be ready by October, but Schroeder believed it would more likely be November. Since Naka and the rest of his Sonic Team had begun working on the game at the Sega Technical Institute, she'd been driving out to Palo Alto once or twice a week to check on its progress. The game was obviously in its earliest stages, just character sprites and background art, but she could tell that it wasn't going to be simply a hastily made rehash of the first game. That had been her biggest fear, and the biggest problem with sequels in general.

Even Nintendo had experienced this conundrum years earlier. After the release of Super Mario Bros., the game's creator, Shigeru Miyamoto, got to work on a follow-up. After nearly a year of work, he delivered a game that many felt was too much like the original, just much more difficult. Nintendo of Japan released the game in 1987 to mixed reviews. Nintendo of America, however, disliked the game so much that they postponed the U.S. release and decided instead to find a different, already completed Japanese game and simply revise it in Mario's image. They ended up settling upon an Arabian-themed game called Doki Doki Panic, which follows a family of four on a perilous quest to rescue kidnapped kids from Wart, an evil frog king who has an intense allergy to vegetables. Nintendo of America then made superficial changes to the game, like swapping out the original characters of Mama, Papa, and children Lina and Imajin for, respectively, Luigi, Toad, Princess Toadstool, and Mario. When their redevelopment was complete, Nintendo of America released their version of Super Mario Bros. 2 to generally positive reviews but a lingering popular sentiment that the game felt "off."

Schroeder didn't know which scenario was worse: a sequel that felt too similar or one that felt too different. But despite concerns about sequel-itis lingering in the back of her mind, everything she had seen thus far from Sonic 2 looked very promising. The game was a bit faster and a bit more colorful, and she had recently heard it might even feature a sidekick for Sonic. Plus, this time around the developers were located mere miles away, which would make any concerns, disputes, or cultural disagreements that much easier to resolve. She was cautiously optimistic about the sequel, and that filled her and Nilsen with a desire to develop a marketing program that was as good as they hoped the game would be.

"Maybe we should send Sonic to retailers a few months before the release," Schroeder suggested. "And he can show off demo versions of the game."

"I like that," Nilsen said. "Add it to the list under in-store promotions."

They'd spent weeks putting together a preliminary marketing plan, which they'd be presenting at the next day's senior staff meeting. The strategy they'd come up with looked like a shinier and more pervasive version of how they had launched the first Sonic game: magazine coverage, radio promotions, and maybe even a mall tour around the Christmas season. There was nothing wrong with what they had put together, but as Nilsen looked over the list he felt greatly underwhelmed. He shook his head, angry with himself for doing what had already been done (even if it had been he himself who had done it the first time).

"This isn't good enough, Mad," Nilsen said. "It just isn't."

"Which part in particular?" Schroeder asked.

"All of it," he said, still shaking his head. "This just isn't good enough."

"You were okay with it ten minutes ago."

"Maybe we should start over from scratch."

"You're joking, right?" she said, not at all amused. "This is weeks of work—half of it yours, by the way."

"I don't care," he said. "It's not good enough. It's not big enough."

"What do you want, Al? A monument?" she asked, getting annoyed. "A blimp?"

"I don't know," Nilsen said, trying to piece together his thoughts.

"I have an entire list of ideas right here. This is what we're going to present tomorrow and, trust me, everyone's going to love it. ."

"But it's not Sonic-worthy, Mad. It just isn't. It's Mario-worthy."

"I don't know what you want from me."

"I want you to help me come up with something better."

"There is nothing better! This is everything we've got—the kitchen sink!"

"Fine," Nilsen said. "Then I want the whole kitchen!"

Their conversation soon escalated into a shouting match, with Schroeder yelling at Nilsen and throwing reason and rational thinking at him, and him refusing to listen or offer any better suggestion. In two years of working together, this was the first time they had ever raised their voices to each other. Though they'd disagreed before, there'd always been the common bond of logic and reason to keep heads cool and smiles on their faces. This time, though, the bickering got so loud that employees stopped outside their office and peered inside to make sure everything was okay. Nilsen saw them watching, their eyes filled with a concern that in the moment he interpreted as taking joy in his failure. He wanted to shout at them to look away, but he knew that he wasn't really mad at them. He wasn't even really mad at Madeline. He was mad at himself, mad that his brain had stopped cooperating, mad that he had failed to come up with an elegant solution to the impossible question.

"You're not listening to a single thing I have to say!" Schroeder shouted. "Maybe it would be better if we each present our own plan and let them choose."

"Oh, yeah?" Nilsen asked, his face reddening by the moment. "Maybe a better idea would be if—" He stopped midsentence, and a maniacal smile began to form.

"You figured it out, didn't you?" she asked, her voice returning to a normal level.

Nilsen nodded slowly. People outside the office were still looking at him. But now he welcomed their watchful eyes.

"So?" Schroeder asked. "What is it?"

"We're going to start in Japan, move to Europe, and then end in the U.S. No trucks, no boats: everything will be delivered by plane exactly one day before. We're going to have a street date, Mad, we're going to have the world's first global launch, and in the process we're going to break every single sales record. What do you think?"

"What do I think?" she mused. "I think a global launch is Sonic-worthy."

And just like that, it was decided. They scrapped their previous plans and started planning the videogame world's first-ever street date. The ideas started flowing.

"The faucet of ideas was turned on full blast and their minds melded into one. Movies come out on Friday, but this should be different."

"Definitely—how about Thursday? Or Saturday?"

"No, it should be Tuesday, and we'll call it Sonic 2sday!"

"That's perfect! Wow, you're 2 smart!"

"Aw shucks, you're 2 kind!"

As they came up with each pun, they started typing it on Nilsen's computer, printing it out, and taping it to the wall. Soon enough, the office walls were covered: *2 Fast! 2 Rad! 2 Day!* They kept going, the pages pouring out of the printer until there was no more space on the walls.

"Now what?" Schroeder asked. "2 bad for us?"

Nilsen got up and went into Van Buskirk's office next door. "Do you mind if Mad and I start taping things to the window outside your office?"

"That's a new one," Van Buskirk replied. "No, I don't mind, but I would not mind even more if I knew why you were doing this."

"Of course," Nilsen said, waving for her to follow him back to his office. "We're going to do the first international global launch. We're going to shatter every record, and it's all going to happen on Sonic 2sday."

Van Buskirk entered his office and saw all the puns taped to the walls.

"For the record," Schroeder said, "we're just a little insane. But not clinically, or anything like that."

Van Buskirk looked at Schroeder, then back at the wall, and then moved to Nilsen's computer. She typed up something quickly, printed it out, and then taped it to the window outside her office: *2 Cool*. With three crazy people hanging up puns everywhere, it didn't take long for their happy insanity to spread. Employees started migrating over from all around the office, coming up with their best puns and hanging them on the window.

Nilsen took a step back and watched the impromptu collage come 2gether, 2 impressed 2 speak. He then got over himself and went back 2 work.

Bill White had suddenly become a very popular guy. And he didn't like it one bit.

"No comment," White said into the phone, and he said it again twice more before the reporter gave up and thanked him for his time. He looked around his messy office, shook his head in disbelief, and wondered how many minutes it would be until the next call came in. Since the press conference the day before, Nintendo's marketing director had been flooded with calls from reporters around the country. What did Nintendo hope to gain from owning a baseball team? Would the team be renamed the Seattle Super Marios? And if so, was there any concern that the new name would sound too similar and possibly upset the local pro basketball team, the Seattle Supersonics? White tried to clarify that it wasn't actually Nintendo buying the team, but rather Nintendo's owner, Hiroshi Yamauchi. He would own the Mariners the same way he owned a variety of other businesses in Japan. And he was only putting up 60 percent of the money, with the other 40 percent coming from local investors who wanted to keep the team in the city. But those facts didn't make for good headlines, so the reporters didn't care.

And, honestly, White didn't care either. He'd thought this was a bad idea from the beginning, and it was even worse now that baseball's commissioner, Fay Vincent, had weighed in. Less than twenty-four hours after Yamauchi's bid was submitted, Vincent cited a "strong policy" against foreign ownership and coldly deemed the deal "unlikely." The folks at Nintendo were disheartened by his quick response, but White didn't understand how they could expect any less. After all, only a year before, some Japanese businessman had inquired about purchasing an available minority stake in the New York Yankees. Any prospect of such a deal was nixed immediately, with Steve Greenberg, baseball's deputy commissioner, explaining that baseball policy prohibited non–North American ownership.

That was all before Japan balked at providing aid to the Gulf War effort and U.S. politicians took to blaming Japan for America's current recession. Since then, things had only gotten worse. Just two days earlier, the Los Angeles County Transportation Commission had been nationally applauded for voting to cancel a $122 million rail car deal with the Sumitomo Corporation, a Japanese firm that had won the contract over a company in Idaho. Videogames were no longer the fastest-growing medium of entertainment; that honor now went to xenophobia, with Japan-bashing being the most popular game.

It also didn't help that Yamauchi, the man at the center of this story, was considered to be a frigid, enigmatic corporate tightwad. Only two years earlier he had responded to inquiries about Nintendo's noticeable lack of philanthropy by saying, "I believe a business can contribute to society by growing and making money and paying more taxes. That's how we contribute to society." Given that previous stance, it was unsurprising that Americans would suspect an ulterior motive behind this out-of-the-blue act of supposed charity. Of course they would think: Here comes Japan, this Godzilla-sized bully—they show up in our country and take over our precious national pastime. As irrational as that point of view might be, the nation was angry, looking for someone to blame, and Yamauchi wasn't doing himself any favors by skipping the press conference and avoiding the American media. The only public statement he made on the matter was to Japan's *Yomiuri* newspaper, saying simply, "This is a kind of social service, not a business activity intended to make a profit."

Another call to Bill White. No comment. No comment. Thank you for your time.

All of this couldn't have come at a worse time for Nintendo, which had recently lost the very American ace that had been up their sleeve for years: Howard Phillips. Nintendo would often parade Phillips, with his curly red hair, boyishly thin freckled face, and gaudy bow tie, in front of the press to provide a familiar-looking face for a mistrusted Japanese company. He was the go-to guy for interviews, the physical embodiment of Nintendo in most of the population's minds. Officially Phillips was Nintendo's Game Master, a tribute to his ability to select hit titles and beat any game, but over the years his job had morphed into serving as the company's unofficial mascot and a role model for a generation of young gamers. Phillips spoke in a fun, fresh parlance that resonated with kids of all ages, and he talked constantly about how he had the coolest job in the world, but in March 1991 he had unexpectedly left Nintendo. White didn't know all the details behind his departure, but he wished that Phillips were around right now to help cushion the blow.

Without Phillips in the picture, Yamauchi's bid for the Mariners had reopened a greatest-hits list of problems from Nintendo's past: the lawsuits with Atari, Tengen, and Galoob; a discrimination case from 1990 alleging that Nintendo had repeatedly failed to hire and promote qualified African Americans; and a failed venture in Minnesota that would have allowed individuals to purchase lottery tickets through the NES, which the press was using to paint Nintendo as a proponent of gambling. This festering negativity would eventually cut into Nintendo's bottom line, which was already suffering as a result of the recession, lack of a must-have title, and the presence of additional options in the marketplace. Less than a month after the holiday season, in addition to dropping the price of the Super Nintendo from $199.95 to $179.95, it had also dropped the price of its handheld Game Boy from $99.95 to $89.95.

Another call, this one from White's wife. "Will you be home for dinner tonight?"

"No comment," he replied.

"That kind of a day, huh?"

"You can't even imagine," he said. "I should be home by nine, provided Nintendo doesn't go ahead and buy a football team."

White understood the good intentions behind purchasing the Mariners, but surely Arakawa couldn't be naive enough to believe that the country would see the situation from his point of view. He wanted to give his boss the benefit of the doubt, but that was becoming harder to do as it became more and more clear that he and Arakawa suffered from irreconcilable philosophical differences. In short, the issue was that White was a classically trained student of advertising who believed in the divine power of marketing, while Arakawa was a master of product development who believed that marketing was pretty much a complete waste of time. This was particularly troubling to White in light of the fact that he was, you know, Nintendo's director of marketing.

Though Arakawa's perspective may have seemed shortsighted, there was an anecdotal basis to his outlook. Over the past decade he'd heard from focus groups that the NES would never sell in America, that the Legend of Zelda was much too confusing, and that an Italian plumber made for a terrible hero. Experiences like these led him to believe that the fundamental problem with marketing was its reliance on the past. It looked backward, not forward, and failed to take into account innovation, trends, or cultural shifts in taste. At the end of the day, the only true predictor of future success was the quality of a product. In short: the reason a game sells is because it's a good game, not because Bill White tells people to buy it.

Though the differing philosophies occasionally produced friction over the years, it was never anything that didn't evaporate in a day or two. After all, why bother blowing gaskets or holding grudges when things were going so well? But as Nintendo shifted to 16 bits and the era of good feelings came to a close, blind faith was gradually replaced with muted skepticism. Fingers that once had seemed to have that Midas touch were now being pointed to allocate blame. And as the company began to lose market share, White found himself disagreeing more and more with Arakawa and the direction that Nintendo was headed. Ultimately, though, his concerns didn't matter much. Arakawa was Nintendo of America's president, and his opinion was the magic wand that enchanted everyone and everything into action. Nothing exemplified the growing divide greater than how Nintendo responded to the emergence of Sega.

For the past year, Sega had been going directly after Nintendo; naming names, trashing games, and making all kinds of bogus claims. At first it was almost cute, their yippity-yapping about how "Genesis does what Nintendon't," like a toy poodle barking in the face of a Great Dane. It was a nuisance, yes, but White knew it wasn't worth the energy to fight them off. But then Tom Kalinske took over, and that poodle turned rabid. Sega lowered their price, signed up third-party developers, and painted Nintendo as nothing more than a cutesy kids' company. Their bark was still much greater than their bite, but it was loud enough to warrant a response. Now was the time to put Sega in their place, either by launching a campaign that bit them right back or by feeding them some scraps and sending them away from the table. But this was not the Nintendo way. They refused to stoop to unsavory levels and negotiate with marketing terrorists.

Sega realized that they could do or say whatever they wanted, without fear of retribution. At the Consumer Electronics Show, Sega had thrown Mario into a footrace with their too-cool blue hedgehog. Nintendo did not respond. Then they took their parlor trick on the road, setting up their side-by-side comparisons around the country. Still Nintendo did not respond. And then over the holiday season, Sega took things to the next level, airing negative ads, puffing up their numbers for the press, and continuing to find ways to exploit their new mascot, who was nothing more than a Mario ripoff in fast shoes. Once again, Nintendo did not respond.

Looking back, White couldn't believe that Nintendo had let this happen. Sega was nothing but a one-trick pony who had fooled the world into believing their hype. If only his bosses had allowed him to bark back, then they could have sent Sega to the woodshed before they became an actual threat. But now it was probably too late. White would never really know for sure, and that was the worst part. Even after everything Sega had done, Arakawa still refused to fight fire with fire. Though he had opened Nintendo's purse strings and allocated $25 million to market the SNES, that money was earmarked for a wholesome, kid-friendly, gee-whiz campaign, which played right into Sega's hands. But Arakawa wasn't too concerned. He would forever be a slow-and-steady-wins-the-race kind of guy, patiently hoping that the nineties would turn out to be as much of a fairy tale for Nintendo as the eighties had been.

Only time would tell, but there was one key difference between now and then. The reason that Nintendo had pulled off the impossible in the eighties was because they had fought for every inch. Now, though, they were sitting back and letting others take their precious real estate. Nintendo had underestimated Sega, undervalued backward compatibility, and underrated the importance of Howard Phillips's freckled face. They were slightly wounded, but still very much the market leader. If they wanted to remain king of the jungle, then they should be spending their time searching for that same fighting spirit, not buying baseball teams. It was time for Nintendo to get back to work.

What was done was done, however, and White was willing to adopt the key tenet of Arakawa's philosophy: look forward, not backward. The bid for the Mariners had already been made, and the only thing they could do now was damage control. For the deal to go through, it would have to be approved by the commissioner, the ownership committee, and at least 75 percent of the twenty-six major league baseball teams. At this point, White didn't know which of the two evils would be better: suffering through months of bad press only to watch the deal fall apart, or miraculously receiving ownership of the team and exposing the company to continued attacks. He started to give this question some serious thought, but was interrupted once again by a ringing phone.

26.

ORIGIN STORIES

Kalinske didn't mean to stare. It was against his nature to gawk at others, but he simply couldn't help himself when glancing at Howard Phillips, whose neck was noticeably lacking a big, bright bow tie. The post-Nintendo Phillips looked older, wiser, and less two-dimensional. It was as if Charlie Brown had upgraded from his everyday yellow shirt with the black zigzag to a special-occasion navy sport coat. It was only a small change, perhaps, but it represented a redefinition of character. Howard Phillips was all grown up now, and the only thing that epitomized that maturation more than a revised fashion sense was the fact that he had come to San Francisco for a celebratory dinner with Kalinske and Toyoda.

"For the sake of decorum," Phillips said, speaking with a delicate cadence that seemed to expertly bend the English language, "I would like to express right off the bat that I have nothing but the utmost respect for Nintendo and my former colleagues."

From across the table, Kalinske and Toyoda nodded in unison. This was too fine a restaurant and too fine an occasion to let things devolve into a gossip session. Tonight was about one thing and one thing only: finalizing the deal for Nintendo's former Game Master to come work at Sega.

"Of course," Toyoda said. "We will not speak ill of our competitor."

"The truth is that we have tremendous respect for Nintendo as well," Kalinske said. "It's just that, unlike you, we also despise them."

Phillips chuckled. As he did, Kalinske couldn't help but notice a slight disconnect between the simplicity of his bulging smile and the complexity in his eyes. Perhaps Phillips harbored more of a grudge than he let on.

"Speaking of the devil," Kalinske said, "how'd you even wind up there?"

"It all began a long, long time ago," he said, his eyes suddenly shining. It was undoubtedly a tale he'd told many times before, but also undoubtedly one that he loved even more with each retelling. When he spoke it was like a superhero revealing his origin story, the legend of how it all began. Over the next decade, there would be many tales filled with the POW! BAM! ZAP! of success, but nothing would ever rival that initial experience. "At the time, I was just a student at the University of Washington. It was 1982, and one day my good friend—my roommate actually, Don James—got a job with this small, nondescript company that had set up shop in the southern district of Seattle."

"Nintendo?" Toyoda asked.

"Bingo," Phillips said. "They had been importing these giant, refrigerator-sized arcade cabinets from Japan until they realized it would be cheaper to simply send over the parts and have someone put the things together over here. So Don got hired on to do that job. And then after a couple months of him putting these things together, they decided they wanted someone to keep track of these things. So they asked me if I wanted a job, and just like that I became the warehouse manager." Phillips shook his head. "There couldn't have been more than ten of us back then. Myself, Don, Mr. Arakawa, and a few other helping hands. Who could have predicted what would happen next?"

As Phillips told the story, Kalinske couldn't help but feel a sense of compassion for Nintendo. It was like hearing about Goliath's early years, when he was just a skinny kid who liked skipping rocks and making macaroni art. The fact that Phillips had been there since the beginning made Kalinske feel even better about bringing him to Sega. It would hurt Nintendo that much more.

"Here am I," Phillips went on, "just some kid working in the warehouse. And one day Mr. A. comes to me, shows me this game with a funny name, Donkey Kong, and says, 'What do you think about this?' He wants to know because the company isn't doing so great and they want the next arcade game they bring in to be a big smash hit. So I turn it on and start playing the game, and a minute or so later I just blurt out, 'Mr. Arakawa, we gotta bring this to the United States!'

"Now, obviously I'm not saying that I'm the reason that they chose to import Donkey Kong, but I will say that the version of the game I played that day was not the same one that we sent out for people to play. Through trial and error I made some tiny tweaks to adjust for difficulty, timing, number of lives; that sort of thing. From then on, Mr. A. would always come to me and ask if a new game was cool or if it sucked. What needed to be done to make it better? Why wasn't this as fun as it should be? It was a dream come true—I was like a focus group of one."

This was the real reason that Sega wanted Phillips. Yes, it would be satisfying to steal away Nintendo's former mascot, but by this point Sega was past high school pranks and focused on going from good to great. Phillips, for whatever reason, had an unquantifiable, superhero-like ability to determine the anatomy of great, and Sega needed this now more than ever. That was why Toyoda had been wooing him for many months. For a long time, his overtures had been respectfully declined, but as Sega continued to trend upward, Phillips couldn't help but seriously consider the opportunity.

After leaving Nintendo, Phillips did not find the grass to be any greener at Lucasfilm Games. Less than a year later, he moved on to THQ, whose grass also wasn't tinted to his liking. By this time, Sega's offer to lead a development team and produce games was looking pretty good. In fact, it looked great, but Phillips was trying his best not to see it that way. As wonderful as the opportunity appeared to be, he couldn't wrap his brain around joining the dark side. The games that Nintendo made were treasures that could be enjoyed by the whole family. No sex, no gambling, no violence. But Sega did things differently. Although they hadn't specifically set out to publish unsavory content, by transforming themselves into the anti-Nintendo, Sega had become the primary outlet for more mature content. Onslaught, Streets of Rage, Fatal Labyrinth—just look at the titles of these Genesis games! Even with Sonic The Hedgehog, a great game for sure, why did they have to give him such a naughty attitude? What kind of values did Sonic teach to kids? Rowdiness? Rebelliousness? Perpetual impatience? Now, with only 16 bits, perhaps this lack of censorship was no big deal, but as the technology of videogames progressed, the differences in philosophy between Sega and Nintendo would only become more pronounced.

And yet, even though Phillips felt this way, he had still agreed to join Sega. He resented himself for saying yes to Toyoda and cringed a little bit while driving to that night's dinner, but working for them would be better than his current position at THQ, and there didn't appear to be any other options. It was a bitter pill for him to swallow, but he promised himself that he would effect change from inside Sega, act as a moral compass, and stay true to himself.

"With the success of Donkey Kong we became the largest shipper in the port of Seattle," Phillips went on. "And every week we'd receive a delivery of about a hundred or so forty-foot containers. We never quite knew what to expect. Not even Mr. Arakawa. Sometimes it was just more Donkey Kong units, sometimes it would be arcade cabinets for a different game, and sometimes it would be a new toy or electronic device that Mr. Arakawa would have to decide if he wanted us to start selling."

"That sounds kind of exciting," Kalinske said.

"It was!" Phillips exclaimed. "It was like Christmas morning every week! And then one week it was like the best Christmas of all time. We were still pretty busy with Donkey Kong, I remember, and were getting a lot of that, but one of the containers was full of nothing but a bunch of boxes for Mr. Arakawa. And inside one of them was this thing called the Famicom. It was a goofy, very toy-ish-looking thing. White plastic, maroon trim, and a little cartridge of Donkey Kong that just slides right in. So we hooked the thing up to the TV and it was just un-frickin'-believable."

"Un-frickin'-believable." That was a word Kalinske had never used before, and he felt confident that he'd end his days without it ever coming from his mouth. It wasn't a matter of taste, but a function of age. His daughters could get away with it, maybe even Karen, but not himself. Words like that were part of the vernacular for a different generation. It's funny how that works, the hidden truthfulness of language. Kalinske could throw his back out playing basketball and convince himself that it was a one time thing, and he could discover a gray hair and feel the distinguished pride of wisdom, but there was no way that he could ever say "un-frickin'-believable" without feeling like he was a hundred years old. But in a strange way, this disconnect was part of the reason he loved Sega so much. By going after teens and young adults, a demographic that he had never targeted before, he felt a small link to a youthful world of hope, change, and irony, one that his daughters would come into as they grew and which they would inhabit for many years.

"This is so very fascinating," Toyoda said. "What an interesting time."

"And it just kept on getting more exciting," Phillips said, now speaking with a noticeable degree of nostalgia. "After Mr. A. decided that it was time to try to bring back the videogame industry in America with the NES, the question became which games we would sell here. By this time, they already had fifty games selling in Japan, but we were only going to release sixteen over here. So I played each of them, all the way through, and provided an analysis of which ones I thought were the best. After that, our head count grew from a hundred to a thousand, sales hit a billion dollars, and, well, things kind of went batshit crazy for a couple of years."

Kalinske and Toyoda laughed. The sensation that Phillips described, the sheer insanity of riding the roller coaster, had recently become a popular topic of discussion amongst Sega's top executives. They didn't want to jinx Sega's success, but they felt a need to be prepared for the best. When it happens, that exponential expansion, it happens fast. And if the right pieces aren't in place, companies can easily collapse under the weight of their own success (an inability to fulfill increasing orders, ill-advised partnerships, failure to adapt with changing technologies, etc.). So to avoid the pitfalls of newfound profitability, Sega of America had brought in a trio of talented veterans: Doug Glen, Joe Miller, and Ed Volkwein.

Doug Glen was a tall, bald, tech-savvy MIT grad who came on board to run business development. It was borderline impossible to spend a minute with Glen and not step away convinced that he'd missed his calling as a college professor. Instead of pursuing a career in education, however, he'd opted for a combination of Silicon Valley, Madison Avenue, and *je ne sais quoi*, arming himself with a sophisticated fluency in technology, advertising, and a multitude of Romance languages. With a finger in so many pies, he was the perfect fit to partner Sega with other companies on the cutting edge. At the top of Glen's list, however, were launching a CD-based hardware system and exploring the futuristic concept of making videogames available to download directly to a player's television. Beyond these diverse talents, Glen's arrival also signified a superstitious victory of sorts: he had a reputation for joining companies that were about to become the next big thing.

If Glen could be considered the kitchen manager responsible for selecting which ingredients ought to be part of Sega's recipe for success, then Joe Miller would be the chef responsible for slicing, dicing, and mixing everything together. Miller was an engineer by trade and a perfectionist by reputation. People tended to see him in one of two ways: as pretentious and pompous, or as a genuine visionary. But regardless of which view people took, they always looked at him with an underpinning of reverence. When it came to engineering, Miller knew what he was doing and had the résumé to prove it. He had spent the past decade bouncing between gaming outfits (like Atari and Epyx) and computer companies (like Koala Technologies and Convergent), which made him familiar with a wide spectrum of software and hardware. He had initially been brought in by Sega's head of product development, Ken Balthaser, to set up Sega's new multimedia studio, where the company hoped to record top-notch musical acts and film live-action movie scenes to be used in games for the CD system they hoped to launch in late 1992. Around the time Miller finished building the studio, Balthaser decided to leave and start a software company with his son. Kalinske, Rioux, and Toyoda all believed that Miller would be the perfect successor, a nimble-minded guy who could handle consoles, peripherals, and software for the next generation. Miller agreed with this assessment but didn't know if he wanted to accept the undertaking. He confided to Kalinske that, unlike other candidates, he had the benefit of several months spent observing Sega from the inside and did not appreciate the constant pressure that Sega of Japan exerted on product development. At the time, Sega of Japan insisted on reviewing every single R&D dollar spent, and when it came to projects that they didn't fully support, SOJ had a habit of "delaying" discussions until they evaporated. So Kalinske told Miller to write up a list of everything that he wanted changed. The following week, Miller submitted a list, and a week after that Kalinske returned it. "It's all taken care of," Kalinske said. "So you're out of excuses. Welcome to Sega."

With Glen in charge of looking down the road and Miller responsible for building the cars to get there, there was still the intangible issue of how drivers should feel when they were on Sega's highway. What were the images, sounds, and emotions that consumers should experience when they saw the name Sega? It all comes back to marketing, and Kalinske was finally ready to pull out all the stops. Between launching Sonic, targeting teens, and adopting a fearless head-to-head strategy, Sega's marketing team had done an incredible job of positioning themselves against Nintendo. Now, however, they needed to find and finalize their own identity. Internally, there was no uncertainty amongst employees about how to define Sega. It represented freedom, revolution, and the next stage in entertainment evolution. And while some outside Sega's office walls may have already known that the coup was under way, it was time to bring on a new advertising agency that would make sure that the rest of the world was fully aware of the new world order.

Now that Steve Race was out of the picture, Kalinske needed someone to run the marketing department, a strong vice president who could manage the increasing head count and lead the agency review process. A part of him wanted to offer the job to Nilsen, rewarding him for being a one-man wrecking crew these past couple of years, but another part of Kalinske felt that promoting Nilsen would be nothing short of cruel.

Nilsen was clearly a master of ideas big and small, someone who always saw the extraordinary in an ordinary world. The problem, though, was that what the organization needed right now was, quite literally, organization. And that was not Nilsen's strong suit, at least not in the conventional sense. In truth, he probably had a better grasp than anyone else when it came to product development, marketing milestones, and industry trends, but instead of maintaining this information in reviewable files and folders, he had it all filed away inside his head. It was the difference between the kid who takes a math test and aces it but doesn't show his work, and the kid who gets most of the questions right and fills the page with detailed equations demonstrating how he arrived at the answers. This is not to say that Nilsen lacked management skills, but right now Sega needed someone who could invite others into his head and show them what needed to be done. They needed someone who was as much a teacher as he was a commander. They needed Ed Volkwein.

Volkwein was a balding, graying journeyman marketer who radiated affability and acumen the way a beloved pediatrician did. He'd risen up through the ranks at General Foods in the 1970s, where he excelled as a product manager for desserts and dog food. After eight years of adhering to their strict by-the-book marketing, he turned the page and went to Chesebrough Ponds to focus on new products and popularize their Ragú spaghetti sauce. The challenge with Ragú was that although they made a fine sauce, it rarely found a spot inside the American cupboard beside tried-and-true condiments like ketchup, mustard, and mayonnaise. To change this perception, Volkwein's team redefined the product by launching a comprehensive national campaign that simultaneously positioned Ragú as the sauce that true Italians prefer and also implied that Americans could easily bring home the exotic taste of Italy with this low-cost product. Each ad ended with the line "That's Italian," which quickly became synonymous with the product. Though Volkwein would go on to craft many more campaigns in his life, for everything from Prince tennis rackets to Funk & Wagnalls encyclopedias, his work with Ragú represented what Kalinske hoped to do at Sega: redefine a product, elevate it above the competition, bring it into homes that had never purchased that kind of product before, and, ideally, come up with a kicker to end each commercial, something that would make the ad memorable and epitomize the Sega experience.

Shortly after becoming Sega's vice president of marketing, Volkwein began the ad review process. He scoured the landscape, looking for agencies that were big enough to take Sega national and bold enough to go after Nintendo without backing down. Basically, an agency that swung for the fences and actually hit home runs. At this point, Sega still couldn't afford the game's heaviest hitters, but Volkwein was confident that the company's recent momentum would attract an agency looking to get into the videogame business. In order to find the right agency and keep the process as transparent as possible, he leaned heavily on six people: Kalinske, Rioux, Nilsen, Van Buskirk, Adair, and Tom Abramson, a rowdy hustler he had recently hired to handle Sega's promotions. Together, they would make the most important decision in Sega's quest to reach that next level.

Kalinske believed that getting there was simply a matter of time, a process that was already being sped up by the additions of Glen, Miller, and Volkwein—the professor, the perfectionist, and the pediatrician. And now it was time to add another fresh face to the mix: that of the man who had almost finished sharing his Nintendo origin story.

"By this point, Nintendo is completely exploding," Howard Phillips explained. He spoke with an infectious, I-can't-believe-this-really-happened-to-me enthusiasm that Kalinske and Toyoda loved. "It's utterly amazing, but there's still this fear that one day, out of the blue, we'll fall apart faster than Atari. How do we avoid that? Quality, quality, quality; not just with the games, no, but with every part of the gaming experience. So Mr. A. comes up with an idea to keep our players up to speed and taps Gail Tilden to launch this little eight-page newsletter. The *Nintendo Fun Club News*, it was called. She wanted to root it in reality and put a face on it to help broker the kids into the gaming experience. So she asked me if I would be willing to do this, and I said, 'Sure, why not.'"

"Wow," Kalinske said. "Did you have any idea what you were getting into?"

"Big responsibility," Toyoda added.

Phillips chuckled. "In retrospect, I didn't have any appreciation for what Gail was asking me to do," he said with a playful shake of the head. "A couple months after I said okay, she comes to me again and says it's not going to be a newsletter anymore, but a full-fledged magazine, *Nintendo Power*. Do I still want to do this? Well, how could I say no? There are so many messages out there surrounding videogames that it's important for me to step forward and present a positive message. Play itself is very rewarding for children. Maybe in some respects there's a little too much play, but play is still important. And there's this whole social opportunity, this currency of interaction, with friends, family, and even strangers. That kind of attitude extends beyond the playground, so I was really happy to be expressing that aspect of games."

"That's beautiful," Kalinske said. Ever since his conversation with Heidkamp about the increasing violence in videogames, he occasionally felt twinges of doubt regarding what he was selling. They were tiny ones, buried deep below the surface, but still, they were there, and anything that could keep them quiet was much appreciated. "Well put, Howard."

"I don't regret what happened next, but it certainly came as a surprise," Phillips said. After *Nintendo Power* became the number one kids' magazine, the bow-tie-wearing Howard Phillips persona went through the roof. His face became instantly recognizable and synonymous with Nintendo, and he spent most of his time traveling around the country promoting the company. "My persona had really started to take off, and they took it to the bank: the freckled Anglo face on the Japanese company. This was no evil plot by them, but it became a bit oppressive because it came at the expense of product development. If I was on the road doing promotion, then of course I couldn't evaluate the games. It got to the point where I was evaluating games that I hadn't even completed, which was just inconceivable from my perspective."

"That's awful," Kalinske said. "It's like a musician going on tour and not having time to work on new songs."

"Nintendo's rock star!" Toyoda exclaimed.

"That's exactly how it felt," Phillips confided. "I couldn't go into public places without being recognized. Sometimes it was fun, but sometimes you're in a rush and someone really wants to co-opt your time. And you feel like you shouldn't let them down, so it gets challenging. Then it started getting to the point where it was getting a little awkward with the moms—the looks they gave me and things they would say."

Kalinske laughed. "And how did Mrs. Game Master feel about all of this?"

"That was the worst part!" Phillips said. "The moms started pulling her aside and asking what it was like to be married to the Game Master. 'Is he a master in the bedroom? Is his manual dexterity really that good? Does he wear the bow tie all the time?'"

When the laughter subsided, Kalinske said, "Well, the good news is, we just removed Sega's mandatory bow tie policy, so you should be in the clear." Phillips grinned.

"When is good for you to start?" Toyoda asked. "We are flexible."

Phillips opened his mouth to answer, but suddenly the words were no longer flowing out the way they had when he was telling his story.

"The thing is . . ." Phillips began, in a way that could mean only one thing. He didn't really want to say what would come next, but what choice did he have? After all, every superhero only gets one origin story. "I greatly appreciate the opportunity to work for Sega, but on second thought it's perhaps not the best fit."

"We understand," Kalinske said, forcing himself to say something other than *Check, please.* He nodded several times, trying to add some truth to his words.

"Maybe another time," Toyoda said without any hint of emotion.

"I'm very sorry. I feel terrible. I should have said something earlier."

"Nonsense," Kalinske said, again suppressing an instinct to ask for the check.

"Maybe another time," Toyoda repeated.

"Thank you," Phillips said with a strange sense of relief.

Kalinske, Toyoda, and Phillips then proceeded to finish their meal, despite the awkwardness. Kalinske held no ill will against Phillips, but that didn't make the minutes go by any smoother. He was stung by the turn of events and felt foolish for not having anticipated this outcome, but he knew that in the long run it wouldn't really matter. With Phillips or without him, Sega was moving forward at a million miles per hour.

"Miles Prower?" Nilsen asked with shock, horror, and disbelief. "Really?"

"Yes," Schroeder said, standing glumly in front of his desk.

"No."

"I know," she said. "But yes. They're really committed to this."

First name: Miles. Last name: Prower. That was, apparently, the moniker of Sonic's sidekick, a splashy new character who would be heavily featured in the sequel. As with the mascot competition from 1990 that had resulted in the creation of Sonic, SOJ once again held an internal design competition, this time to create a partner in crime for their hedgehog. The winning entry came from Yasushi Yamaguchi, Sonic Team's primary level designer. Yamaguchi submitted an orange fox with stylish bangs and a distinct pair of tails, inspired by a mythical creature from Japanese folklore. According to legend, certain Japanese red foxes, or *kitsune*, magically grow an additional tail for every thousand years they have lived. Each tail provides unique powers, like the ability to create illusions, manifest in dreams, or rub two tails together and create fire. Though legend implies that *kitsune* are notorious tricksters, Yamaguchi's creation was much more domesticated. His fox featured a jovial smile and a can-do attitude, and his tails could spin fast enough to let him fly through the air. Nilsen loved the new character. He looked cool, but not in a steal-the-spotlight kind of way. He skewed younger, but not to the depths of Disney. And his lackadaisical flying ability and Boy Scout persona were strong complements to Sonic's speed and bad-boy attitude. He was perfect, except for the name he had been given.

"Miles Prower?" Nilsen said again.

"It's a play on 'miles per hour,'" Schroeder said. "Get it?"

"He sounds like a Bond villain."

"I was thinking porn star."

Nilsen shook his head. He liked puns as much as the next guy (probably more, even), but this just went too far. What if Batman's understudy were called Marshal Arts? Or if Mario's brother had been named Pie Zano? It made a caricature out of an indelible character in a fully realized universe. Plus there was another reason. "The only Miles I know," Nilsen said, "is this kid from grade school. And I really didn't like him."

"You have my empathy," Schroeder said. "But as I mentioned before, the developers are really committed to this."

Nilsen sighed. The last time the developers had become committed to a name, Sonic's archenemy suffered from a case of multiple personality disorder. In the character bible for Sonic The Hedgehog, Sega of America named his main adversary Doctor Ivo Robotnik. Sega of Japan, however, had taken to calling the evil scientist Doctor Eggman. Both sides were unwilling to compromise, so Sonic's foe ended up with one name in the East and another in the West, creating an international nomenclature incident that took years to be resolved. Nilsen refused to let this happen again. Miles Prower had to go. "Then I guess it's up to us to show them that they're wrong."

His first stop was Kalinske, who agreed that they could and should do better (though there was admittedly some sweet relief in the fact that this time around the biggest development controversy was a matter of names and not personality, aesthetics, fangs, or rock bands). Still, this wasn't an ego contest between SOA and SOJ; rather, it was about doing what was best for Sega and the fictional Sonic universe. Sega of America decided upon the name Tails, and Schroeder took this to the developers, who were unwilling to budge. Toyoda then tried to mend fences, perhaps even find a compromise, but there was no progress to be made. With common sense and politics failing them, Nilsen and Schroeder resorted to Kalinske's favorite asset of all: story. After they finished crafting a compelling backstory, Nilsen drove out to Palo Alto so he could share the tale.

Toyoda joined him at the Sega Technical Institute, where it was clear that the Sonic Team was less than thrilled to see them. In their minds, this was another case of the Americans imposing their will purely for its own sake. Nilsen knew they felt that way, and he was aware that there was little he could do to prove to them that this wasn't the case. All he had was his short story, which he hoped would be enough. He cleared his throat, ignored the glares, and began reading from "The Renaming of Miles Monotail."

> This is the story of Miles Monotail. Miles was your average four-year-old fox. He loved to play with his friends, but his friends weren't really his friends. Whenever they saw Miles they laughed and made fun of him.
>
> Why? Well, Miles wasn't like all of the other foxes. Miles

Monotail had two tails. And as kids tend to do when someone is different, they make fun of him. It didn't help that Miles sometimes tripped over his second tail and went rolling down the hill. Coordination was not one of Miles's virtues.

Because of the rough time that his friends gave him, Miles became very depressed.

One day he was walking along with his head hanging down when a blur and a whoosh crossed his path. There's only one person who could be moving that fast, and that is Sonic The Hedgehog.

Miles thought Sonic was the greatest person in the world. Miles wished that he could be as cool and coordinated as Sonic was. And most of all, he wanted to meet Sonic.

This was his big chance. Miles took a deep, deep breath and at the top of his lungs yelled out, *"Sonic!"*

The blur that was Sonic turned around and stopped in front of Miles. "You called?" said Sonic.

"Oh, Sonic, you're my hero," exclaimed Miles as he ran around and around Sonic. Well, you can guess what happened next—Miles tripped over his second tail and fell down. Tears came to his eyes.

"Hey, cheer up, little fellow. What's the matter?" said Sonic.

"Sonic, I want to be just like you, but I'm a freak. I've got two tails."

Sonic leaned over to Miles and said kindly, "You're no freak. You're more special than anyone because you have something that everyone else doesn't have. And you can do things that they can't. If anything, your friends should be jealous of you."

"But I can't do anything special," cried Miles.

"Oh yes you can," Sonic said. "I'll show you. You're about to enter Sonic's special training camp."

Well, Miles couldn't be happier. His hero took him under his wing and started teaching Miles how to use his two tails to their best advantage. He showed Miles how to curl his tails up under his body so that he was like a very aerodynamic ball and could do Sonic's famous Supersonic Spin.

Sonic then taught Miles how to use his two tails as a helicopter rotor so that Miles could fly around. Even Sonic couldn't do that.

Needless to say, Miles was ecstatic. He was special, and when his friends saw what Miles could do that they couldn't do, they became very jealous, but also every single one of them wanted to be Miles's best friend.

But Miles had a new best friend. Someone who believed in him. Someone who was his hero. And that friend was Sonic.

Sonic was happy that he could help his buddy gain new confidence and new abilities. "See, Miles, you *are* special because you have two tails. And because of that, I'm going to give you a very special nickname. From now on I'm going to call you Tails because you should always be reminded that you are special because you have two tails."

So from that day forward Miles Monotail became known as Tails.

When Nilsen finished, he folded the paper and slid it into his pocket. The members of Sonic Team, who'd been planning on hating every word out of Nilsen's mouth, found themselves entranced by the story. One developer was even moved to tears.

After allowing a moment to let it sink in, Naka walked up to Nilsen and said, "You may call him Tails." Yet despite Naka's proclamation, the issue was still not resolved. Although moved, several of Sonic Team's other members remained unconvinced. Sensing that the momentary camaraderie was about to fizzle away, Toyoda blurted out a compromise: "How about his real name is Miles Prower, but Sonic calls him by the nickname Tails?" This suggestion managed to do the trick: the fox would go by Tails, but Sonic Team would take solace in the fact that in a fictional filing cabinet somewhere, there was a birth certificate with the name Miles Prower (although they would eventually decide to make this less fictional and graffiti the name "Miles" throughout the game).

But in this moment, on this day, they were all part of the same story. And they would be writing the happy ending together. For now, at least.

27.

SOMETHING BEYOND VIDEOGAMES

On April 6, 1992, names were on Tom Kalinske's mind. But for the first time in months, Tails, Miles Prower, and the PhD-holding Robotnik had nothing to do with it. On this occasion, the only characters who mattered were his daughters, Karen, and the just-born boy nestled in her arms. A crowd of family members gathered around her hospital bed, inspecting the tiny creature and beginning the process of loving him unconditionally. With so many eyes fixed upon him, he flapped his elbow, as if waving hello, and wowed the room. His father was certain that he already had the Kalinske charisma. He was perfect in every way, except that he still needed a name.

"Brandon," Karen declared, finalizing the creation of this tiny being.

Later that evening, eight hundred miles to the north, there was another cause for celebration. And though it lacked the sentimental drama of childbirth, this event was sentimental and dramatic to those in attendance and would perhaps lead to a playoff berth one day soon. At 7:05 p.m. the Mariners opened the 1992 baseball season, and to the delight of fifty-five thousand cheering fans, they did so in Seattle. The heroes responsible for preventing the team's relocation enjoyed the game from the Boeing luxury suite on the third-base side. There, Arakawa, Lincoln, Senator Gorton, and a handful of minority investors commemorated the occasion by passing around a bottle of Chateau Ste. Michelle, a local wine befitting the local royalty.

Yet despite feeling like kings and having submitted the highest bid on the table, the Seattle Baseball Group had not yet had its $125 million offer accepted by Major League Baseball. This was partly due to the molasses of bureaucracy, but mostly it was the by-product of a national backlash that had been piling up against Nintendo ever since their fateful press conference. Headlines like "Japanese Bid for Seattle Team Gets Baseball's Cold Shoulder" (*New York Times*, January 24) and "Buy American Cry Spreading Across Nation" (*Boston Globe*, January 25) described the attitude of a nation that had interpreted Nintendo's noble motives as sinister machinations. The controversy shifted from national concern to national outrage on February 15, when the commissioner of Japan's Nippon Professional Baseball league, Ichiro Yoshikuni, stated that "Japanese baseball is for the Japanese and Japanese fans would try to exclude the possibility of foreign-country involvement." Following this, Nintendo simply became a metaphor for "them" in the "us vs. them" prism that came to define America's relations with Japan. On television, in newspapers, and inside bars across the country, denouncing Nintendo was socially acceptable—and often even socially expected.

At this point, it would have been easy and even understandable for Nintendo to simply give up. They had tried to do a good thing, but others did not see it that way. And while it's admirable not to base decisions on the perceptions of others, in this case those others were also Nintendo's customers, and the bad press was hurting sales. But to Arakawa (and also Lincoln and Yamauchi), giving up wasn't even an option. For better or worse, Arakawa believed in staying the course and doing right by those he cared about most—usually that was the gamers, but in this case it was the people of Seattle. Obstacles are a part of life, but still, it's always better to be the tortoise than the hare.

Instead of raising the white flag, Nintendo chose to cloak themselves in the red, white, and blue. In early February, Yamauchi said that he hoped to move Nintendo's global headquarters to Seattle. In late February, seven new American-born minority investors were brought into the Seattle Baseball

Club. And in March 1992, for the first time in Nintendo of America's history, they hired someone to handle corporate communications. That someone was Perrin Kaplan, a PR vixen whose skills of persuasion were so ferocious that she claimed to have once convinced an Israeli soldier to smuggle her across the Lebanese border. Between Kaplan's finesse and some restructuring of the ownership group, the backlash began to dissipate. The nation was still disgruntled, but at least they weren't as vocal about it as they once were.

And by opening day of the 1992 baseball season, the only voices that really mattered were those of the 55,918 fans cheering as the umpire shouted "Play ball!" and the still-in-Seattle Mariners took the field.

Since bringing his son home from the hospital, Kalinske had developed a talent for answering phone calls before they reached the second ring. He figured this was the least he could do, what with Karen having handled the whole pregnancy, labor, and delivery side of the equation. And tonight his full range of dexterity was on display as he swiped the kitchen phone from its cradle in the middle of its first ring.

"Tom? Hello, friend. It's Olaf," Olafsson said with the sly but sincere friendliness he had perfected. "Forgive me for calling at this hour."

"No problem," Kalinske said. "It's not too late here with the time difference."

"Ah, yes, saved by the Pacific time zone," Olafsson replied with some relief. "I must admit that I'm traveling so much I often lose track of the hours."

"I know the feeling all too well. Anyway, what's going on?"

"Well, my friend, I am calling to offer my congratulations."

"You spoke with Japan?" Kalinske asked, unable to hide his excitement. After months of loose discussions, he and Olafsson had spent much of the past couple of weeks trying to formalize a partnership between Sega and Sony, an alliance based on shared respect and mutual need. As the two of them had become friendlier and grown more willing to speak more openly, it became clear that they were a natural match. Olafsson confided to Kalinske that, above all else, his goal was for Sony to get into the videogame business in a major way and stay there for years to come. This seemed like a no-brainer, but Sony's old guard was still resistant to spending the time and resources to make that happen. In terms of software, they already looked down their nose at Olafsson's Sony Computer Entertainment division, which had been unable to obtain the money and internal support to attract top-flight developers. And when it came to hardware, they were looking to pull the plug on Kutaragi's work, perhaps as early as June. Olafsson was convinced that he could still make Sony a player in the videogame industry, but he needed more time. An alliance with Sega would buy him that time by ensuring that Sony would remain in the videogame business at least as long as that partnership lasted.

For Kalinske, the upside to a relationship with Sony went beyond videogames. He envisioned the possibility of not only creating software together, but also working on hardware, music, and one day even movies. With Sony, Sega could become the intersection of technology, entertainment, and pop culture. With Sony, Sega could own the future. "So what exactly did Japan say? Were they interested in us working together?"

"Tom, Tom, Tom," Olafsson playfully chided. "You must get your priorities in order. I meant congratulations on the baby boy. Brandon, is that his name?"

"Oh, that congratulations," Kalinske replied sheepishly. "I feel like such a fool."

"Don't. Japan has been very much on my mind as well," Olafsson admitted, which made Kalinske feel better. "I suppose your jumping to conclusions is a testament to the type of guys we are and, perhaps, also a sign of the times we live in."

"That sounds like a good excuse. I'll take it," Kalinske said, making them both laugh. "But seriously, thank you so much for calling about the baby. We're all very excited. And it's nice to finally have another guy inside this house of girls."

"I can only imagine," Olafsson said. "Anyhow, I should leave you to more important matters. I'll call as soon as I hear anything new. It'll likely be a slow burn, but I think we'll be able to make it happen."

"I look forward to hearing from you," Kalinske said.

"Fantastic," Olafsson said. "Until then, enjoy the new dynamic."

Kalinske hung up and returned to his new dynamic: a happily exhausted wife, three very curious daughters, and a teeny-tiny son.

There was certainly a sense that things were changing, not just at the

Kalinske home but also at the office. For starters, only a few months after moving into the new office in Redwood Shores, Sega had moved again, to accommodate its rapid expansion. Whereas the previous building felt like a place for transition, this new location felt like one meant for transformation—from good to great, from challenger to contender, from also-ran to front-runner. The six-story, 113,000-square-foot building at 255 Shoreline Drive, with its series of elegant fountains in front and placid lagoon out back, served as a daily reminder of Sega's legitimacy.

The new office represented Sega's metamorphosis, but the overall sense of changing dynamics went beyond a change in location. Ever since Sega's coming-out party in Vegas, the company and its employees were being treated with new respect. While this was certainly good for the ego, it was even better for the bottom line. Two years ago, Toyoda had approached Warner Bros. about licensing some of their characters for Genesis games. Back then, their licensing department wasn't even willing to have a conversation about the matter, for fear that it would get back to Nintendo. Now, two of Sega's most promising games in the pipeline were Batman Returns, based on the upcoming Warner Bros. movie, and Taz-Mania, based on the WB-owned Looney Tunes character. In addition to Warner Bros., Sega was also having high-level discussions with Disney, Universal, Twentieth Century Fox, and of course Sony.

With Sega beginning to take Nintendo's world by storm, Kalinske thought it was time to start giving back. While Nintendo considered purchasing a major league baseball team to be an act of charity, Kalinske thought that Sega could do something more creative and effective. So in early 1992 he formed a charitable trust called the Sega Youth Education and Health Foundation, whose mission was to fight diseases plaguing children and to sponsor a variety of educational endeavors, particularly those with an emphasis on pairing learning and technology. At Mattel, philanthropy had always been a top priority, so Kalinske relished the chance to get back in the corporate giving spirit. An individual can donate money to help support a cause, but an ambitious company can actually cause a shift in the global conversation.

Beyond the social benefits, launching a Sega foundation also served to placate those worried about the social or education merits of videogames. Rarely was Kalinske himself overcome by such thoughts, but doubts were cropping up more frequently now. Maybe it was because as Emil Heidkamp had mentioned the games did appear to be quickly drifting toward increased violence. Maybe it stemmed from the sort of personal reflection that happens when you've brought a new life into the world. Or maybe his concerns came from the growing whispers of parents who were concerned that videogames might be detrimental to their children's development. A minority of moms and dads believed that Nintendo (and now Sega) were irresponsibly raising a generation of "vidiots." If Kalinske had wanted to face these concerns head-on, perhaps he could have better understood their origin. But that wasn't something he wished to take on, at least not right now. There was just too much to do, and no time to waste thinking about it. Besides, even if there was any merit to these concerns, they would surely be canceled out by the good that would come from Sega's new foundation. Yes, he realized that this type of thinking was how a drug kingpin with a penchant for charitable donations might rationalize his lifestyle, but that didn't necessarily make it wrong, right? No, of course not. Sega was already making a difference, and for proof he need not look further than a recent conversation with Nilsen.

A few months ago, Nilsen had ambled into Kalinske's office. "I just got out of a meeting with Cheryl from KIIS." This was Cheryl Quiroz from the radio station KIIS-FM, whom Nilsen had continued to work with ever since the "Sixteen Weeks of Summer" campaign. She was one of the few outside Sega who truly bled Sonic blue. "She flew up here to talk about a summer concert that they want to put on. And she brought along the station's program director, a guy named Bill Richards."

"Don't know him," Kalinske said. He was slightly confused, but unable to hide his eagerness to find out how this all fit together. "That's kind of unusual, that she would bring him along, isn't it?"

"It is," Nilsen said. "I kept wondering why they'd spent the money to fly him up here, but then it hit me: they're desperate. Turns out that Bill, the PD, had this crazy idea to bring a bunch of hot acts together for a benefit concert. You know, something to raise money and awareness for a good cause. Except they can't find a single sponsor, because that cause is pediatric AIDS."

It's hard to accurately describe the national sentiment toward AIDS during the late eighties and early nineties, particularly when it came to children who were infected, but the case of Ryan White in Kokomo, Indiana, goes a long way toward explaining the emotional tug-of-war between fear and sympathy.

In 1971, when he was only three days old, Ryan White was diagnosed with hemophilia A. As treatment for this disorder, he was given weekly transfusions of a blood-clotting protein called factor VIII. This enabled him to live a relatively normal life throughout most of his childhood, but that changed in 1984 when the thirteen-year-old was rushed to the hospital with symptoms of pneumonia. Following a partial lung transplant, he was diagnosed with AIDS, which he had acquired through a transfusion. White was given only six months to live, but after beating those odds and regaining some of his strength, he wanted to try to resume a normal life. A large part of that normalcy entailed returning to school, but when community members learned of his intentions, they protested.

Fearing that he might be contagious, fifty teachers and over a hundred local parents signed a petition to ban Ryan White from Western Middle School. Even though the health commissioner of Indiana informed the school that White posed no risk to other students, he was expelled from the school. The White family challenged this decision and turned to the legal system to get their son readmitted. Over the next year, White remained at home as his case went through various courts and appeals until finally, in August 1986, he was allowed to return to school for eighth grade. Although this appeared to be a major victory, White was generally unhappy upon returning to classes because he had few friends and was often accused of "being a queer." Meanwhile, his family received threats on a nearly daily basis, and after a bullet zinged through their living room, they decided to withdraw their son from the school.

From then until his death in 1990, White became a national spokesman for the disease, appearing frequently on *The Phil Donahue Show* and participating in charity benefits. His life even inspired a television movie on ABC. Although his efforts helped to significantly raise awareness, the cultural perception of the disease didn't change a great deal. The mention of AIDS was still intrinsically toxic, and that's why no corporations were willing to support the concert that KIIS wanted to put on.

"All they're asking for is fifty grand," Nilsen explained. "And for that, because of the stigma, they're offering two hundred thousand dollars' worth of promos and ad time. An amazing cause and a fantastic deal." Nilsen shook his head again, this time with some anger. "So I told her that Sega would be more than happy to sponsor the event. And that's the good news."

When the information hit, Kalinske leaned forward. "Wait," he said. "Let me get this straight: without my approval, you already agreed to commit a sizable amount of money to a press-heavy event for a disease whose mere mention makes people look the other way?"

"Yes, I did," Nilsen said tentatively, but with no hint of embarrassment.

Kalinske slapped his desk. "Al," he said, "I like you more and more each day. This is fantastic!" Kalinske then went on to tell him about Anique Kaspar, the family friend that the Kalinskes had run into at Disneyland. He explained to Nilsen what the resilient young girl had been going through and how he'd been trying to help her in whatever small ways he possibly could. "Paul Newman opened this great organization in Connecticut, the Hole in the Wall Gang Camp. It's a place for very sick children to enjoy the summer camp experience with others in the same position."

"That's incredible," Nilsen said.

"It is, it really is," Kalinske said. "And Anique deserves all the incredibleness she can get. So Karen and I are going to pay for her to go there this summer. Wait, I hope it doesn't conflict with the concert. Do you know what day KIIS has in mind?"

On April 25, 1992, just days after the birth of Kalinske's son and months after Sega's charitable foundation had been formed, Sega and 102.7 KIIS-FM presented the first annual "KIIS and Unite" concert at the Irvine Meadows Amphitheatre. The groundbreaking eight-hour event supporting the Pediatric AIDS Foundation included appearances by Céline Dion, Kid 'n Play, Eddie Money, and many more. Kalinske didn't recognize all of the performers, but luckily he had brought Anique with him, and she was proving to be a tiny encyclopedia of pop music.

"That's Color Me Badd." Anique pointed out the group as she and Kalinske hung out backstage. She was dressed in a cartoon-print long-sleeved shirt and a slightly oversized white summer hat with flowers tucked into its brim. "They sing a song called 'I Adore Mi Amor.' That means 'I adore my love.'" Kalinske knew Anique had good days and bad days, with more bad than good lately. Today, however, she was bright as could be and that filled Kalinske with gratitude.

"Is the song any good?" he asked.

"Yup," she chirped. "It's great!"

"Then let's go talk to them." Anique nodded, and they began to move toward Color Me Badd, but Kalinske could see how weak she was. "Actually," he said, "why don't I bring them over here?"

Color Me Badd happily came over and treated Anique like a princess. At an event filled with celebrities, she was the star. She deserved it many times over, for an infinite number of reasons, Kalinske thought, but most of all for her smile. How did she do it? he wondered. No matter how she felt, or how the world felt around her, she always had at least the hint of a smile on her face—a smile that was both irresistible and contagious. It really was a gift, and Kalinske wanted to share it with everyone who was there. So when Rick Dees, KIIS-FM's number one DJ, called Kalinske up to the stage, he made sure to bring Anique and that smile of hers.

"Ready?" Kalinske asked her when Dees called his name.

She nodded, and he proceeded to lift her up so that she wouldn't have to exert herself. She was so unexpectedly light that Kalinske nearly dropped her; she must have been forty pounds, forty-five tops. He raised her onto his shoulders, and together they slowly walked onto the stage to the sound of fifteen thousand people clapping.

Kalinske was handed the microphone and made a short speech, but even as he spoke, he knew that the words didn't matter. The girl on his shoulders and the smile below her flower-trimmed white hat were what mattered. The people in the audience who had replaced fear with concern and the celebrities backstage who were so happy to be a part of this—they were what mattered. The $211,069 that Sega and KIIS-FM raised for the Pediatric AIDS Foundation—that was what mattered.

Two months later, Kalinske would receive a call to let him know that Anique had passed away, curled in her mother's arms on the way back from Paul Newman's camp. But that was still several weeks away. On this sunny April day, Kalinske looked out at the cheering crowd that had packed the

Irvine Meadows Amphitheatre to publicly support children with AIDS—an event that he and his team had helped make possible. And then he looked up at a smiling, wonderfully brave girl being showered with the love she deserved, and he knew he would carry her image with him forever. At least there would always be that.

28.

BOCA

Walking onto the stage at the Irvine Amphitheatre, hosting the hottest party at the Consumer Electronics Show, sharing a dinner with Sony's top executives at an exclusive top-floor dining room—there was something addictive about the power and pleasure of stepping into the spotlight. It validated all those invisible hours of hard work Kalinske had put in. And he craved more of these moments, many more, which seemed inevitable with his company now controlling 25 percent of the videogame market and seemingly going nowhere but up. The rocket ship had blasted off, loudly and irrevocably, and what lay ahead was a thrilling space race with Nintendo.

But even as the stars seemed to align, there was still one factor with enough gravitational force to potentially send Sega crashing back to earth: the company's lack of a proven track record. This was less a criticism than an undeniable and unavoidable fact. It was the equivalent of watching a rookie ballplayer smack a home run in his first big-league at bat and lamenting that there was no way such a pace could be sustained. But still, there was an element of validity to the concern. And it hinted at the larger question on everyone's mind: was Sega the next Nintendo, or merely a one-hit wonder? That was what secretly scared retailers, and what less secretly scared Sony.

With everything that Sega of America had in the works, Kalinske was certain that his company had staying power. But the retailers didn't necessarily know that. All they had to rely on was the past, the present, and a tiny glimpse into the future gleaned from a few hours at the Consumer Electronics Show. Part of the reason Kalinske had been brought in was that his long track record could substitute for Sega's short one. But in the booming videogame industry, his personal reputation could only go so far. Since Kalinske had taken over, the retailers had quietly supported Sega's coup, but now he wanted them to take that next step and loudly herald his company's claim to the throne. What he needed was an opportunity to pull retailers behind the curtain, dazzle them, and prove that Sega was here to stay. Lo and behold, after much deliberation, there did appear to be such a way, and that way went straight through Boca.

"Welcome to sunny Florida," Kalinske announced from the front of a large, sparkling, red-carpeted ballroom. The room was teeming with retailers for the opening-night cocktail party of Sega's newest innovation. "More specifically," Kalinske continued, finding the pulse of the room, "welcome to the Boca Raton Club, where you are all invited to spend the next three days golfing, fishing, drinking, and getting to know the men and women who have made Sega a stunning success. And don't worry, I've instructed them not to talk business on the golf course. Well, not on the front nine, at least."

Laughter ricocheted from wall to wall. There were about four hundred retailers in all, sporting smiles and the varying hues of recently acquired tans. And it wasn't just senior buyers and CEOs in attendance. Sega wanted its friendship with the retail community to be a long-lasting one. So they also invited midlevel executives, junior associates, and even a few assistants. In the best-case scenario, these would be tomorrow's decision makers; in the worst-case scenario, they'd talk up Sega by the watercooler.

The event formally began on May 11, 1992, but Kalinske and his team had arrived in Florida a couple of days earlier. Though guests were promised fun and sun, for Sega's employees this was a Broadway performance designed to make retailers forget about Nintendo. Nobody in the videogame industry had ever planned anything like this in advance of the trade shows, and so there was a lot on the line. Typically, retailers placed their big Christmas orders at the June CES, which put Sega in the position of competing directly against Nintendo. Since buyers only had a finite budget to spend on videogames, an unexpected dazzle from Nintendo would indirectly frazzle Sega. An event in May, however, removed this element of surprise and gave

Sega an opportunity to tempt retailers into blowing their budgets before they even saw what Nintendo had to offer. "So if you need anything, anything at all," Kalinske said, gazing out into the packed hall, "don't hesitate to speak up. We're here to make sure you have a good time. You deserve it. And if it should happen that by the end of the trip you've forgotten about our friends in Redmond, then all the better!"

Kalinske finished speaking and sat down to an ovation. But that kind of applause came easily when there was an open bar involved. The real challenge would be the next few days. Boca presented an opportunity unlike any other, and there was little margin for error. If we ever want to pass Nintendo, Kalinske thought, then even perfection today isn't good enough; we have to offer a guarantee of perfection tomorrow as well. In the fairy tale that Sega was writing, there was no room for too hot or too cold—everything had to be just right. Somewhat neurotically heeding his own dictum, Kalinske found himself looking over his shoulder to make sure nobody was watching as he used the end of a tablecloth to quickly polish a spoon whose shine was mildly dulled.

"Not afraid to get your hands dirty?" Nilsen asked, catching Kalinske in the act.

Kalinske smiled. Busted. "Whatever it takes, right?"

"Mind if I pitch in?" Nilsen asked.

"I would not object to that," Kalinske said, happy to have a partner in crime.

For a short, strange moment, the two men stood in silence, polishing pieces of silverware that were already acceptable and whose shine nobody would ever notice.

The absurdity of the moment was heightened by the fact that this was likely something neither man would ever do at home. Kalinske probably hadn't helped Karen with the dishes in over a decade, and Nilsen gave the impression of someone who circumvented such busywork by using plastic utensils. But this was different—this was Sega—and the rules of normal life didn't apply.

"Out, damned spot," Nilsen playfully said.

"Channeling your inner Lady Macbeth?" Kalinske asked. "Does that mean that we've already slain the king?"

"We've been number one in 16 bits for three years running!"

"Well," Kalinske said with a hint of amusement, "it's certainly easier to be number one when you're the only game in town."

"I do not deny that fact," Nilsen said. "But still, numbers don't lie!"

"Good one," Kalinske said. "That's exactly the kind of propaganda I'm looking forward to hearing tomorrow." Tomorrow, like today, would include an itinerary of golf, tennis, and lounging by the pool. But unlike today, tomorrow would also include formal presentations by Kalinske (general), Burns (sales), Adair (Game Gear), and Nilsen (Genesis). Inside the Trojan horse that was Boca, this was the carefully coordinated Greek invasion. As Kalinske thought more about that, he relaxed a little. A successful event in Boca hinged on tennis, golf, and persuasive speeches from himself and three of his most trusted employees. And like a pitcher going into the sixth inning without having given up a walk or a hit, he started to believe that he might really have a chance at pulling off a perfect game after all. Kalinske put down the spoon and looked out into the bustling room. "Actually," he said, "how would you feel about tapping into that propaganda right now? I see a room full of folks sitting on the fence who need to be pushed over once and for all. Shall we go help them make the right decision?"

"Of course," Nilsen said, putting down the silverware. "It's what I do best."

"Me too," Kalinske said, surprised by how true the words felt coming off his lips. "Me too," he said again before the two men moved into the crowd to do what they did best, and prove to the world that Sega was no one-hit wonder.

"In 1992 hot advertising is going to be very, very important," Nilsen boomed from the edge of a twenty-foot-long stage that had been constructed in the ballroom. In a matter of hours, the venue had transformed from a place designed to dole out tiny hors d'oeuvres to a place designed to bring in big orders. Nilsen spoke beside a wooden podium with a large projection screen behind him and hundreds of retailers jammed into seats in front of him. "So we've searched high and low trying to find the perfect spokesperson for Sega Genesis."

Kalinske watched from the front row, toggling his focus between Nilsen's words and John Sullivan's eyes. Sullivan, sitting beside him, was the buyer for Toys "R" Us. Ever since 1985, when the chain's founder, Charles Lazarus, and executive vice president Howard Moore had taken the leap and become the first national chain to gamble on Nintendo, Toys "R" Us had developed a reputation as the community's tastemakers. That torch had been passed to Sullivan, who was so dead set on having the largest quantity of the best games available that he often couldn't even wait for product to be delivered to his stores. Instead, he would routinely send out a fleet of trucks to pick up the inventory as it entered the country, and then funnel it out to the eight hundred Toys "R" Us locations nationwide. Though Wal-Mart, Kmart, and Target each had more stores, it could easily be argued that Toys "R" Us was the most important account of them all.

"The perfect spokesperson has to be knowledgeable about videogames and also possess unimpeachable integrity," Nilsen said. As he readied the projection screen for the big reveal, Sullivan tilted his head toward Kalinske. "Could it be?" he asked under his breath. "Someone even better than Sonic?"

Kalinske nodded with confidence. "Two someones, actually."

"Right now," Nilsen said, "I'd like to introduce you to the first of our new spokespeople." As soon as he finished the sentence, a large photo of Nintendo's Howard Lincoln popped up on the screen. Below the portrait of Sega's competitor was a statement that Lincoln had made about Sonic during the Tengen trial: "They came up with a darn good game. They're going to be a very strong competitor."

After a brief moment of are-we-allowed-to-laugh-at-this silence, everyone gave in and erupted in laughter. Before the chuckling could dissipate, Nilsen surfed forward on the wave of momentum. "And to help promote our third-party licensees, I'd like you to meet our second new spokesperson." This time, a large photo of Nintendo's Minoru Arakawa popped up, with an equally incriminating quote of his praise for EA's John Madden Football for the Sega Genesis.

Kalinske noticed Sullivan bending forward to try to contain a bout of belly laughter, and he couldn't help but think back to the baseball analogy from last evening. The perfect game was still in progress, but now he felt like he was headed into the ninth with only the bottom of the order standing in his way. Kalinske had started off day 2 with an inspiring speech. He'd wanted to get everyone as excited about peeking into Sega's future as he had been a couple of years ago when Nakayama took him to that R&D lab.

When Kalinske felt like his words had done the trick, he passed the baton to Burns, who happily bombarded the audience with sales figures and forecasts. More than anyone else, he spoke their language, and there was no translator needed to see how much the retailers liked what he had to say. After Burns described how much there was to be made, it was up to Adair and Nilsen to pull back the curtain and prove that this was possible. Adair spoke first, using Sega's colorful Game Gear as a metaphor to describe everything that Nintendo's black-and-white Game Boy could never be. And then finally, to close the deal, Nilsen took over.

Nilsen announced that Sega was dropping the price of the Genesis to $129.95 and then disclosed Sega's war plan for 1992, which consisted of having the "hottest games," the "hottest promotions," and the "hottest advertising backed up by Sega's largest advertising budget ever." Though Kalinske loved a good game as much as the next guy and took great pride in clever promotions, it was this last bit that excited Kalinske the most. Since bringing on Ed Volkwein, Sega's search for a new ad agency had been narrowed down to five choices, including the front-runner and his personal favorite, Wieden+Kennedy, an Oregon-based firm that had become famous for their work with Nike during the past decade. While Kalinske would readily admit that he missed Steve Race, there was a poetic justice to losing the man behind Reebok and gaining the men behind Nike. Kalinske was thrilled that Wieden+Kennedy had joined the field of competitors, and he was looking forward to watching them win the account next month, when each agency would formally pitch Sega. He knew that a company like Wieden+Kennedy would help him finish Sega's reinvention, and he wanted the new national campaign to veer memorably outside the box. If Sega could do that and also get the retailers on their side, there would be no stopping them. But Kalinske knew that he was getting ahead of himself. One step at a time, he reminded himself. Wind up and pitch. Catch and throw. Finish the perfect game.

"After all is said and done," Nilsen said with folksy charm, "what is it that kids want? Great games. It's that simple. And in 1992, we've got the must-have games that kids will be clamoring to buy in your stores."

Kalinske nodded to Nilsen, who was controlling the room magnificently. "Starting off our new titles for the second half of 1992 is David Robinson Basketball." Nilsen paused to play a tape featuring some of the gameplay, then glided back into his speech. "We're talking a full-on, full-court running game here. With two dozen incredible moves that were digitized from videotapes of real basketball action, David Robinson Basketball includes all the elbow-pumpin', board-crashin', ball-stealin' excitement the floor can dish out."

Nilsen had a fast-talking, strangely enticing style that somehow made it seem like he was making fun of specific games, of videogames in general, and even of himself, while simultaneously expressing reverence for all those things, including himself. It was hard to say how he did all this, but he did it well as he continued through the highlights of Sega's upcoming roster.

There was Taz-Mania, which featured "seventeen levels of postcard-quality Tasmanian scenery, including whirling waterspouts and perilous quicksand." And there was also Evander Holyfield's Real Deal Boxing, a game centered around the fighter who had once knocked out Buster Douglas. "It's 360 degrees of nonstop action in the ring!" Nilsen boasted, before moving on to Super Monaco GP 2 ("The ultimate fantasy for anyone who's ever been thrilled by the smell of burning rubber"), Batman Returns ("Holy bat smoke! It's Batmobile versus Duckmobile as Batman hurls into acrobatic action against the death-dealing Penguin"), and TaleSpin ("You'll circumnavigate the globe while Don Karnage and his air pirates try to thwart your progress").

Nilsen was on a roll, and the crowd loved it, Kalinske most of all. Nilsen was throwing strikes left and right, showing off the full arsenal. Fastball. Curveball. Slider. Even a knuckleball from time to time. "In October we'll be rereleasing some of our previous best-sellers, like Michael Jackson's Moonwalker, under the banner of Sega Classics, and they'll be available to customers at only $29.95." Kalinske could feel the momentum building. Perfection might be an abstract concept, but it had a specific feeling, and Kalinske was sure that everyone in the room must have felt it.

Then Nilsen introduced the next game—and something went horribly wrong. "In November," he announced, "comes a game with the most unusual goal. For the first time ever, you want to die," Nilsen said. Seconds later, the power went out. The air-conditioning cut out, the projection screen went black, and though slivers of sunlight trickled through the window shades, the hundreds of retailers who would decide Sega's fate suddenly found themselves in the dark.

29.

AFTER THE BLACKOUT, BUT BEFORE THE SURGE

Shit.

Kalinske tried hard to think of any word besides this, but it was no use. The world had gone dark right in the middle of Sega's most important presentation. Shit, shit. After the momentary shock wore off and he realized that this was a power outage, he turned to John Sullivan, whose eyebrow was ominously raised. Shit, shit, shit.

"Well," Kalinske said, mining for some kind of silver lining, "I guess Sega has become so darn powerful that even the state of Florida can't handle us."

"Hey, if they're looking at the same sales figures that I have," Sullivan said, slapping Kalinske on the shoulder, "then I wouldn't be surprised." It was then that Kalinske realized Sullivan's eyebrow had arched not with condemnation but with curiosity. And he was not alone. There was a wide range of reactions from audience members, including shock, disbelief, and incredulity, but no evident annoyance. If anything, they seemed pleasantly surprised by the turn of events, and happy to have a story to spread around back at the office. "To be completely honest, Tom," Sullivan said with another slap to the shoulder, "I figured this was all part of the show. You guys are kind of crazy like that. Good crazy."

Good crazy. Kalinske liked the sound of that, and gained a further appreciation for the description as Sullivan continued to explain while the club's staff ushered them out of the ballroom and through an equally unlit hallway. "There's a kamikaze-like aggressiveness to everything you guys do. You're not afraid to shake things up," Sullivan said as he, Kalinske, and hundreds of other retailers were led out into the parking lot.

"As long as we're helping to put money in your pocket," Kalinske said, "then I'm a happy man."

"You are," Sullivan said. "Believe me, you are." They continued walking beneath the zealous sun until they found a comfortable spot in the shade where they could continue the conversation. "Can I trust you with a story?"

Kalinske nodded and took a half step closer. He had a feeling this would be good.

"Okay," Sullivan started. "Coming out of this past holiday season, we'd sold a ton of goods. You know this already, I know, but the point is that we sell out of almost everything except some older 8-bit stuff. We ask you to mark down the goods, you guys oblige."

"Of course," Kalinske said. "We don't want that stuff taking up your shelf space."

"And I thank you for that sentiment. As I did with Electronic Arts when they agreed to mark down some excess inventory we had on hand. But I'm sure you can guess who didn't feel the same way."

"It's their policy," Kalinske said, rolling his eyes. "They don't mark down."

"Yup," Sullivan continued. "So sometime in January, I'm walking through one of our stores with Charles. Well, there are all these unsold games piled up, and he's not a happy camper. I mean, the man is seventy years old—he's our founder, for God's sake—but Charles Lazarus loves selling more than anything else, so this gets him up in arms. He picks up one of our inventory scanners, checks the SKU, and out pops the sell-through information with the vendor's name: Nintendo. He tells me that I need to get them to budge, and I try, but it's no use."

"Not good," Kalinske said. "When Charles wants something done—"

"Believe me, I know. But they're steadfast in saying that it was our problem for not selling the inventory. They refuse to take it back and refund what we paid, and they refuse to let us discount the goods. So what am I supposed to do?"

"You're supposed to ask how high," Kalinske said, "when Nintendo says jump."

"I keep talking to Peter Main and Randy Peretzman, telling these guys that something needs to give. Finally, they agree to come out, and they say they're going to bring Arakawa. I take that as a good sign, since he doesn't travel much. So it's Charles, myself, and Howard Moore, and then Peter, Randy, and Arakawa."

"Wow," Kalinske said. "Everyone's bringing out the big guns."

"A good old-fashioned Mexican standoff, right?" Sullivan said. "But they're not Mexican, they're Japanese. Or Arakawa is, at least. So I warn Charles in advance that dealing with them is a little different than he's used to. When the Japanese say yes, it means 'Yes, I understand,' not 'Yes, I agree.' So if you tell Mr. Arakawa you want markdown money, he will say yes, but that doesn't mean he's going to give you any dough."

"The quirks of Japanese business!" Kalinske belted. "It is slowly becoming the story of my life. And, my friend, it's a horror story if ever there was one."

"Not getting along with Nakayama-san?"

"Of course I am. It just takes way more time and energy than it ought to."

"That sounds about right," Sullivan said. "So they come in and we go through all the usual pleasantries. I think we even exchange gifts with Arakawa, each giving the other some trinket destined for the garbage someday. Anyway, we get down to business. Charles lays out the importance of a partnership, then cuts to the chase. For this relationship to keep functioning at a high level, things need to start changing around here. There's a long silence, and all eyes go to Arakawa, who finally nods and says yes. But Charles remembers what I told him earlier, and so he says, 'Yes? What kind of yes? Yes, we're going to work together on this, or yes, you understand, but you don't care?'"

"Can I interrupt for a moment?" Kalinske asked. "I just want to tell you that I don't even care how this story ends. You've already made my day."

"Just wait," Sullivan said with noticeable excitement. "After another long silence, Arakawa once again does that nod and says yes. Charles is starting to get cranky and keeps asking what that's supposed to mean, until Peter finally pipes up and says that it means that what we want is not going to happen."

"For the record, you've now made my week."

Sullivan rolled his eyes and continued. "So Charles has had enough. He stands up and says, 'In my mind, this is a partnership. Now you can either try to meet us halfway, or this meeting is over.' Charles stomps out of the room and doesn't come back. As you might imagine, there is a pretty big awkward silence there. So we fill the void by going back and forth with niceties. I mean, it's a couple million dollars we're talking about here. That's a lot, but not the end of the world. We don't want to make enemies, so we talk about the wife and kids and send them back to Redmond."

"I apologize for interrupting again," Kalinske said. "But I would like to amend the record to let you know that you've now made my year."

"Oh, yeah?" Sullivan said. "Then I'm about to make your life. After I walk out of the meeting, Charles comes up to me and asks if they changed their mind. Of course not, I tell him. So he says, 'Fine. Even if they're not going to give us the money, I want you to do the markdown anyway.' I tell him that's silly. Now we're not only going to lose money, but we're going to piss them off and remove any incentive they might have to change their mind. But Charles doesn't care. 'Look,' he says. 'We're going to play a chess game. You might be right and they may not give us the money. But if we mark down their products, there isn't going to be a single retailer in the marketplace who doesn't think that Nintendo gave us the money to do it.' So I do exactly what he wants, and the next day I get a call from Peter and he's flipping out. I explain that we're doing the markdown anyway and that Charles has a message he'd like to pass along: 'We'll take the loss this time, but we won't ever forget that we did.' Well, not long after that, Nintendo agrees to give us the money, and they start a corporate markdown program."

"Wow," Kalinske said. "Wow, wow, and wow. I don't know what to say."

"Say 'You're welcome,'" Sullivan replied, "and I'll say 'Thanks.' Because, I tell you, that game of chicken doesn't turn out the way it did without Sega in the mix. So thank you, Mr. Kalinske. From myself and Charles. Since the day you took over, it's been clear to me that Sega wants nothing more than to beat the crap out of Nintendo. And so far, you and your team are doing a damn fine job."

"Thanks, John," Kalinske said, genuinely touched. He was overjoyed to hear such unfiltered thoughts. Before today, he had assumed that everything he'd just heard from Sullivan was in fact reality, but there was nothing better than having that perception confirmed straight from the source. And for Kalinske, the best part was Sullivan's acknowledgment of the work done by his team. Almost all of them had been handpicked by Kalinske, Rioux, or Toyoda and thrust into a job they'd never had before. By hook or by crook (but never by the book), they got the job done, and the industry was taking notice.

Kalinske looked around the parking lot, now flooded with a mixture of his employees, retailers, and perplexed members of the Boca Raton Club wondering who'd turned out the lights. Even in this bizarre moment, he could see that his guys were finally getting the respect they deserved. Nilsen was being congratulated like he'd just delivered a nation-changing inaugural address, Adair was being treated with the same kindness she always showed to everyone else, and Van Buskirk was telling a story about the Sega LPGA Tour event and naturally drawing in anyone who happened to catch a word. "After a month of trying to coordinate this thing I realize that I'm stuck in the middle between Sega of Japan and the LPGA," she recounted. "Only problem is, due to the time difference, whenever Japan's in the office, the LPGA isn't, and vice versa. So I cut through the bureaucracy and call the tour myself. We get the thing done in a matter of days. Great, right? Or not," Van Buskirk said, and paused briefly to shake her head.

"Oh, man!" someone said—just a random club member cheerfully eavesdropping.

"This guy knows what I'm talking about," Van Buskirk said with a wink. "Anyway, I fly down to this fabulous event with John Carlucci, who helped put this whole deal together and, by the way, runs the best sports marketing company in the business. So shortly after we arrive, SOJ's best and brightest arrive and we get summoned to an inquisition. I had no idea what was coming. I don't know if anyone else knew it was coming. But man, we got taken down. We were asked to stand in front of the table. We were not asked to discuss anything. We just got a verbal barrage: 'You really messed up, you embarrassed us, and this is an abomination to how we do business.' We both walked out with our eyes as big as the moon. I've heard of these meetings where the guy in the center is so full of authority. We left the room shaking. Literally shaking, I swear. But it wasn't twenty-four hours later, at some dinner event, that the same people—the same guy in the center and all of his lieutenants—line up to get their photos taken with me because I'm tall and blond." Van Buskirk playfully ran a hand through her hair. "By the end, it occurs to me that I'm just a blow-up doll to them. Knock me down, deflate me—heck, even use me to ride in the carpool lane—and then just blow me back up when you need something."

Van Buskirk had the crowd surrounding her in stitches, including Kalinske and Sullivan. "She's a keeper," Sullivan said.

"Yup," Kalinske added, watching Van Buskirk transition into another anecdote. Even when the power went out, the presentation was halted, and the script had to be tossed out the window, the show went on at Sega. "But the thing is, John," Kalinske said, "they're all keepers. Otherwise they wouldn't be here, and neither would I."

About ten minutes later, the power came back on, and the parking lot broke out in applause. Kalinske, his employees, and the retailers made their way back into the ballroom to finish the presentation. After everyone took a moment to feast on the glorious invincibility that only air-conditioning can provide, there was an overwhelming sense that maybe it would be better to skip the rest of Nilsen's speech and trade in business casual for swimwear. Nilsen could sense this, and knew he'd have to work to get them to listen to him.

"Greetings," Nilsen said, readjusting to the stage. "Thank you for bearing with us through that pesky disturbance. Unfortunately, we'll never know what caused the blackout, but I can't help think twice now about those rumors that Luigi was spotted earlier today hanging out by one of the generators."

The audience laughed, but wasn't yet convinced to sit back down.

"He claimed to be working on the plumbing," Nilsen explained, "but with the recent popularity of Sonic, I wouldn't be surprised if he and his brother were looking into other lines of work. Though it seems electrician might not be the best fit for his skill set."

This time more laughter, and some started to take a seat.

"Now, it occurs to me," Nilsen said, "that we were just out under that big yellow sun, and some of you might prefer to be there rather than here. Particularly those of you from the artic tundra of Minneapolis," he said, glancing at the contingent from Target. "And I will be the first to admit that

I can't compete with tropical weather. So if you would prefer to sit around by the pool with a cold piña colada in your hand instead of sitting around here and listening to me yap away, then I encourage you to do so." He paused to allow the flock of retailers a momentary thought of paradise. "But if you decide to leave now, you're going to miss learning about the most important event in the history of videogames."

Kalinske loved it, the way Nilsen and all of his employees ran through obstacle courses with a smile on their face, no matter how difficult or peculiar the challenges were. This was what Sullivan had been talking about, the good kind of crazy, and Sega personified it through and through.

"So the choice is yours," Nilsen declared. "And to those of you who are going out by the pool, save me a piña colada!" But there would be no tropical drinks, because everyone returned to their seats excited to hear about the next big thing. "Wonderful," Nilsen said, and then took up where he had left off. "Crooks beware!" he proclaimed, describing the Home Alone game due out in November. "Kevin's back, and he's got a houseful of clever tactics to defend the entire neighborhood, keep the robbers on the run, and keep the gamers hollering for me."

Kalinske proudly watched Nilsen get back in the groove, introducing The Little Mermaid ("The whole family will fall hook, line, and sinker for Ariel and, as a result, you'll be reeling in the profits"), X-Men ("Disaster, riddles, muscle, and endless play arrive in November as the top-selling comic book comes to the Genesis"), and Streets of Rage 2 ("This isn't some wimpy imitation. The characters are 25 percent bigger and 100 percent tougher"). The longer Nilsen spoke, the more Kalinske realized that he had been wrong earlier. There was no need to pitch a perfect game. Sega had already pitched a perfect game these past two years, and now it was just about continuing to throw hard.

The retailers furiously scribbled notes as Nilsen took them further and further behind the curtain, their anticipation growing for what he'd called "the most important event in the history of videogames." Nilsen could see their minds salivating, but Nilsen continued to expertly stretch out their anticipation. "This next product needs to remain so top-secret that we can't even show you what it looks like," he began, introducing NFL Sports Talk Football Starring Joe Montana. "The cast is off, and in October he'll be back on the playing field. The Genesis playing field, that is!"

After that, Nilsen finally arrived at the moment they had all been waiting for. "Here we are, folks," he said, adopting a hushed tone. "I have the pleasure of giving you a preview of a game we're really excited about. It won't be out until 1993, but we're so proud of the graphics, sound, and control that we wanted to give you a sneak peek at the next generation of Genesis games."

Kalinske noticed retailers leaning forward in their chairs, hoping as much as he did that Sega had the goods to keep going after Nintendo. He studied Sullivan, who looked like a child about to unwrap an unexpected Christmas gift. "It's an adventure game set in the not so distant past, filled with puzzles, unforgettable science fiction elements, and a quest to restore order in your civilization. No, I'm not talking about a Zelda knockoff. This is something completely different."

Nilsen stepped off the stage and walked into the audience, his passion so overwhelming that he wanted them to see it firsthand. "Even if you don't like to play videogames, I urge you to pick up the control pad and try out this game, Dolphin, when I'm finished speaking." This was the game that Ed Annunziata had fantasized about more than a year ago, over their Italian meal during CES. "The feel of controlling this magnificent mammal," Nilsen continued, "is unlike anything you've ever experienced before. This game is the next step in videogaming." At that, he returned to the stage and played a short clip of the game. It really was unlike anything that had ever been done before, a game as smooth and beautiful as a book or album. Maybe it wouldn't be the most important event in the history of videogames, but it was certainly proof that Sega had no intention of remaining second-best.

When the clip finished, the retailers appeared noticeably impressed, which made what was about to happen that much more fun. "Hey, Al," Kalinske said, loud enough for all to hear. "Don't make them wait any longer. Show them the big thing already!"

Nilsen smiled. "Sure thing, boss." The performance wasn't over, which surprised the audience almost as much as the blackout had. "But wait, there's more," Nilsen said, clicking on the next slide. "Peripherals, an important part of your profit picture!" Following a collective groan, he raced through descriptions of an innovative wireless remote (the Cordless Elite), an intimidating plastic bazooka (the Menacer), and a Sega Genesis cleaning system headed to stores in September. "And that," Nilsen said, sticking his notes into a folder, "is all I have to tell you today."

Just then, through the ballroom doors, Sonic The Hedgehog himself entered, accompanied by Schroeder. "Okay, Sonic," Nilsen said with a pinch of feigned exasperation. "I'm sorry. I'll tell the whole world about your new game. But would you mind if I brought Madeline up here to help explain why Sonic 2 is going to be the number one game of 1992 and the most important event in the history of videogames?"

This was the moment that Nilsen had been waiting for, where he got to finally reveal Sega's big plans for Sonic 2. But he'd have to wait a moment longer, because as he opened his mouth, Sonic received a thunderous standing ovation. The retailers were on their feet, all of them, shaking the room and causing a tiny earthquake. These people weren't even cheering for Sega's sequel plans, since Nilsen had yet to reveal the X's and O's. No, they were simply cheering for Sonic; the character, the icon, the emblem of the war against Nintendo. And after this week, the army would be much bigger and stronger, especially after what Nilsen and Schroeder had to say next.

"Sonic's back and he's better than ever!" Nilsen proclaimed as the lingering applause quieted down. After that, he and Schroeder launched into one of their lightning-quick, Abbot-and-Costello-esque banter sessions.

"That's right, Al. And this time he's faster."

"Yup, Mad. And he's cooler!"

"He's got more enemies!"

"More levels!"

"More attitude!"

On and on they went, ping-ponging superlatives, introducing Tails, and welcoming a costumed orange fox to join them onstage. But this was all an appetizer for the entrée, which was Sonic 2sday, the videogame world's first-ever global launch.

"The entire marketing program," Nilsen said, "is designed to build awareness for Sonic 2sday, the street date of Sonic 2."

"And on that date," Schroeder explained, "the game won't just be available at retail in the U.S., but also in Europe, Japan, Australia, and the rest of the world!"

"Yup," Nilsen confirmed. "We're working to make Sonic 2sday the biggest international event since the fall of the Berlin Wall."

To someone outside this room, the idea of a coordinated worldwide release might have seemed interesting but irrelevant. But the point of a global launch wasn't to dazzle with concept; the point was that the concept created connection. Normally, with games released at different stores on different days, customers couldn't help but feel like these things sort of fell out of thin air. But to know the exact date that something would be arriving, to have it circled on the calendar ahead of time, gave the gift of anticipation. For kids, there's no feeling better than excitement, and for parents, there's no feeling worse than bursting that bubble, which made it every family's duty to celebrate Sonic 2sday. It was a marketing ploy, yes, but it worked in the same self-fulfilling way as a blockbuster film did. They're not called "blockbusters" just because of their budgets; rather, it's because of the event-like, don't-be-left-out way that they are marketed, which makes people rush to the theater for the opening weekend, which then makes more people rush to the theater when they hear how big that opening weekend was. The art of the blockbuster is that it popularizes something before it ever even exists, and though Sonic 2 was still months away from completion, Sonic 2sday gave Kalinske and company an opportunity to unleash the biggest blockbuster the video-game world had ever seen.

"Anyway," Schroeder said, "you probably want to know when Sonic 2sday is. Well, right now all I can tell you is that it'll be one of the Tuesdays in November."

"Oh, come on, Mad," Nilsen pleaded. "Tell these nice buyers the date."

"I can't do that," she explained. "Sonic won't let me!"

Actually, Sonic's creator wouldn't let her, because he himself didn't know. As the game neared completion, Naka's notorious perfectionism became even more pronounced. The game had already been pushed from October to November, and Sega of America was desperately hoping it wouldn't be postponed again. That would ruin Sonic 2sday, the Christmas season, and Sega's plans for a new world order.

"When it comes to the challenge of leading the 16-bit revolution, we're up to it," Nilsen said as he neared the end of his presentation. "But we need your help."

The retailers were the conduits to the consumer, which made them gardeners capable of planting the seed. So they had to decide whether this was a garden worth growing. Did they want to bet big on Sonic 2? Cover their windows with images of Sonic and Tails? Plaster posters everywhere and hang a giant sign out front with a countdown to the game? Presell the game, track those customers, and reward them with a promotional T-shirt (which would be free to buyers who preordered but would cost stores $2.50 each)? Most important, were they willing to alter their operational and financial procedures so that Sonic 2 could be shipped directly to each of their individual stores? This last proposition was unprecedented—Sega was asking Fortune 500 companies to bypass their long-standing distribution systems for what was literally a fly-by-night delivery. But this was the only way that Sega could ensure a coordinated global release. For this to work, it required a partnership with the people in this room. So it all boiled down to one question: were they in or out? Ultimately, this was what Nilsen's presentation was asking, but the question wasn't posed that simply. In the tradition of Sega's "good crazy" spirit and the torrential pun-derstorm that was Sonic 2 (official tagline: "2fast, 2cool, 2day"), it was issued as a challenge.

"I have just one question for you," Nilsen said slowly, measuring the full weight of his final words. "Are you up 2 it?"

30.

JUST DO IT

Smiles widened, glasses clinked together, and inside a chic French restaurant in San Francisco, toasts were made to celebrate revolutionizing the way videogames were advertised. Tom Kalinske then shot a quick, happy glance at Shinobu Toyoda and Ed Volkwein before taking a long sip of an oaky Bordeaux and savoring the significance of this dinner with Dan Wieden and David Kennedy. If someone had told him two years earlier that Wieden+Kennedy would be interested in becoming Sega's agency of record, Kalinske would have either laughed, rolled his eyes, or assumed that Dan and David's siblings had started their own ad shop. After all, these were the guys responsible for building the Nike brand into what it is today, coming up with the famous "Just do it" tagline and iconic sneaker campaigns like "Bo knows" (with Bo Jackson), "Mr. Robinson's Neighborhood" (with David Robinson), and "Spike and Mike" (with Spike Lee and Michael Jordan). It just seemed wonderfully inconceivable that the Oregon-based agency responsible for transforming Nike into the *it* sneaker company now wanted to do the same for Sega in the videogame space. But here they all were.

"So, what happened next?" Wieden asked. "Come on, Tom, don't leave us in the dark. Did the retailers rise to the challenge or not?"

"You tell me," Kalinske said, resuming his recap of Boca. "After the presentations, our first meeting was with Toys."

"Toys 'R' Us," Volkwein clarified.

"Right, Toys 'R' Us," Kalinske continued. "One of the biggest guys in the marketplace. The tastemakers, if you will."

"Also very close relationship with Nintendo," Toyoda piped up.

"Yup, Toys and Nintendo are peas in a pod," Kalinske said. "And we've got the big pea himself, Charles Lazarus."

"And let me guess," Wieden said. "He didn't accomplish all that by acting like a gunslinging risk taker."

"Bingo," Kalinske said. "Which is why we schedule the Toys meeting first. If they're in, then we'll know right away that this is going to happen. If not, then at least we've got three days in paradise to shake it off."

"Go big or go home," Kennedy said. "Gotta love it."

"So the meeting starts, and it's myself, Paul, Al, Diane, this guy," Kalinske said, nudging Toyoda, "and Richard Burns, our VP of sales. First thing Mr. Lazarus wants to know is if others are on board. Now, we've all been down there a few days, so they don't know that we haven't met with anyone else yet. I'm about to tell them this, hoping that the compliment of making them our first might carry some weight, but Burns cuts me off and says that everyone is already on board and we've been on with Japan all night to see if they can accommodate this demand. Lazarus looks around the table, each of us trotting out our best poker faces, and he says, 'Looks like we don't have any choice. Count us in and make sure we get the biggest order.'"

"That is positively outstanding," Wieden declared.

"I can't even describe the size of the weight that fell off my shoulders," Kalinske said. "Not only that, but Charles then goes on to instruct his people to work closely with us every step of the way so that both companies can set records with this thing. Amazing, just amazing, and yet . . . that wasn't even the biggest triumph at Boca."

Toyoda and Volkwein crinkled their faces, unsure of what exactly their boss had in mind. "Are you talking about the comment from Griffiths?"

"No, but that's up there too," Kalinske said, in the mood to brag and turning to Wieden and Kennedy to offer an explanation. "Jeff Griffiths is the buyer from Electronics Boutique, and he said something that reinforced how we hoped everyone in the industry was feeling. Toward the end of Boca I asked him how he felt about the event, and without the slightest hesitation, he said, 'Empowered.' He then said, and I quote, 'After years of being lectured to by powerless drones at Nintendo, whose only apparent talent was an annoying ability to make us feel eternally beholden to them for the pitifully small allocations and onerous payment terms, we finally find ourselves sitting face-to-face with people willing to listen to our ideas and act on them.' His words, not mine," Kalinske said proudly.

"Right on," Wieden said. "We're honored that you're even considering us as the guys to help you take down those drones."

"Oh, please," Kalinske said. He appreciated the humble words, but their guests from Oregon were far and away the front-runners for the account. After months of conducting the agency review, Sega had narrowed down their list to three candidates: Wieden+Kennedy, the gods of Nike; Foote, Cone & Belding, the masterminds behind Levi's 501 jeans; and Goodby, Berlin & Silverstein, a boutique agency whose biggest claim to fame was how close their office was to Sega's headquarters (a fact they emphasized often, via spontaneous drop-bys). The guys at Goodby, Berlin & Silverstein had the creative chops, no doubt, but with a limited track record and the recent departure of Andy Berlin, one of the small firm's partners, the competition for Sega's business was pretty much narrowed down to two. And in the next few weeks, both suitors would be pitching and Kalinske would finally have the agency that would help him bring down Nintendo. "I think your work speaks for itself," Kalinske told them. "And I honestly can't wait to see what you come up with."

"Can you give us a sneak preview?" Volkwein asked.

Wieden and Kennedy glanced at each other, smiled, and shook their heads.

"Not even a hint?" Kalinske asked, goading. "Just a tiny little glimpse?"

"I could tell you, but then I'd have to kill you," Wieden said. "And then you'd miss the presentation, so it would all kind of be for nothing."

"But I will say," Kennedy explained, "that you wanted something revolutionary, and this campaign will be your rallying cry."

This got Kalinske even more excited. He couldn't believe that it had been only a year and half since that big presentation in Japan where he had received approval to revamp Sega. Hiring a new ad agency and investing in a company-redefining campaign had always been the final step in Sega's makeover, and it was almost time to try to win this beauty pageant. Things were coming together even quicker than Kalinske had imagined, and not just with advertising and retailer relations, but in other spheres as well. Plans for a Sonic The Hedgehog cartoon were reaching the point of reality, distributing Genesis games over television cable wires was almost possible, and the crown jewel of progress was an alliance with Sony. A few days after Boca, Sega of America and Sony Electronic Publishing announced an alliance to make games for the Sega CD. The power play involving these companies shook up the videogame industry, and it even made waves in the mainstream press. On May 21, 1992, the *New York Times* ran an article by Adam Bryant that explained the crux of the deal: "The broad alliance announced yesterday calls for Sony to tap its full stable of recording artists, actors and movies to create games for the new Sega machines, which will use compact disk technology." Although Kalinske and Olafsson had wanted this for ages, getting approval from both parent companies had been no small task. After months of respectful pleading, a turning point came when Sega helped Sony out with an unsettling situation.

In the 1980s, there had been a growing fascination with full motion video (FMV) games. With real actors and hours of prerecorded footage, FMV games were essentially movies where the viewer, or in this case the player, interacted with the content (i.e., shooting monsters, avoiding obstacles) in a way that dictated how the story played out. This intersection of Silicon Valley and Hollywood led many to believe that this represented the next step in the evolution of videogames. One of those believers was Sony's Michael Schulhof, who in 1990 acquired the rights to four FMV games that had originally been created for Hasbro's NEMO console but became available when the toymaker pulled the plug on that system. His thinking was that these games, whose files were much too big for 16-bit consoles, would be perfect for Sony's CD venture with Nintendo. But when that deal fell apart, Sony was left with costly assets and no place to exploit them—until the Sega CD came into focus. This would seemingly be the right home for Sony's FMV games, but upgrading just two of the games, Night Trap and Sewer Shark, to make them playable on the Sega CD was going to cost nearly $5 million.

Sony was already lukewarm on the videogame industry and not sure that it was wise to throw more money at the FMV problem. Kalinske and Toyoda, however, really wanted to make a partnership work, not to mention that they needed more software for the upcoming Sega CD release. Given the situation, it seemed like a no-brainer for Sega to offer to foot the bill;

this would forge an alliance with Sony, and also keep Sega on the cutting edge of technology. This was a dream scenario, everything Kalinske had wanted, until he took a closer look at the games in question and realized how violent they were. By movie standards, Night Trap and Sewer Shark were no worse than a schlocky horror film, but compared to the typical videogame, they may as well have been snuff films (particularly Night Trap, in which players were tasked with saving scantily clad women from a vicious flock of vampires). As Kalinske watched one of the vampires attack a defenseless girl in the shower, that pang of conscience he hoped had been vanquished returned once again louder than before. Was it really a good idea for Sega to release this kind of game? Probably not, Kalinske thought, but then was reminded of another thought that he'd had years earlier. In the tiny kitchen of Sega's old office, he had been reading that *New York Times* article about Peter Main and came to the conclusion that if Nintendo represented control, Sega must offer the freedom of choice. Neither of these games were Kalinske's cup of tea, but who was he to stand in the way of choice? And so after he decided to get out of his own way, Kalinske, with Toyoda, arranged for Sega to foot half of Sony's bill for these two games and future FMV development costs, which ultimately helped push through an alliance between the companies.

After becoming partners on this, they decided that they would divvy up the slate of FMV games, with Sega publishing some as first-party titles and Sony publishing others under their Imagesoft label. As a courtesy, Sony gave Sega the opportunity to choose first. Well, Kalinske thought, if I've already decided to go down this road then I may as well ride it all the way, and he selected Night Trap, which was definitely the more violent of the two, but it was also the one he thought would sell better. The game would be released on October 15, 1992, along with the launch of the Sega CD, officially marking the beginning of a relationship between Sega and Sony that Kalinske hoped would grow much deeper. For better or worse, Sega was all about choice. That was the backbone of their videogame revolution, and it was why an agency like Wieden+Kennedy was excited to work for Sega.

"Wait," Glen interrupted, as everyone at the table began to discuss favorite Super Bowl commercials from over the years. "What is the biggest triumph?"

"What biggest triumph?" Kalinske asked.

"Earlier," Glen explained. "You had mentioned a sizable triumph that was neither Toys 'R' Us nor the comment by Jeff Griffiths. Now I'm curious."

"Oh, that," Kalinske said with a chuckle. He put his hand on Toyoda's shoulder and said, "The biggest triumph of Boca was Shinobu catching a massive sailfish."

Everyone at the table laughed and turned to Toyoda, who nodded bashfully.

"How big was the fish?" Wieden asked.

"I will show you," Toyoda said, and then pulled out his wallet and showed off a photo of him proudly standing beside an enormous sailfish. "On the first day of Boca, Gary Kusin, who is the founder of Babbage's, caught a big eighty-inch sailfish. To see this, I thought: wow, very amazing. But then the next day I got one that was an inch bigger and it was an incredible moment."

As everyone gushed over Toyoda's eighty-one-inch sailfish, it felt like the perfect end to the evening, and the beginning of a wonderful new relationship. Except that during the conversational ride off into the sunset, a waitress came by to drop off several expensive bottles of champagne wrapped with a bow.

"What's this?" Kalinske said, delighted by the liquid generosity. "What a nice gesture," he said, turning to his guests. "But you know we can't let you pay for this."

Wieden and Kennedy turned to each other with a puzzled look, but before they could say anything Kalinske picked up a note attached to the bottles. "Well," Kalinske said after an echoing laugh, "it appears that we have an unexpected benefactor."

"Oh, yeah?" Volkwein asked. "And who's that?"

Kalinske answered by passing around the note, which read:

Enjoy the champagne. We eat here all the time.

Sincerely,

Goodby, Berlin & Silverstein

After everyone finished laughing, Toyoda asked the question on everybody's mind. "Do you think we should drink it?"

Wieden and Kennedy both shrugged and left the decision up to their host.

Kalinske reread the note and slowly shook his head. It really was a nice gesture, deserving of bonus points for cleverness. But unfortunately, a bold move and offices nearby weren't going to be enough to get Sega's account. Maybe there was a way to throw them some business for the handheld, but the upcoming ad campaign was Sega's big shot, and they couldn't afford the risk of a less experienced agency.

"I'm tempted to say no," Kalinske said. "But I can't help but think that this move demonstrates a little bit of Sega spirit. So, what the hell?"

Once again smiles widened, glasses clinked together, and a toast was made inside a chic French restaurant in San Francisco. This time, however, it was hard to take a sip and imagine anything other than Jeff Goodby smirking somewhere in the night and thinking to himself, "Just do it."

Shortly after dining with Wieden and Kennedy, Kalinske and company headed to Chicago for the summer CES. With the agency pitches fast approaching and more and more retailers signing on for Sonic 2sday each day, Sega went into the show with more momentum than ever before. And having already accomplished in Boca much of what would typically need to be done at CES, Sega was in the unique position of playing defense more than offense. But just because they would be playing defense didn't mean that they'd act passive; as any football fan knows, there are tons of strategies, schemes, formations, and blitz packages. With so many permutations, the key to winning on defense is all about identifying the offense's play as soon as possible and then calling the right audible to shut it down. That was exactly what Kalinske's team set out to do, but after nearly two days in the Windy City, all anyone heard was that Nintendo had a big announcement planned. So Kalinske gathered the troops together in his suite at the Sheraton for an emergency meeting.

"They're getting smarter," Kalinske said, shaking his head. All of Sega's top executives were crammed into the suite, as was David Rosen and Dai Sakarai, one of Nakayama's top lieutants (someone who, in the pseudo-war between SOA and SOJ, Kalinske and the others trusted; plus he knew the lay of the land from his very brief term as interim president of SOA). "I hate them for it, but I definitely respect it."

"Nobody's giving me anything," Rioux said. "I just keep hearing 'big move' over and over. What am I supposed to do with that?"

"Could it be a new Mario game?" Kalinske asked.

"That is not their style," Toyoda replied. "They like to wait a few years in between the Mario titles."

"Realistically, what could it be?" Nilsen asked, half closing his eyes, as if by doing so the answer would be psychically revealed to him. "We should be able to figure this out," he said with frustration, upset at himself for not being able to do so.

"My money is on a theme park," Van Buskirk said. "The Amazingly Incredible Mushroom Kingdom. Complete with real-life Koopa Troopas!"

"I'd welcome that," Kalinske said. "Another thing to keep them distracted."

"What about a color handheld?" Adair suggested. "They can't just ignore how well Game Gear is doing, can they?"

"I don't know," Burns said. "They've gotten quite good at burying their heads in the sand, haven't they? One might even call it a forte."

"It must be a partnership," Glen suggested. "It's hard to imagine a world in which they don't consider our alliance with Sony to be a slap in the face. I would venture to guess that they slap back in some fashion."

"I doubt it," Rioux said. "They'd never let anyone in their sandbox."

"What about a new PR agency?" Sega's PR agent, Brenda Lynch, suggested. "The Mariners backlash had to have woken them up."

"You may very well be right," Volkwein said, "but with all due respect, I don't think a new PR agency qualifies as big news."

"If it truly is big news," Dai Sakarai mused, "then perhaps they have decided to give up on 16 bits and move on to a 32-bit system."

"I like that theory," David Rosen put in. "Do to us what we did to them."

Back and forth everyone went, yipping, yapping, and playfully snapping at each other. Kalinske loved the energy and enthusiasm of the team he had built. He loved that they believed in the cause and weren't afraid to act. And it was with this last part in mind that he suggested they take things into another gear. "Exactly," he said. "It could be anything. So we're going to have to figure this out ourselves. Does anyone object to going undercover?"

Nobody in the room objected. In fact, they thought it was about time. Their favorite part about working at Sega was the whatever-it-takes mentality and the shenanigans that often came about as a result of that. There were six industry parties scheduled for that evening, so Kalinske broke everyone up into teams and sent them around Chicago to see what they could dig up.

Adair and Van Buskirk were paired together and sent to the Electronic Arts party, where their efforts to solve the mystery initially amounted to nothing. It wasn't that they weren't the top-notch spies they fancied themselves to be; rather, it truly seemed as though nobody really knew what Nintendo had planned. That changed a little, however, as the night rolled on and drinks continued to be served. Finally, in tandem, Diane and Van Buskirk cornered an EA producer with close ties to Nintendo.

"Come on, I know you know," Van Buskirk said.

"Yeah, honestly, just tell us already," Adair added. "It's almost midnight, anyway. How could it possibly make a difference at this point?"

"Okay," the producer said. "What difference does it make?" He then informed them that the next day Nintendo would be dramatically dropping the price of their hardware. Without even thanking him for the intel, off they ran into the night.

They reported the news back to Kalinske, who couldn't hide his disappointment. Nintendo was dropping the price of their deluxe SNES package (bundled with Super Mario World) to $129.95, the same as Sega. In addition, they'd also be selling an SNES without any game for $99.95. The latter piece was particularly crushing. Up until this point, Sega's consistent advantage had been a lower price than Nintendo, but by tomorrow afternoon the Big N would be stealing headlines and beginning to undo all the work that Sega had done.

Shortly after midnight, everyone returned to Kalinske's suite to digest the news and brainstorm possible responses. Some were sleepy-eyed and some were clearly drunk, but they all shared a manic insomniac energy and excitement. They not only worked for Sega but had become Sega, and took a strange, wonderful pride in doing whatever it took to help defeat Nintendo.

"Thank you all for being here at this hour," Kalinske said, showing genuine gratitude. "As some of you may have already heard, Nintendo is trying to outprice us. Before we even talk about a response, I want everyone to take a moment to think about how much of a thorn in their side we must really be. They refuse to acknowledge us, they treat us like second-class citizens, but it's because of us and only us that in the span of only eight months they have dropped their price nearly 50 percent. They may still control the market, but I'd say that counts as proof that they are no longer invincible."

Though it was late and the news about Nintendo was disappointing, the truth behind Kalinske's words filled everyone with a second wind. Look at how much they had done in such a short time; there had to be something, then, that they could do in the few hours before the announcement.

"I'm going to open the floor," Kalinske said, "because I want whatever we do to be done together. What is everyone thinking?"

"What about leaking a story?" Lynch suggested. "Something about how bad finances prompted the move. Company in trouble. The end is near. Et cetera, et cetera."

"Do we have something like that?" Kalinske asked.

"No," Lynch said. "But we can get creative and make them waste time responding to our storyline, instead of writing their own."

"Hang on a second," Burns cautioned. "Maybe this isn't as bad as it looks. They'll have a small price advantage, sure, but we've got Sonic 2sday."

"I love Sonic 2sday more than anyone," Nilsen said, "but, believe me, Nintendo's move is as bad as it looks."

"There's only one thing we should do, and we all know it," Adair said. "We need to fire back and match them. Paul, what about that idea you mentioned to me before, of selling just the Genesis without any games? Could we sell that for $99.95?"

"You're talking about the Core System," Rioux replied, considering her suggestion. "Yeah, that price is doable, but we both know Japan won't go for it. And even if they did, we simply don't have enough time to deliver a measured response."

"Hey," Van Buskirk said, "haven't any of you ever pulled an all-nighter?"

"You're exactly right," Kalinske said, feeling a sudden invincibility flow through his veins. She was right, and so was Adair. Beneath the mirage of possibilities, Sega really had only one option. And in no way was this option ideal, because it meant somehow cramming weeks of work into a single pressure-packed night, but if Sega wished to really become the company that deserved to decapitate Nintendo, that would revolutionize videogames, that was worthy of working with an agency like Wieden+Kennedy, then they really had no choice but to suck it up, find the fun, and just do it. "Let's do it," Kalinske said. "Come on, we need this. We're on the brink of something here, and we really need this. So it's simple: let's make this happen."

And just like that, everyone was flooded with energy and ready to work through the night. But in order to get to work, they needed something bigger than a hotel room, preferably something with some form of technology to help them pull this off. The hotel had a business center, but it was closed at this hour—or it was closed until Adair found a way to break in and temporarily turn the place into Sega's workshop. They figured out the logistics, developed the strategy, and designed the necessary materials for the presentation that morning that would cut Nintendo off at the knees.

Beside the time crunch, the hardest part of pulling this off was that they had to make it appear completely preplanned. More deadly than being upstaged by Nintendo was the allegation of being reactionary. Everything that they prepared this evening had to look, sound, and feel like it had been planned out for months. Although that wasn't factually true, emotionally it sort of was; the crux of Sega's philosophy hinged on instant adaptability, the art of embracing opportunities in the blink of an eye. So although they only had hours to create something from scratch instead of months, if anyone could do it, it was these guys. And they did, finely crafting all of the following materials before the preshow sales meeting in the morning:

- Two hundred press kits
- Updated fall/winter price lists
- Comprehensive sell sheets for the new $99 Core System
- Price stickers and display photos to adorn Sega's booth
- Updated pocket cheat sheets for Sega's internal sales force
- Talking points for the press, retailers, third parties, etc.

The impossible task of putting together an entire campaign in one night was the easy part compared to changing Nakayama's mind. Luckily, for that task Kalinske got help from Rosen and Sakarai. All through the night they strategized about which buttons to push and how hard to push them. At last they took the request to Nakayama, bracing themselves for that pivotal now-or-nothing moment.

"You're being impulsive," Nakayama said. He felt strongly that this would be something they would regret in the morning.

"Not at all. We've actually been considering this for quite some time and had all the materials ready just in case," Kalinske said, speaking quickly so as not to get caught in a lie. It was true that Sega of America had been seriously considering a cheaper, game-free system, though not one at $99.95. But Nakayama didn't need to know that. "Please trust me," Kalinske pleaded. "Remember, you hired me to make decisions like this."

There was a long pause.

"It's the logical move," Kalinske continued. "If not now, then we'll do this in six months. But what's six months of some additional profits compared to all the sales we'll make by doing this now and taking Nintendo down in the process?"

And then a longer pause.

Kalinske looked at Rioux and Toyoda, his brothers in arms, then glanced over their shoulders at Nilsen, Adair, Van Buskirk, and the rest of his devoted employees. This had to work out; they just had to be given permission to try to pull off this miracle.

Finally Nakayama broke the silence. "Okay, Tom. If you think this is best."

As soon as the line went dead, the team erupted with joy. In just a few hours they would return to being serious, focused, business-suit-wearing executives. In just a few hours they would match blows with Nintendo and look for new and unexpected ways to tilt the scales. But that was all a few hours away. Right now they got to stay up late and enjoy just being kids stuck in adult bodies.

31

TOO HOT, TOO COLD, AND JUST RIGHT

"That sucks," some random teenager said, and he said it with such visceral disdain that it was as if he'd been forever traumatized by the level of suckitude he'd been forced to witness. "It just, you know, sucks," he said again, this time earning dismissive nods of agreement from the other kids in the room. There were about a dozen of them in total, all hired by Goodby, Berlin & Silverstein to review the commercials that the agency had prepared for their upcoming Sega pitch.

"He's right," another said. "Like, we already know our parents don't know how to play videogames. Tell us something new!" More dismissive nods, and this time some high fives of agreement from around the room. Although none of these kids had ever met before, and likely never would again, they were fully united in their distaste for the commercials they had just seen.

Jeff Goodby, the agency's leader and cofounder, watched all this from behind a two-way mirror. Goodby was a Harvard-educated ad man with the renegade mentality of a pirate philosopher, the lumbering physique of a friendly yeti, and the rare gift of being able to sport a ponytail and make it work. Once upon a time, he'd considered advertising to be the lowest form of writing, nothing but a nuisance to those in its perpetual line of fire. But after leaving his post as a city hall reporter for a newspaper in Massachusetts and moving to San Francisco with his wife, he needed a job quickly and wound up working for the ad agency J. Walter Thompson. There, his mind was blown by the art of advertising, and in the process he learned that he had a penchant for blowing minds himself with his creative work. After a stint at Ogilvy and Mather, he left there in 1983 to start his own agency with Rich Silverstein, his creative partner, and Andy Berlin, a brash entrepreneur.

Throughout the 1980s, GB&S grew into one of the industry's top boutique shops, able to compete with the big firms by coming up with bigger ideas, like the Electronic Arts "We See Farther" campaign (1984), which personified the young ad agency almost as much as it did the young computer company. They also distinguished themselves from the competition by developing a distinct style (a cinema verité high-concept, low-production-values technique), promising clients a more hands-on experience (one of the founders would personally head up every account), and placing a forward-thinking emphasis on account planning (the initial consumer research and messaging that inform the creative process). After a decade of impressive growth, during which the agency had evolved into a firm of nearly fifty people, it seemed like only a matter of time before Goodby, Berlin & Silverstein reached the next echelon of success. That inevitability was derailed, however, when tensions among the partners caused Andy Berlin to leave and start a new agency (Berlin Cameron) in early 1992. To Goodby and Silverstein, this changed nothing, but to the outside world there was a skeptical feeling that this signified the beginning of the end. As a member of an industry where perception is said to equal reality, Goodby knew he needed to find a way to show that the Berlin-less agency was stronger than ever. And the way he planned to do that was by winning Sega's business, but the shitty feedback that he was receiving from the focus group didn't bode well for the agency's chances.

"All right, then," Silverstein said to his partner and the other creatives on their side of the glass. "Welcome to our worst-case scenario."

Goodby nodded, keeping his eyes on the teenagers. "I'm tempted to celebrate the fact that their response is so overwhelmingly unanimous," he said, "until, you know, I take into account the fact that they're basically telling us to quit our jobs and go jump out the nearest window."

With no silver lining in sight, Goodby, Silverstein, and the others returned their attention to the focus group.

"That footage you showed wasn't even from Sonic 2!" said one of the teens. "It was from the original Sonic, and it was from level two, which isn't hard at all."

From another: "Those commercials look like they were written by adults."

And from a third: "The guys who made it weren't even good enough to get to the difficult parts."

The kids were right. The guys who'd made the commercials weren't very good at the games, but they'd hoped to compensate for that knowledge gap by spending time with people who were good at them. To truly understand what it meant to be a gamer, the agency sent Jon Steel, their director of account planning, and a small team of his best planners around the country to spend time with this demographic. As the head researcher on this project, Irina Heirakuji arranged for a two-week tour through the country during which she and the other planners would visit with boys between eight and twelve, invite over members of their friendship circle, and observe how they played videogames. In addition to these in-home observations, the planners also studied the children's bedrooms, closets, and any other area that might offer an insight into their overstimulated minds.

What the account planners discovered was that these kids, just like those in the test group, had their own language, customs, and rituals; in a sense, they had their own secret society that adults could observe but never quite understand. Whereas secret societies in the past have met in lodges, taverns, and dimly lit alleys, this new generation's meeting spot was the virtual world of videogames. And also unlike the covert gatherings of yesteryear, where furtive glances and secret handshakes were needed to gain entry, the only password into this world was up-up-down-down-left-right-left-right-B-A-start. Playing videogames wasn't even so much about being good at them as it was about understanding them in a fundamental way that adults never could. For the first time ever, the kid who could never beat his father on the baskctball court or beat his mother in an argument now had a place where he could reign supreme.

Goodby, Berlin & Silverstein recognized the value of this new dynamic, and for the pitch they created a series of commercials highlighting the concept of the kid as king. One of the ads featured images of adults-only luxuries, like sports cars and scantily clad women, playing over an unseen and unemotional narrator's description of the power of adults. "They can drive cars," he says. "They can go to R-rated movies, and they can also decide when to go to bed." As he speaks, a barrage of semi-related buzzwords—"speed," "babes," "midnight"—blink on and off the screen. "But," the narrator forcefully says as the screen cuts to game footage from Sonic The Hedgehog, "it will be a chilly day in hell before an adult gets this far on Sonic 2."

In another ad, a kid named Mitch is seen playing Sonic 2 at home, desperately trying to reach the seventh level. As he nears this goal, his boring father tells him to stop playing and hit the books. Undeterred, Mitch continues playing. Eventually his father delivers to Mitch what he believes to be the most shattering of insults: "If you keep playing videogames, you'll never grow up to be like me." Following these harrowing words, the audience is told that Mitch's new goal in life is to reach the eighth level.

Both ads ended with a tagline that had been cooked up by Dave O'Hare, the chief creative on the account, with a major assist from Jon Steel. After spending those weeks with gamers around the country, Steel informed the creatives that this videogame world is all about speed. It's all about being able to get through a level in order to move on to the next one. The longer you spend on a level, the slower you are as a player, the less competent you are as an individual. These kids don't just want to win; they want to win quickly. He also told O'Hare that although the majority of kids owned and played Nintendo, those who had tried both systems believed that Sega's was superior. One kid even said that after playing Sega there was "no going back." With comments like these, Steel paraphrased that going from Nintendo to Sega was like getting called up to the big leagues. O'Hare considered this, and after a minute of blending it all together, he suggested that the gamer version of this analogy would be getting to that next power, that next zone, that next level. Seconds later, he came up with the phrase that somehow incredibly summed it all up: "Welcome to the Next Level."

Between the tagline, the account planning, and the scattershot we-get-you feel of the campaign, Goodby thought that he and his guys had nailed it. They had infiltrated the secret society and learned how to speak directly to this demographic. Or so they thought, until the focus groups' constant allegations of suckitude.

"The more I think about the commercial," another kid said, "the more it sucks."

"Plus Sega shouldn't be insulting our parents!"

How could they have been so wrong? Goodby wondered. When had they gotten so out of touch with America's youth? And why had he wasted money on those bottles of champagne for an account his agency had no chance of winning?

"I wish this was a commercial for Nintendo!"

"Oh, yeah, did you see the new one for Mario All-Stars?"

Goodby and his colleagues watched the onslaught continue, still searching for some kind of silver lining. But eventually it became clear that there was none, and, more important, there was no time to look for one. The presentation was in a week and they didn't have shit. Game over.

It's all about cool; that's the holy grail. You're born, you die, and in between you spend a bunch of years searching for it—looking cool, sounding cool, buying cool, and, no matter what, *not* being *un*cool. That right there, that's the secret formula. It's addictive, it's enlightening, and it's goddamn recession-proof. In a world full of too many people shouting too many things, it's the only adjective that really matters: "cool."

Tim Price knew cool, or at least as much as any mortal could claim to know it. Price was an award-winning copywriter whose passion for wild, restless, high-velocity advertising could be traced to his love for off-road racing. He joined Foote, Cone & Belding as a creative director in 1978 and quickly caught fire with his team's edgy work on the Levi's Youthwear account. In the process of transforming the esteemed jeans company from a cute brand into a cool one, he found his voice (a sort of counterculture romanticism), he met his wife (the VP director of event marketing at Levi Strauss and Co.), and his firm won part of Nintendo's business (they got the Game Boy, Leo Burnett got the NES). Price was psyched by the chance to help define this relatively new and fast-moving industry of videogames, to do for Nintendo what FCB had done for Levi's and appeal to an older, hipper audience.

Unfortunately, this was not at all what Nintendo wanted. They weren't interested in branding, expanding their demographic, or producing memorable, high-concept ads. All they wanted was cartoony, easy-to-digest commercials that showed a lot of game footage. If that was what they wanted, that was their prerogative, of course, but it raised the question of why they'd hired Foote, Cone & Belding to do this. Price tried to talk Nintendo into taking bigger risks, and whenever he did, he'd receive positive reactions from the NOA marketing team. Peter Main liked clever ideas, Don Coyner loved sophistication, and Bill White wanted to take over the world. But shortly after getting excited about big ideas, reality would sink in; either they'd decide that Nintendo was going to stick with what had been working thus far, or they'd bring the plans back to Minoru Arakawa and get a thanks-but-no-thanks. Fun, happy, and title-driven advertising was Nintendo's bread and butter, and they didn't need FCB's jelly. Within the next year they moved all their business to Leo Burnett, which was not entirely surprising, though it stung Price nevertheless. But now, two years later, he had the chance to right that wrong by helping Foote, Cone & Belding win Sega's business.

In terms of marketing philosophies, Sega of America was Nintendo of America's polar opposite. They craved bold ideas, sky-high concepts, and aggressive, in-your-face branding. Sega represented FCB's chance to let it all hang out, which led to the creation of a wacky, high-octane campaign centered around the tagline "Make your brain sweat." Surrender to madness, give in to the insanity, and let Sega take your brain into overdrive—this was the hyperactive sensation that the campaign wished to impart. To help hammer home the point, Price wanted to bring an actual brain to the presentation. Obtaining a real one turned out to be harder than he'd anticipated, but he did manage to track down a medical supply company that sold brain-shaped molds, and he figured that just might work.

On the day of the pitch, Sega's key marketing members arrived at FCB's offices not knowing quite what to expect. Neither did Price, whose wife had yet to arrive with the key to the presentation. The night before, she had used the mold to make a life-sized green Jell-O brain, complete with veins of red licorice. It was gorgeous, it was disgusting, and, for the moment, it had been stopped in the hallway by someone at FCB who didn't recognize Price's wife and didn't quite understand why she was carrying something that looked like it would be evidence of a murder committed in Candyland. Eventually she managed to sweet-talk her way past the evil Lord Licorice and get the noggin to her husband. After a thousand thank-yous and a kiss on the cheek, Price then explained to Sega how Foote, Cone & Belding would make consumers sweat their brains for Sega.

Back at Sega of America's headquarters, Tom Kalinske, Ed Volkwein, Al Nilsen, Diane Adair, Doug Glen, Tom Abramson, and Ellen Beth Van Buskirk met in the conference room to discuss the agency review process.

"Thoughts on Foote Cone?" Kalinske asked, opening up the floor.

"Not as many as I would have hoped," Nilsen said. "A bit unmemorable."

"But the brain, Al," Van Buskirk gravely reminded him. "The brain!"

"All right," Abramson said, "I'll just come right out and say what we're all thinking: the human brain has to be the ugliest organ in the body, right?" Chirps of laughter fluttered around him. Tom Abramson was the newest member of Sega's inner circle, and he fit in perfectly. His intellectual absurdity pleasantly turned every conversation into banter, but it was his ability to will promotions into existence that earned him a seat at the table. With a background in event marketing for the Ice Capades, the Harlem Globetrotters, and Walt Disney World, he had a certain savvy for nontraditional marketing, which at Sega propelled him to do things like hire student reps at colleges and send out Sega Shuttles to easily transport the newest games anywhere, anytime. Plus the guy was just plain fun. "Now, I know there are plenty of repellent organs out there, certainly more for the menfolk than the ladies, but when you factor in the weight and those creepy little folds . . . well, it's gotta be the brain."

"Maybe not the best presentation," Kalinske said, "but you have to admit that it's a cool line: 'Make your brain sweat.'"

"Eh," Volkwein said with a quick shake of the head. "Good, not great."

Adair nodded. "Net-net, I thought it was cool, but did anyone else think that maybe it was too cool?"

"Too far out there?" Kalinske asked. "I can definitely see that."

"Yeah," Van Buskirk said. "Cool we want, but that was frostbitten."

"Now I can't help but wonder about other species," Abramson blurted. "Dogs? Cats? Are there brains as objectionable? And what about koala bears?"

"I agree," Glen added, his typical subdued enthusiasm even more noticeable in contrast to Abramsom's gleeful musings. "I should clarify: I agree with Ellen Beth's comment regarding the frostbite, although I'm admittedly intrigued by the tangential brain curiosity."

"Bottom line," Kalinske proclaimed, "we can do better. And I have a feeling that Wieden+Kennedy will find a way to strike the right balance."

"Agreed," Glen said. "I think we can do better."

"Absolutely," Kalinske said. "I have a feeling that Wieden+Kennedy will find a way to strike the right balance."

"I'd be shocked if they didn't knock it out of the park," Nilsen said. "Shocked."

"They do fine work," Volkwein said. "No denying that."

"Wieden will deliver," Glen said, "but let's not forget about Goodby just yet."

"Nobody has forgotten about Goodby," Nilsen said. "And if we ever did, I'm sure you'd be there to remind us."

"What specifically should I infer from that comment?" Glen asked.

"Come on, Doug," Kalinske said. "They sent us champagne during our dinner with Wieden and Kennedy. How do you think they knew where we were eating?"

Glen blushed a little and then smiled. "Because they have great market research!" Laughter broke out around the table.

Suddenly Kalinske pointed at the window. "Did anybody else see that?"

"What did you see?" Adair asked.

"Something small, moving quickly," Kalinske said. "Looked almost like a golf ball." Everyone froze and stared out the window, but nothing else appeared. "I guess not," Kalinske said. "Hm. Strange." And then the discussion about Sega's marketing plans resumed.

Jimbo Matison had just puked his guts out, and he was certain that another round of gastric fireworks would be coming very soon. Until then, there was nothing he could do but curl up on the couch and hope that the distraction of crappy daytime television might be enough to briefly postpone the inevitable. The twenty-six-year-old had the flu, and sick days had stopped being fun about ten years ago.

Not long after failing to find something watchable, he received a phone call from a producer at Colossal Pictures, the commercial production company where he'd been doing grunt work for years. "I know you're sick or whatever," she said, "but do you think you can come in for just an hour or two?"

"I'm sick," he said. "Like, legit sick."

"It's a voice-over thing," she said. "I think you'd be good."

"Oh," Matison said, suddenly feeling a little better. He'd been trying to break into the voice-over business for years. "What's it for?"

"We're doing a bid for this thing called Sega."

"What's that?"

"Just come in."

"What's in it for me."

"I'll buy you lunch."

He thought over the offer. "And?"

"And what?"

"And you'll help me get my SAG card."

"Yeah, whatever," she said. "Just come in, okay?"

Matison waited for the next wave of vomiting, and as soon as it passed he jumped on his bike and rode over to the production house. When he got there, it was just his producer, a sound guy, and some dude from an ad agency. The dude seemed pretty cool—he wasn't wearing a suit, at least—and he asked Matison to scream the word "Sega" as loud as he possibly could.

"Okay," Matison said skeptically. "What's this for?"

"Come on, Jimbo," the producer said. "This isn't rocket science. Do you want that free lunch or not?"

For the next hour, Matison repeatedly screamed the word "Sega" at the top of his lungs. The producer, the sound guy, and the dude from the ad agency were loving it, having him shout it from different parts of the room; faster, slower, faster, slower, over and over, screaming "Sega" as many ways as the four-letter word could be shouted.

When he was finished and the agency dude had thanked him for a job well done, he asked how they planned to use his scream.

"We don't know yet."

"Gotcha," Matison said with a nod. "Because I was thinking: do you remember those old Quasar commercials, how they used to say 'Quasar' at the very end? It really stuck in your head, didn't it?"

"Hmmm," the agency dude said. "That's not a bad idea."

"Cool," Matison said, and got on his bike, hoping not to puke on the way home.

Inside a former Gothic-style church that had been converted into a chic nightclub, Wieden+Kennedy employees decked out the holy space in futuristic fonts, pale neon colors, and the unusual feeling of a science experiment gone right (think rock-and-roll Frankenstein). It was like *A Clockwork Orange*, but for teenagers. It was as if someone had ripped the pages out of *Brave New World* and glued them into *The Catcher in the Rye*. It was *1984* meets *The Breakfast Club*, new-speak meets teen-speak meets Sonic The Hedgehog. And it was called "vidspeak," the concept of new, hip, future-is-now language that Wieden+Kennedy had invented as the backbone of their campaign. The agency provided a sampler from the vidspeak lexicon, which included the following words, terms, and phrases:

Gearlets: The vidspeak word for gamers. Also known as gamelets, gameys, whoossies, vidiots, speaklets, bossaroos, and cluelets.

Hedgy wedgy: Anything pertaining to Sonic The Hedgehog, or to any fan of said Sonic The Hedgehog. Also that cute little way Sonic has of stamping his foot when he can't believe you're so slow and stupid. (See Slow geezer trying to play the game)

Whammy jammy: The way you feel when playing a good game.

Gobble-degoop: Running wild. Running fast. Running all over the place without time to say "Hasta la vista, baby." (See Sonic The Hedgehog)

Mobile mover with wings: What gamers will call the Game Gear. Also referred to as the a-to-go-cup, a minda-rama, a home away from home, and a great thing to wrap your knozzles around.

Master blaster: What you do when you start playing the Sega Menacer Master Module. Or, how to be a real sure-shot full-tilt accu-sight kind of guy.

I was Brahms: I was drunk with power. I was mad with passion. I was blitzed with energy. I was actually able to reach the next damn level.

It was a little stranger than Kalinske had expected, a little more to swallow than "Just do it," but there was a beauty to the chaos, and if anyone could make it pop into pop culture, it was Wieden+Kennedy. And, like Nike's iconic slogan, the agency had created a tagline that was easy to connect with. "You are here," Dan Wieden said. "You are here," he repeated, and then elaborated on many ramifications of this seemingly simple phrase. "It's a tagline that means: You are in. You are hip. You are cool. You are not there, which is where everybody else who is not here is. You are with us. You are smart, cunning, and extremely creative. You are inside the game, inside this new world, inside another reality. You are so good you don't need a glossary to explain any of this stuff to you."

When Wieden finished, he was pelted with applause, both from Sega's marketing team and from his awestruck colleagues. He waited for the clapping to energize the room, and then he began to discuss a possible programming schedule. As he started to explain how his agency could get airtime at unmatchable discounts, Kalinske subtly elbowed Nilsen. "What do you think so far?" Kalinske whispered.

Before whispering back, Nilsen quickly flipped through the vidspeak sampler. "Mostly whammy jammy, but every now and then a bit hedgy wedgy."

And just like that, Al Nilsen became the first gamer to express himself in vidspeak.

Before Sega of America could officially select Wieden+Kennedy, they needed to sit through one final pitch. It was all but a formality at this point, but they owed Goodby, Berlin & Silverstein the courtesy of at least allowing them to believe that the competition was wide open. Besides, what was the worst that could happen? They liked some tiny aspect of the Goodby campaign and passed it along to Wieden+Kennedy? Thoughts like these swirled around Kalinske's mind as he walked into the lobby of the Foster City Crowne Plaza. Like W+K, the guys at GB&S had opted to pitch outside a dull office environment, though it's hard to say that the Crowne Plaza engendered feelings of joy and enthusiasm. But any doubts about the venue instantly evaporated when Kalinske and a dozen colleagues entered the ballroom.

"Welcome!" Jeff Goodby shouted, ushering the Sega employees into what had once been an unspectacular ballroom with frumpy velvet curtains and asparagus-colored carpeting. "Welcome to the Next Level."

The breathtaking what-the-huh sensation of swinging open those ballroom doors felt slightly on par with stumbling through a closet and suddenly falling into Narnia. In the center of the room, sixteen giant television sets had been multiplexed together to create one mega-sized screen. This beautiful monstrosity looked like what you might expect Transformers to use if the Autobots and Deceptions decided to momentarily put aside their differences and watch the Super Bowl together. Standing in front of the TVs was a twelve-year-old boy suavely playing Sonic The Hedgehog on the IMAX-sized screen. Sounds from the game blared through a booming sound system that had been installed by the roadies for the Grateful Dead. The speakers, the boy, and the tower of televisions were all on top of a massive stage that had been fashioned by George Coates Performance Works, a San Francisco–based ambient art company renowned for developing innovative types of theatrical performance. Part of this performance required a large ensemble cast, which explained the rows of stadium-style seats up against the walls. These seats, all of them filled, were occupied by employees from Goodby, Berlin & Silverstein. As the team from Sega was shown to their own seats, Kalinske concluded that this had to be greatest presentation he had ever seen in his twenty years of corporate meetings. And it had only been going on for about fifteen seconds.

"Thank you all for coming to our humble presentation," Jeff Goodby said. He spoke from atop the stage, joined by Rich Silverstein, Jon Steel, Irina Heirakuji, and Harold Sogard, the agency's director of account management. They were each wearing a school letterman's jacket they had made, featuring a patch of Sonic on the shoulder.

Silverstein nodded. "We're psyched to show you a little bit of what we've cooked up," he said. "And we hope that you're prepared to be blown away."

"We've spent the past month traveling around the country and living with gamers," Steel said. "We got to know them and understand what they want."

"To search for any clue about how their minds work," Heirakuji said. "We even raided their backpacks, bedrooms, and closets."

"It's true," Sogard said. "Jeff even turned our conference room into a ten-year-old's bedroom. Right down to the dirty clothes!"

As laughter emanated from the Sega crowd, Goodby gave an unapologetic shrug. "Hey, I've never had a problem with getting my hands a little dirty," he said. "But before we get into the campaign, I wanted to first show you how committed to Sega our agency really is. So in preparation for the pitch, I went around the office and assigned everyone a Genesis game to master." Goodby took a step forward and pointed to his employees in the stadium seats. "Over there, we've got an expert on every single game that you guys make. Go ahead and ask them any question about any game. I'm totally dead serious."

When the Sega folks realized that Goodby was, in fact, totally dead serious, Nilsen was selected to come up with some brain busters. "In Phantasy Star III: Generations of Doom, what is the name of the main character in the First Generation?"

The agency employee assigned to Phantasy Star III stood up. "That's a tough question," he said, making Goodby sweat for a second. "But only because there are six playable characters: Rhys, Lyle, Mieu, Wren, Lena, and Maia. If I had to narrow it down to one, though, I'd go with Rhys, the Crown Prince of the Orakian Kingdom of Landen."

"I couldn't have put it better myself," Nilsen said amidst applause for the Phantasy Star III expert.

After his employees correctly answered a few more of Nilsen's questions, Goodby moved on to the campaign. "Welcome to the Next Level," he said. "That's not just the tagline, but the entire essence of our campaign. 'Welcome to the Next Level'—that says it all. It's a badge of honor; it's the ultimate challenge; it's an invitation to join the revolution. It represents the only thing that a player cares about when he's locked into a game. Shut off the real world, dive into the game world, and just keep going-going-going at all costs until you get there. It means wake up and let's get started. Put away your toys, put away that juvenile Nintendo, and go get a Genesis if you want to find out what life is really all about. Welcome to the Next Level. It means that you have finally arrived. And just in time, because we've been waiting for you."

Kalinske loved everything that Goodby had to say. The guy got it—he just totally and completely got it. But as spot-on as his words continued to be, they were nothing more than that. Beyond the theatrical presentation, was Goodby's "Welcome to the Next Level" really that different from Wieden+Kennedy's "You are here"? And if it came down to a matter of execution, Wieden+Kennedy would get the benefit of the doubt based on their track record. By this point, however, Kalinske wanted Goodby to win the account. He wanted to give the business to a bunch of guys as smart, scrappy, and subversive as his own. In order to do that, however, he needed proof that "Welcome to the Next Level" could be more than just words. He had to know that this would be the weapon used to assassinate Nintendo. As if reading his mind, Goodby screened some commercials that the agency had prepared in advance.

The ads shown at the pitch were unrecognizable as those the focus group had seen. Following that barrage of atomic suck-bombs, Goodby and Silverstein had decided to take a step back. As they reflected upon what had gone wrong, they concluded that the agency's account planners had conducted incredible research and the creative teams had written great copy, so the problem had to be a matter of execution. The dots were all there, but they needed a better way to connect them. After further examination, however, what they realized was that they shouldn't even connect the dots at all. Beneath the focus group's insults and criticism, there was a lesson to be gleaned: kids notice everything. For example, when one of the respondents had said, "That footage you showed wasn't even from Sonic 2!" what he really meant was, "To you, all game footage looks the same." And then when he'd said, "It was from the original Sonic, and it was from level two, which isn't hard at all," what he really meant was, "Unlike you, I can see a split-second frame from a level, identify it, identify with it, and also have an emotional reaction to it."

These kids, their minds just operated at a much faster speed than anyone gave them credit for. In the same way that dogs can hear things that humans cannot, this generation of kids could see things that adults couldn't even process. Not only that, but they remembered them too. So the best way to speak to them was to invent a dog whistle that played something only they could hear. This was kind of what they had been trying to do with the semi-subliminal messaging ("speed," "babes," "midnight"), but that was too obvious. That was like an old man wearing baggy pants and a backward hat.

Kids were too smart for that; they could spot a poser from a million miles away. With this in mind, the agency had their editor, Hank Corwin, go back through the material and create some kind of beautiful chaos that spoke up for kids and, for once, not down to them.

What Goodby unveiled at the Crowne Plaza was unlike anything Kalinske and his colleagues had ever seen before. Quick cuts. Crazy zooms. Wild camera angles. It felt less like watching a regular commercial than like fast-forwarding through one on the VCR. Loud punk music. Intense lens flares. Aggressive close-ups. It looked sort of like a music video, but only if that music video was suffering from manic-depression and had just ingested a cocktail of heroin, cocaine, and speed. Weird lighting, unpretty actors, nonlinear storytelling—the whole thing was off-putting, migraine-inducing, and offensive to the senses, but it was absolutely incredible. And to tie it all together, at the end of every spot some maniac shouted, "Sega!"

"And just remember," Goodby said as the video presentation came to an end, "we're only a short drive away." He then played a short video clip of himself, Silverstein, and a few other guys whacking golf balls off the roof of their office building. Except whenever they hit the ball, the real reaction shot was replaced with footage of golf balls hitting Sega of America headquarters.

During the ground-shaking applause that followed, Nilsen subtly elbowed Kalinske. "What did you think?"

Kalinske blinked for a second, then replied, "I think vidspeak just became a dead language. Sorry, hedgy wedgy."

He was practically in a state of shock. This was it—everything he had wanted. The tone was edgy, but not too sharp. It cut, but only deep enough to leave a cool scar. It was sex without a condom, smoking two packs a day, and watching the speedometer break a hundred miles per hour; and the best part was that none of it hurt because it was only a videogame.

If there had been any lingering doubts that Goodby, Berlin & Silverstein was the right choice, they disappeared when the Sega people got back to their office and saw that the entire parking lot was covered in chalk markings. At first it looked like teenage vandalism, but upon closer inspection it turned out to be a single message written over and over: "Welcome to the Next Level."

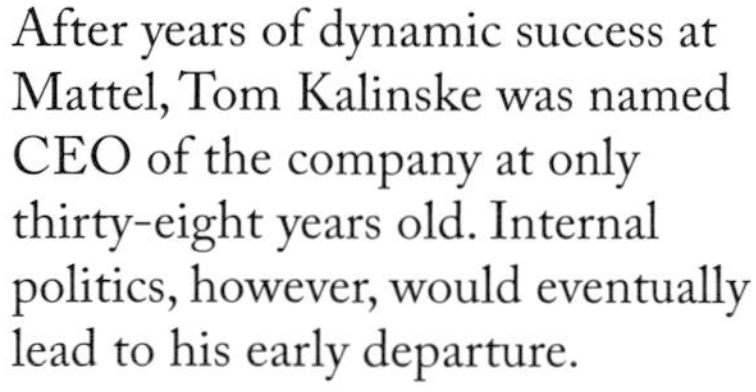

After years of dynamic success at Mattel, Tom Kalinske was named CEO of the company at only thirty-eight years old. Internal politics, however, would eventually lead to his early departure.

Photograph courtesy of Tom Kalinske

While between jobs, Tom Kalinske and his family took a trip to Hawaii in 1990, but that vacation was cut short by an unexpected guest . . .

Photograph courtesy of Tom Kalinske

Hayao Nakayama, the president of Sega Enterprises, tracked down Kalinske in Hawaii and whisked him away to Japan to show off what his company had in the pipeline. Impressed, Kalinske agreed to become the CEO of Sega of America and take on mighty Nintendo, who controlled 95 percent of the videogame market.

Photograph courtesy of Tom Kalinske

Nintendo's massive booth, often referred to as the "Deathstar," at the 1989 Consumer Electronics Show.

Photograph courtesy of Consumer Electronics Association

To dethrone Nintendo, Kalinske relied heavily on his right-hand man, Shinobu Toyoda (center), and Sega's marketing maestro, Al Nilsen (left).

Photograph courtesy of Shinobu Toyoda

Al Nilsen prepares for Sega's innovative mall tour, which traveled around the country to pit Genesis against Super Nintendo . . . months before the SNES was even released.

Photograph courtesy of Ellen Beth Van Buskirk

Shinobu Toyoda poses with Joe Montana (center) who headlined one of Sega's early hits, and David Rosen (left), who founded Sega back in 1965.

Photograph courtesy of Shinobu Toyoda

Shortly after Kalinske took over, Sonic mania swept through the nation. The blue blur's cool-dude attitude quickly infected men, women, and children of all ages (including, of course, the entire Kalinske family).

Photograph courtesy of Tom Kalinske

Before dominating the videogame world, Nintendo of America was just a tiny arcade company run by brilliant visionary Minoru Arakawa.

Photograph courtesy of Howard Phillips

Howard Lincoln, an attorney in the Seattle area, helped Arakawa put out some early fires and then became Nintendo of America's senior vice president.

Photograph courtesy of Howard Phillips

Throughout the early eighties, Arakawa and Lincoln led Nintendo to greatness with the help of hit arcade games like Donkey Kong, Punch-Out, and Mario Bros. (pictured).

Photograph courtesy of Howard Phillips

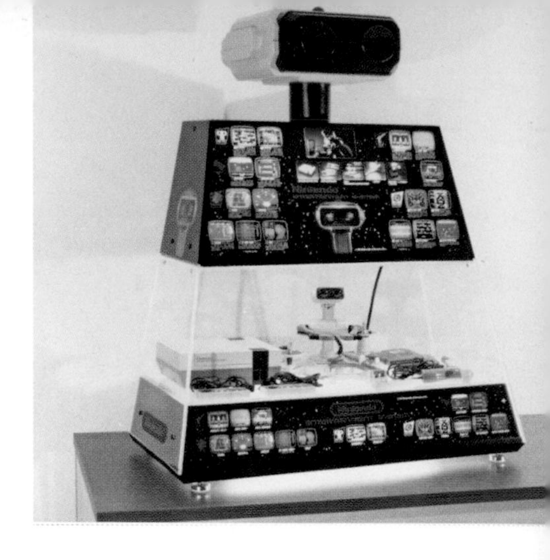

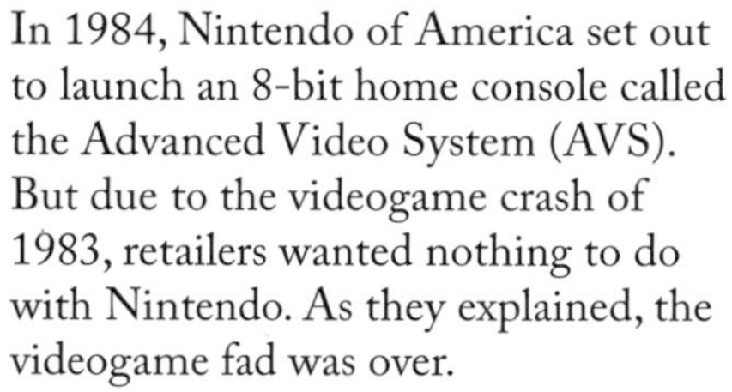

In 1984, Nintendo of America set out to launch an 8-bit home console called the Advanced Video System (AVS). But due to the videogame crash of 1983, retailers wanted nothing to do with Nintendo. As they explained, the videogame fad was over.

Photograph courtesy of Howard Phillips

One year later, the Nintendo Entertainment System (NES) was released in New York. To avoid the stigma from the crash of 1983, Nintendo used R.O.B. (Robotic Operating Buddy) to position its NES as much more than just a home console.

Photograph courtesy of John Sakaley

In 1988, *Nintendo Power* became the fastest magazine to reach one million paid subscribers and also personified Nintendo's relentless commitment to providing an unparalleled user experience.

Photograph courtesy of Howard Phillips

Nintendo's game master, Howard Phillips, the "man who played videogames for a living," became so popular that in 1990, his Q-rating was higher than Madonna, Pee Wee Herman, and the Incredible Hulk.

Photograph courtesy of Howard Phillips

Phillips became a celebrity, but not even his star power could compare to that of Michael Jackson, whose 16-bit game Moonwalker was an early hit for the Genesis. Here, the legendary King of Pop poses with Sega's King of Marketing.

Photograph courtesy of Al Nilsen

After watching Kalinske's group take a bite out of the market, Nintendo's "main man," executive vice president Peter Main, proudly coordinated the launch of the Super Nintendo (SNES) in 1991.

Photograph courtesy of Peter Main

To blunt the impact of Nintendo's 16-bit SNES and further Sega's entertainment revolution, Kalinske relied heavily on the dazzling PR work of Ellen Beth Van Buskirk.

Photograph courtesy of Ellen Beth Van Buskirk

Kalinske aspired for Sega to be more than just a game company and in 1992 established the Sega Youth Education Health Foundation.

Photograph courtesy of Cheryl Quiroz

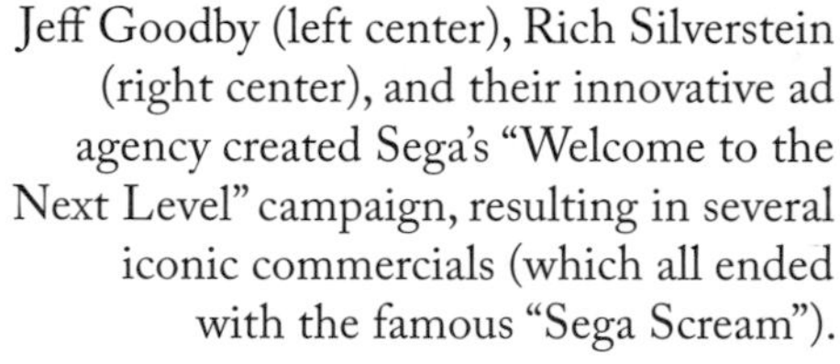

Jeff Goodby (left center), Rich Silverstein (right center), and their innovative ad agency created Sega's "Welcome to the Next Level" campaign, resulting in several iconic commercials (which all ended with the famous "Sega Scream").

Photograph courtesy of Jeff Goodby

The relentless work of Diane Fornasier (left) on Game Gear and then the blast-processing Genesis insured that Sega's marketing kept reaching the next level.

Photograph courtesy of Diane Fornasier

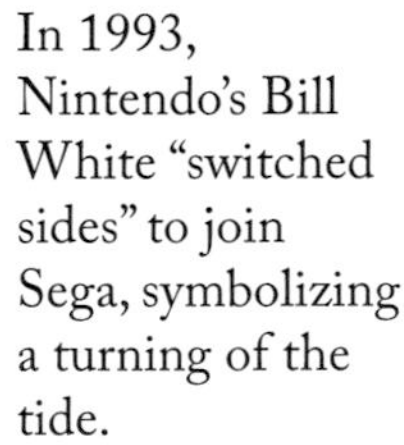

In 1993, Nintendo's Bill White "switched sides" to join Sega, symbolizing a turning of the tide.

Photograph courtesy of Bill White

Despite losing market share, Nintendo remained committed to its core values of making great family-friendly games. Here, Nintendo's director of licensing, Tony Harman, proudly announces the release of Major League Baseball with Ken Griffey Junior.

Photograph courtesy of Tony Harman

While Nintendo fought back with great software, Kalinske focused on the next generation of hardware and tried to broker a deal with Olaf Olafsson, the head of Sony Electronic Publishing.

Photograph courtesy of Johann Pall Valdimarsson

Nintendo followed up its baseball game with the earth-shattering success of Donkey Kong Country. After selling more than seven million copies in only six months, Lincoln and Arakawa enjoy seeing the pendulum swing back in their favor.

Photograph courtesy of Howard Lincoln

Tom Kalinske, Shinobu Toyoda, and executive VP Paul Rioux dress up for a Sonic & Knuckles promotion. But behind the smiles, their wardrobe personifies the new dynamic between Sega of America and Sega of Japan.

Photograph courtesy of Shinobu Toyoda

The final blow to Sega came at the 1995 Entertainment Electronics Expo, when Sony's Steve Race delivered the greatest (and shortest) speech of his career.

Photograph courtesy of Steve Race

Following the triumphant success of Nintendo's N64, Howard Lincoln and Minoru Arakawa race for charity outside of the company's office in Redmond, Washington.

Photograph courtesy of Howard Lincoln

32.

THE KUTARAGI DREAM MACHINE

While Sega was preparing for the next level, and Nintendo seemed to finally be adjusting to this one, Sony was trying to decide if they even wanted to be in the game. To figure this out once and for all, Sony held a pivotal management meeting at their Japanese headquarters on June 24, 1992.

From the outset, it was quite clear that Sony's board of directors wanted nothing to do with developing a console. The R&D costs of creating proprietary hardware were astronomical, and the long development cycles would create an endless money pit. More abstractly, the videogame business was little more than an offshoot of the fad-driven toy industry, and that just did not fit with Sony's brand. The software, however, was a different matter, and the board was okay with this less risky endeavor. In the same way that Sony's Columbia Pictures made movies and CBS Records distributed music, Olafsson's software division would be permitted to keep producing games. In fact, the board had just recently allowed him to work with Sega in this arena. Was software not enough? Why this continued fixation on hardware? If Nintendo were involved, things might be different, but following last year's public humiliation at the Consumer Electronics Show and some recent failed attempts to test those waters once more, that prospect looked highly unlikely. Given all of these factors, plus the aging board of directors' you-will-understand-when-you-grow-up sensibility, they believed it was time to put Sony's PlayStation to rest.

Judging by the many nods of support, this sentiment appeared to be unanimous. These powerful men who once upon a time had shaped Japanese culture, back in the 1960s, '70s, and '80s, were now all ominously agreeing like a fleet of bobblehead dolls in the trunk of a car headed off a steep cliff. Needless to say, not everyone agreed with Sony's elder statesmen, but nobody in that boardroom disagreed with them more than Ken Kutaragi, the father whose baby the board of directors intended to kill.

Ken Kutaragi was one of Sony's top engineers, a sharply charismatic forty-one-year-old man with a proclivity for aggressively questioning authority. This was a rare quality in Japanese culture, one not generally welcomed at large corporations, but Kutaragi's defiant spirit had served the company well in the past and was the primary reason that Sony was even considering videogames in the future. Originally they'd had no interest in videogames (not just consoles, but even manufacturing parts for Nintendo), and neither had Kutaragi. He was not a gamer by nature, but his interest in the art form changed in 1988 when watching his daughter play the 8-bit Famicom. It wasn't so much what he saw on the screen, but the potential he saw for videogames: a way to bring computers out of the office and into the living room. Or more specifically, what he saw as a Sony employee: the mainstream consumer electronics appeal of vast computing power.

Although this revelation changed Kutaragi's perspective, it had not persuaded most of Sony's executives, and this frustrated him. But instead of pleading his case to deaf ears, he decided to do something much more proactive. He knew that Nintendo was in the late stages of developing a 16-bit console, and he also knew that they were worried about the sound capabilities of Sega's soon-to-be-released Mega Drive. Sega had worked with Yamaha to create an advanced audio processor capable of FM synthesis, which led Nintendo to Sony in the hopes of finding something superior. Sony's executives, however, did not want to gamble on manufacturing videogame components and balked at the opportunity. Or so they thought. Without their knowledge, Kutaragi continued the conversation with Nintendo and developed an audio chip called the SPC700. When Kutaragi revealed his secret project, Nintendo's designers were thrilled. This was exactly what they had been searching for. Sony's top executives, however, were much less enthusiastic and wanted Kutaragi fired, which likely would have happened if he had not been protected by Sony's chairman, Norio Ohga. As angry as the executives were about the defiance, they weren't angry enough to start a civil war. And those tempers soon softened even more when Sony received a lucrative contract for Kutaragi's sound chip. Kutaragi had gambled and, amazingly, had won. But more important, his win was really a win for Sony, and this propelled the company to reexamine the videogame industry. If there was this much money to be made with an audio chip, imagine how much could be made with an entire console. This was the logic that had guided Sony into working with Nintendo, first as a supplier for the Super Famicom and then as a partner on the PlayStation. But that logic spontaneously combusted when Nintendo snubbed Sony for Philips.

Now everything was in limbo. That's why this boardroom was overwhelmed with so much frustration. There had been a brief window when these executives could see the potential in what Kutaragi had shown them, but now they were blind to that promise. And as Kutaragi watched this supposed discussion of possibilities evolve into a monologue of foregone conclusions, he realized he could wait no longer to try to open their eyes.

"Having listened to what everyone is saying," Kutaragi began, addressing Sony's board of directors, "I can see three options. First, to continue indefinitely with the traditional, Nintendo-compatible 16-bit game machines. Second, to sell game machines in a format proprietary to Sony. Third, to retreat from the market." Kutaragi paused for a moment to let it all sink in. "Personally, I believe Sony should choose the second option."

The board members looked at him with suspicion. Of course he believed they should choose the second option. This was not news; this was how he had always felt.

Kutaragi knew that his words alone would indeed not be enough to change anyone's mind. That was why he had come prepared with something more than words. Kutaragi looked around the room and smiled with anticipation. As he had done with the Super Nintendo audio chip, he had once again been working in secret, and finally the time had come to reveal what he had been working on.

33.

A QUICK LAP AROUND VICTORY LANE

Nintendo of America's employees felt such head-to-toe reverence for Minoru Arakawa that there was not a thing in the world they would not do for their beloved leader. Or so they thought. In the summer of 1992, however, it dawned on NOA's employees that there was actually one small thing they were not willing to do: go swimming.

"But the water, it's perfect," Arakawa pleaded to a group of guests loitering beside the long pool in his backyard. They had come to his home in Medina for the barbecue that he and his wife hosted every year for Nintendo's highest-ranking employees and their spouses. For people working in a speed-of-light industry, the annual summer celebration provided one day per year when Nintendo's employees were encouraged to step back, relax, and smell the burgers being grilled to perfection.

If burgers didn't do the trick, there was also a full-color spectrum of sushi rolls from one of the city's finest Japanese restaurants, as well as slivers of succulent king salmon that Howard Lincoln had brought back from his recent trip to Alaska. Mountains of food crested everywhere, and drinks could be found wherever the food was not—wine, champagne, and umbrella-topped cocktails, as well as a wide variety of beers from both America and

Japan. There were about a hundred guests in total, all happily eating, drinking, and swapping stories, but even so, none of them were willing to take a plunge into the pool. After failing to entice a different group of guests to let loose and go swimming, Arakawa returned to his post at the grill. Yet he remained patient and undeterred, confident that by the end of the day the pool would no longer be left unrippled.

"Is there anything I can help you with, Mr. A.?" Tony Harman asked, catching Arakawa supervising the grill. Although Harman possessed the physique of a football player and the chilly detachment of a baseball pitcher, his true athletic calling had been soccer, where he'd shined at the collegiate level. In 1988, after earning an undergraduate degree in engineering and a graduate degree in business, he walked into Nintendo's Redmond office and coaxed his way into an interview. He was hired to localize the company's games—basically, tweaking and translating Japanese titles so that they would be more palatable to Western audiences. Less than a year later, Harman's relentlessness caught Arakawa's eye and earned him a promotion.

"Tony!" Arakawa exclaimed with a nearly paternal elation. "Hungry?"

"No, no, I'm not here for that," Harman replied. "I'm here to help. Just tell me what you need."

Technically, Harman's promotion made him Nintendo of America's director of development and acquisitions, but in reality his job was to manage Arakawa's office, coordinate his affairs, and just generally be his "guy." Whether playing the role of caretaker, confidant, or creative strategist, Tony Harman was Arakawa's first call when he needed something done.

"Your wife is here," Arakawa said, shooing Harman away. "Go eat and enjoy."

"I already ate."

"Maybe a swim?"

"Please, I want to help."

"No," Arakawa said. "Not today."

"Okay, fine."

Harman reluctantly obeyed and joined his colleagues amidst the tables scattered throughout the backyard. From the grill, Arakawa watched with a grin.

Wholesome games and Disneyesque characters were what the world saw, played, and enjoyed, but the origin of this sensibility trickled down directly from the company itself. Family first—that was the Nintendo way. No matter how many games were sold or how much money was made, Nintendo rarely felt much larger than a family business. Part of that familial atmosphere stemmed from the fact that a Yamauchi had always run NCL and that Hiroshi Yamauchi's son-in-law ran NOA, but a larger part was due to Nintendo's unique corporate culture. Profitability was always the primary goal, but never at the expense of loyalty, integrity, and family values. That's why most people who joined Nintendo tended to stay there for their entire careers.

Typically, these barbecues were Nintendo-only affairs, but this year's event doubled as an opportunity to celebrate some long-overdue good news. After nearly six months stuck in the center of America's anti-Japanese party, Nintendo's days of being treated like a piñata had finally come to an end. On June 11, 1992, Major League Baseball's owners voted 25–1 to approve the sale of the Mariners to the Baseball Club of Seattle (with the lone holdout being the owner of the Cleveland Indians, who cited a need for more time to review the proposal). Although this decision ended up being nearly unanimous, getting there had required months of concessions, secret meetings, and, ultimately, Nintendo's Yamauchi agreeing to accept a minority voting stake. Although it would be hard to find a logical (i.e., not xenophobic, and not "just because we said so") explanation for why someone putting up the majority of the money should receive a minority of the control, this was one thing on which Nintendo was willing to concede. That's because buying the Mariners had never been about swimming in a stadiumful of money but had always been about demonstrating loyalty, integrity, and family values. And in honor of this recent news, Senator Slade Gorton and everyone in the ownership group had been invited to join Nintendo's barbecue and get to know their extended family.

"It still feels surreal," Chris Larson explained to Gail Tilden and her husband as they all lingered beside the yet-to-be-used pool. Larson was a former computer prodigy who had started working for Microsoft at age sixteen and now, at thirty-three, had just become a proud new owner of the Seattle Mariners. Of everyone in the investor group, Larson was likely the one who loved baseball the most, as evidenced by his extensive collection of memorabilia and his decision to ditch business casual in favor of a throwback Seattle Rainiers jersey for the barbecue. "I grew up collecting baseball cards," he said, unable to suppress a childlike grin. "And now, well, I've got a whole team."

"I'm glad it all worked out in the end," Tilden said, quickly thinking back on all the undue drama of these past few months. "I can't wait to go to one of the games."

"Especially a playoff game," her husband joked.

All three laughed before switching gears to talk about Tilden's experience at Nintendo. "So tell me, what do you do over there?" Larson asked.

"Oh, you know, all sorts of things," she replied. Her eyes lit up in a subtle but special way anytime she talked about working at Nintendo. "But typically I spend most of my days running the *Nintendo Power* magazine."

"That's fantastic!"

"It really is," she said, momentarily appreciating how nice it was to have a fun, challenging, and satisfying job. With a background in marketing, Tilden had never expected to find herself running a magazine, let alone a kids' magazine, but the unexpected nature of the journey was part of what made it fun. There were always fires of assorted sizes to hose down along the way, particularly over the past year. Now that Nintendo had three different systems (NES, SNES, and Game Boy), juggling coverage without alienating readers had become something of an art. While Tilden was gymnastically coordinating this balancing act, there was also the departure of Howard Phillips to figure out. In addition to serving as Nintendo's Game Master, Phillips also had starred in the magazine's comic strip, "Howard & Nester," and reviewed games along with Don James and Tony Harman. In order to avoid another situation in which a single Nintendo employee was elevated to celebrity status, Nintendo didn't appoint a second Game Master, Howard was written out of the comic strip, and Tony Harman assumed a greater role when it came to games. In fact, Harman's role grew into what Phillips had always wished his job had entailed: not just evaluating games but developing them as well. "Every now and again there are some bumps in the road," Tilden said about running the magazine, "but it looks to be smooth sailing ahead."

Tilden's words reflected the renewed excitement in the air. After a shaky start to 1992, the pendulum appeared to be swinging back in Nintendo's favor. Sega may have been the first to put out a 16-bit system, but Nintendo's aggressive price drop (matching Sega at $99.95) at CES allowed the Big N to reclaim any market share they might have lost to their new foe. In addition to the pricing, there were several factors responsible for the recovery, including Perrin Kaplan's PR savvy and John Sakaley's incredible in-store displays, but above all, the resurgence could be attributed to what had always defined the Nintendo experience: good games. Sega still had a larger library, but Nintendo's roster was growing quickly, and unlike their competitor's games, which they viewed as a mixed bag of awful, acceptable, and above average, Nintendo's games were all above a certain threshold of quality. At its core, this distinction personified the difference between Sega and Nintendo. Sega was okay with the fact that some games were subpar, some were weird, some were violent, and some were racy; philosophically, Sega believed in the power of choice and wanted the consumer to ultimately decide what worked and what did not. Nintendo, however, opposed this laissez-faire attitude and took it upon themselves to exert greater control over the process. From their stringent development cycle to their game-centric marketing focus, Nintendo sought to control the creative process the same way they controlled retail sales, operations, and distribution. Although it's difficult to use the word "control" so often and not inspire a sudden appearance by Big Brother, it's important to note that there was nothing sinister behind Nintendo's motivations. Rather, they were more like a lowercase big brother, looking out for a younger sibling and wanting to ensure that when he spent the money saved from birthdays, tooth fairy visits, and couch-cushion expeditions, that little brother received a certain type of gaming experience. Maybe this goal of slowly but surely building trust wouldn't matter to someone craving Sonic The Hedgehog today, but Nintendo believed that after the Sega owners got burned by a few duds, their perspective would change.

That was the long-term strategy, at least, but it was actually a pair of runaway hits that pushed Nintendo forward in the short term. If Mario was Nintendo's Mickey Mouse, then the Zelda series was something like their Donald Duck: offbeat, unexpected, and a little more complex than his cutesy Disney mates. The darker and more sophisticated tone to the Zelda series immediately resonated with American audiences following the 1987 release of the first game. The same, however, could not be said in Japan, although this reaction had less to do with demeanor and more to do with distribution. The original Zelda, along with the first Metroid game, had been the launch titles for Nintendo's ill-fated Disk System, which was essentially a hulking computer drive that connected to the Famicom and played tiny, rewritable diskettes instead of the standard cartridges. The logic was that when someone had beaten a game or just simply got bored with it, they would bring the reusable diskette to a Disk Writer at their local toy store and pay a small fee to erase the old game and get a new one. Although Nintendo appeared to be walking on water throughout much of the eighties, this was one example where they'd ended up getting soaked. The Disk System was poorly received, to the point that NOA chose not to release it in the United States. And because not many Japanese players ever owned it, they lost the chance to fall in love with Zelda in the same way as their American counterparts, who could buy it as a regular game cartridge.

In truth, NOA likely would have preferred to have a Zelda title closer to the SNES launch, but because of the different cultural appetites for the franchise it was a bit of a lower development priority in Japan. Not only that, but the new Zelda game was incredibly complex and wound up taking nearly sixty thousand hours to program. When it was finished, however, and finally came to America in May 1992, The Legend of Zelda: A Link to the Past became an instant classic and presented a compelling reason for consumers to buy a Super Nintendo and not a Zelda-free Sega Genesis. If this was not reason enough, then Nintendo hoped that another exclusive title released in June 1992 would put the SNES over the top: Street Fighter II.

Although the 1990s had not been kind to the arcade industry, Capcom's Street Fighter II bucked that trend with a hot release in early 1991. If the allure of videogames is the opportunity to act out repressed fantasies in a virtual world, then this game's scorching popularity demonstrated that a lot of kids dream about kicking the crap out of each other. Although this wasn't the first so-called fighting game, where two virtual combatants slug it out, Street Fighter II's innovations turned this type of game into its own genre. With distinctive characters (like Blanka, a green-skinned beast-like Brazilian man who had been raised in the jungle), Street Fighter II introduced signature moves for each character (Blanka, for example, can wound opponents with his electrical blasts). The game also pioneered the concept of a "combo move," where players could cleverly combine several attacks in a single sequence that would harm opponents more than if each had been inflicted individually. Ultimately, by virtue of these innovations, Street Fighter II was the first fighting game that was actually based on skill and not luck. That's what made it really click. Well, that plus the fact that it was so goddamn cool to control fighters who possessed the same kind of depth, backstory, and superpowers as iconic comic book characters.

Given the arcade game's runaway success, Nintendo naturally wanted to license a home version for the Super Nintendo. Also naturally, Sega wanted to do the same. But those desires fell short when the close personal and professional relationship between NCL and Capcom in Japan enabled Nintendo to obtain exclusive rights to the game. As a result, in the summer of 1992 Nintendo had two must-have games that weren't available for the Sega Genesis, and there was a third on the way: Super Mario Kart, a fun-for-all-ages go-kart racing game scheduled for August. Part of the beauty was that, for the first time ever, it brought together all the key members from the Mario universe. With the option to play as both heroes and villains, from Mario and Luigi to Bowser and Donkey Kong, it had the strangely wonderful effect of giving players the feeling that this is what life was really like in the Mushroom Kingdom—as if when these characters weren't busy trying to save the world, they all just hung out and raced each other for fun. Sure, they threw banana peels at each other and fired off the occasional heat-seeking red turtle shell, but that didn't change the fact that they seemed to have a special friendship.

Beyond the frivolous fun, there was a technical aspect to Super Mario Kart that had everyone at Nintendo convinced they'd be riding through Victory Lane. It would be the first game for the Super Nintendo to effectively take advantage of the system's Mode 7 capability, which was a graphics mode that enabled background layers to be rotated and scaled in such a way that a three-dimensional graphic could be achieved. Nintendo was well aware that this esoteric feature was not reason enough for someone to buy a game, but if the marketing team could somehow show consumers that technology made a good game great, and in the course of doing so make it clear that Super Nintendo had this Mode 7 and Sega most certainly did not, then that just might be enough to take over the 16-bit market once and for all. The future looked bright, and whether retailers loved or hated Nintendo, the great software available, the greater software to come, and the well-priced hardware were now too much for anybody to resist. Not even Target.

"So I get a call from the buyer at Target," Peter Main explained to a small group of colleagues and spouses huddled around him at the barbecue, "and he says that they'd like to start doing business with us again. Now, I have nothing but the utmost respect for these Target guys, a real classy bunch, so I feel comfortable needling him a little bit. I said, 'That's very interesting to hear, but what about all these philosophical differences that I've been hearing about?' He laughed, a real nice and sincere one, and he says to me that there really are philosophical differences but he realized that between the two, he preferred Nintendo's philosophy."

Everyone chuckled, refilled their drinks, picked at the food, and continued to decline Arakawa's invitations to go swimming. This joyous cycle continued for hours, until the sun began to set and guests took this as their cue to go home. But even as the celebration thinned out, many of Nintendo's longer-tenured employees still remained: Main, Lincoln, White, Harman, and pretty much anyone who could positively identify R.O.B. the robot. These were the people who, no matter what, would forever be bonded by an incredible experience. They hadn't chosen each other, but they had all chosen Nintendo, and that meant that despite their differing personalities they all shared a unified philosophy. For better or worse, they were part of the same family, and thank God these past few months had been veering toward the better.

By the time it happened, nobody remembered whose idea it was, but the men of Nintendo found themselves swimming in the pool. Because of the drinks, the laughter, and the intrinsic nature of barbecues, the last few hours had become a blur. But like the moonlight shimmying through the water, it was a nice blur to be a part of. They weren't quite sure what tomorrow would bring, but they were confident that everything would work out okay as long as they continued to do things the Nintendo way. It wasn't always flashy and it wasn't always popular, but it was a slow journey into what was possible, and they couldn't wait for Arakawa to lead them there.

34.

COPS AND ROBBERS

Toward the end of the summer, Tom Kalinske could feel Sega's momentum slipping. Although Anique's passing had instinctively triggered a desire to take his foot off the pedal and spend more time with his family, that just seemed selfish at a time when Nintendo was beginning to wake up. He owed so much to Sega, to its people and their sacrifices, that postponing personal goals for professional ones was the least he could do. It was not ideal, but he was willing to do what was best for the company; in the long run, Sega's revolution would make the world a better place for him, his family, and kids everywhere.

"Is that how you really feel, or is that just a convenient justification?"

The question hung out there for a bloated second, as Kalinske realized that the words that had just come out of his mouth applied as much to himself as they did to the man on the other end of the phone. He shook off the strangely surreal moment, and in the solitude of his home office, he repeated the question: "Do you honestly feel that way? Or are you just rationalizing decisions made in the past?"

"That's a good question," Nintendo's Bill White said from his home in Redmond, Washington. "But I'm afraid it might be a rhetorical one. What does it matter if I'm happy at Nintendo? I'm here, it's a great company, and the people are generally acceptable. Oh, yeah, and one more thing that I probably should I started with: it's none of your goddamn business!"

Although White had said this last part harshly, both men couldn't help but laugh, because it was the first thing that either one had uttered during the call that even began to resemble sincerity. "Okay," White continued, "now I've got one for you: how did you get my number? And no, it's not a rhetorical question."

"Don't worry about the how," Kalinske suggested. "I'd like to talk about the why. And if it's not immediately obvious from my calling you at home and asking about your current employment situation, it's because I'd like you to come join us at Sega."

Ever since Sega had failed to hire Howard Phillips, Kalinske had been looking to poach someone else from the royal family. But because Nintendo was such a small outfit, and because their employees rarely ever left the company, this was no easy mission. Bill White, however, seemed like the best candidate, and recent rumblings that he was growing increasingly frustrated by Nintendo's neutered marketing approach made Kalinske cautiously optimistic. Plus, there was the fortuitous connection of Bill's sister, Renee; she was the general manager at Sega's old ad agency, Bozell. She had actually been the one to introduce the two of them, and by virtue of that relationship, Kalinske imagined that Bill differed from his Nintendo colleagues and actually had some shred of respect for Sega. Or at least he didn't think they were Satan incarnate. Either way, it was worth a call. Especially during a period when Nintendo was gaining steam and Sega was waiting on Sonic 2sday, the first round of Goodby commercials, and a handful of next-level business opportunities.

Although White had little interest in leaving Nintendo, his ego was instinctively boosted by Kalinske's offer. It felt like a notorious robber asking a decorated cop to join his crew for a big heist; sure, the offer was insultingly presumptuous, but there was also something inherently flattering about it, a suggestion that in this world of black and white, there was something uniquely gray about your character that signified a flourish on both sides of the law. "I appreciate the consideration," White said, "but I don't have any interest in making a move at this time."

The last three words—"at this time"—convinced Kalinske not to end the call. Over the past few years, White had been programmed to fundamentally despise Sega, but there was something about those three words that indicated that perhaps he could be reprogrammed. And so, at this time, Kalinske decided to keep pushing.

"Come on, Bill," Kalinske pleaded. "I can't sit here and genuinely believe that you're happy with things at Nintendo. From everything I've heard and the few times we have met, I can tell that you're not some drink-the-Kool-Aid ad man, but someone who actually appreciates the art of marketing. And I don't know if you get that from your father or your sister, or if it was just something you were born with, but you've got the gift, and I'm confident that Nintendo isn't utilizing it properly."

"Nice of you to say," White said. "But I'm curious how you arrived at that conclusion. It's certainly not from my sister; she and I never talk business."

"No, that's not from Renee," Kalinske said. "I would never even consider exploiting a family relationship like that. My assumption comes from a much more obvious source: Nintendo's advertisements."

"What about them?"

"They're terrible."

"Screw you!"

"Okay," Kalinske corrected. "They're not terrible. They're good. Never great, never groundbreaking, never even memorable. They're always just good enough. And I understand why Nintendo would want to do things that way, but I also understand why, if I were in your position, I'd want to rip my hair out."

"Interesting assumption," White mused. "Although I am quite sure that the last time we saw each other I was not missing a single hair on my head."

"I didn't notice one way or another," Kalinske said. "I was probably too busy thinking about the $65 million generation-defining campaign that we're gearing up to launch."

"That much?" White asked. "I thought it was only $45 million."

"Don't believe everything you read in the press," Kalinske said. "Besides, does it really matter whether it's $65 million, $45 million, or even just $5 million? It's money that will be spent on actual marketing. Messaging. Branding. Influencing culture. People are going to be talking about this so much that I bet the watercooler business doubles."

"Ha," White replied. "And the unfounded self-aggrandizing business might triple, right? Look, you're obviously someone who—"

"What about the movie?" Kalinske cut in.

"What is that supposed to mean?"

"I know a few journalists, and they seem pretty convinced that what's happening in North Carolina is a complete disaster."

"That's a lie," White said, knowing full well that what Kalinske had said was, in fact, the truth. Or at least it had been when *Los Angeles Times* writer Richard Stayton visited the North Carolina film set earlier that summer. The movie in question was *Super Mario Bros.*, the same one that Peter Main had excitedly announced at the Consumer Electronics Show back in January 1991. "Hey, Tom, why don't you do yourself a favor and heed your own advice? Don't believe everything you hear from the press. Especially the lies."

Like every movie ever made, *Super Mario Bros.* started off with the best of intentions. In early 1990, just weeks after *The Wizard* hit the box office, White decided to get serious about making a movie based on Nintendo's incredibly popular plumber. With *The Wizard*, which starred Fred Savage and centered around an autistic videogame prodigy, Nintendo had agreed to license their logos, trademarks, and game footage to Universal Pictures, who produced and distributed the movie. Nintendo was paid $100,000 for the intellectual property, but notably, and contrary to public opinion, they had no creative approval over the final film beyond the initial script and the implementation of game footage. On one hand, this deal could be viewed as a major coup for Nintendo, who was actually being paid to have a ninety-minute commercial produced about its games. But on the other hand, in this case the dominant hand, a company who considered their best quality to be control had effectively parted with that right for a modest sum. This is not to say that Universal kept Nintendo in the dark, because they didn't—White even had an open invitation to visit the set in Reno but it cautioned Nintendo against getting involved in a similar situation, especially after the resulting film turned out to be exceptionally mediocre, fun and watchable but not much more than a ninety-minute Nintendo commercial. Perhaps Nintendo's commercials were guilty of being "just good enough," but everything else the company produced was great, so when White pushed the idea of a Mario movie, Arakawa had been open to the idea as long as the movie could achieve greatness.

Throughout 1990, several movie studios made pitches to Nintendo, with story ideas, production budgets, and potential talent for the feature film. Competition for the film rights was intense, particularly at a time when four of the year's highest-grossest films turned out to be action-adventure family movies (*Home Alone, Teenage Mutant Ninja Turtles, Dick Tracy,* and *Kindergarten Cop*). The suitors each made multimillion-dollar offers, but as Nintendo had learned with *The Wizard*, the money came with strings attached. And this time Nintendo cared less about the money and more about controlling those strings, which led them to bypass the movie studios in favor of a pair of independent filmmakers: Jake Eberts and Roland Joffe.

When Eberts and Joffe presented their creative vision, they were less concerned with their production company's stature and more concerned about their personal track record. Both had made great films, but those films tended to involve death, rape, and the fragility of human life, like *The Killing Fields*, a 1984 Joffe-directed film about Cambodian dictator Pol Pot's state-sponsored genocide of two million innocent civilians. Although this was several shades darker than King Koopa's death-by-Goomba strategy, Arakawa was impressed with the adult-skewed pitch and thought that perhaps a more serious film could attract an even larger audience. Feeling like Eberts and Joffe offered the best chance of something more than mediocrity, Arakawa and Yamauchi sold them the film rights to Super Mario Bros. at a reduced rate in October 1990. And from there, everything that could go wrong did.

Murphy's Law first reared its head when Bill White, Nintendo's point guy for the film, received a call from Joffe letting him know that Dustin Hoffman was extremely interested in playing the role of Mario. White was thrilled by this news, believing that the Oscar-winning actor had the perfect skill set to achieve Nintendo's lofty goals: the dramatic chops to skew older (*Rain Main*), the cartoonish eccentricities to skew younger (*Hook*), and the ability to be taken seriously while acting silly (*Tootsie*). White asked Joffe to set up a meeting and then excitedly brought this news to Arakawa, who didn't share the same enthusiasm. Instead, he just kind of scratched his head and said that Hoffman wouldn't be right for the part. In traditional Arakawa fashion, when he was asked if there was any particular reason why not, NOA's president thought for a moment, squinted ever so slightly, and then quietly said no before moving on to the next thing. By the time this response was relayed back to Joffe, a meeting had already been set up with the actor. As a result, White was sent to New York for a two-hour meeting with Hoffman, where he was forced to reluctantly rain on Rain Man's parade.

With Hoffman out of the picture, White and the producers set their sights on Danny DeVito, who resembled a pudgy, charismatic plumber as much as anyone in Hollywood. Arakawa approved of this new choice, but the feeling turned out not to be mutual when DeVito turned down the role in order to focus on his directing career. Next on the list was a young actor named Tom Hanks who agreed to take the role for $5 million. But before signing the deal, the producers and Nintendo started to have second thoughts. Hanks's recent credits included *The 'Burbs*, *Turner & Hooch*, and *Joe Versus the Volcano*, spurring serious doubts that he could handle a dramatic role. There were also concerns that although Hanks was tremendously likable, he lacked that "it" factor to carry a big-budget movie. And with *Super Mario Bros.* now budgeted at a hefty $40 million, the offer to Hanks was pulled. Hanks wasn't thrilled but quickly got over the loss when his next two films (*A League of Their Own* and *Sleepless in Seattle*) made him the most popular actor in Hollywood, and the next two (*Philadelphia* and *Apollo 13*) won him Academy Awards for Best Actor.

When attaching actors to a film, this game of musical chairs is pretty common. But when it happens with writers and directors, that usually signifies it's time to put your tray table in the upright position and brace for a crash landing in development hell. Although Joffe had directed many movies, he saw *Super Mario Bros.* as the perfect vehicle to launch his producing career; wanting to focus only on that aspect, he hired Greg Beeman to direct the movie. Beeman had a powerful, whimsical style, but because he had directed only one film to date (and because that film was an $8 million teen adventure romp starring Corey Feldman and Corey Haim), no distributor would finance production with Beeman at the helm. Without financing there would be no movie, so Beeman was jettisoned in favor of Rocky Morton and Annabel Jankel, a husband-and-wife directing team based in London. Morton and Jankel were a pair of former music video directors famous for creating *The Max Headroom Show*, a cutting-edge British TV program featuring the world's first computer-generated talk-show host. In keeping with Max Headroom's subversive spirit, Morton and Jankel wanted the movie not only to skew older but also to slant much darker, dingier, and a bit demented.

Although this direction frightened White and Nintendo more than a little, they had approved the hiring of Morton and Jankel, so it stood to reason that they ought to let their new directors do what they did best. In order to pull that off, however, the new directors needed a script that reflected their hyperkinetic vision.

The original script for *Super Mario Bros.* was supposed to be written by Barry Morrow, who had recently won an Academy Award for his *Rain Man* screenplay, but his initial treatment outlined a darker movie that Nintendo had anticipated. Like *Rain Man*, Morrow's vision centered around the complexity of brotherly relationships and even mimicked the relationship between Tom Cruise's and Dustin Hoffman's characters, with Mario serving as something of a heroic guardian to his lower-functioning younger brother, Luigi. Nintendo wanted serious, but this was too serious and, besides, wasn't *The Wizard* already kind of a Nintendo-inspired reboot of *Rain Man*? Instead of revising his vision, Morrow was displeased by its reception, not to mention Dustin Hoffman's being passed over for Mario, and stepped aside to allow someone else to give it a go. As a result, while the press kept printing stories about how an Oscar winner was bringing Mario to the big screen, the first script was actually penned by Jim Jennewein and Tom S. Parker in early 1991.

Jennewein and Parker were an unusual choice given that they had no previous screenwriting credits, but although they lacked experience, they possessed something that Hollywood valued even more: heat. A few months earlier, the unknown pair had sold a spec script called *Stay Tuned* to Morgan Creek for $750,000, and in the period that followed, they became the industry's newest next great writers and got the gig to write *Super Mario Bros.* Jennewein and Parker, who went on to script child-friendly adaptations of *The Flintstones* and *Richie Rich*, wrote a lighthearted modern-day fairy tale that upped the ante for an adult audience with complex characters and tongue-in-cheek humor. The script was praised by Nintendo and the producers, but because it had been written specifically for Beeman to direct, it was tossed aside when Morton and Jankel took over.

The new directors wanted something less fantastical and more steeped in the mythos of science fiction. To give life to this concept, they hired Parker Bennett and Terry Runte, who were best known for writing the outlandish movie *Mystery Date*, about a teenage boy whose dream date with the girl next door turns into a nightmare when he finds a dead body in the trunk and other horrors ensue. Again, White and Nintendo were skeptical of where this was headed, but they felt that Morton and Jankel deserved the chance to follow their instincts; besides, they had the safety of creative approval. To Nintendo's surprise, the sci-fi script was actually not so bad. It had the futuristic flair of a space opera but still contained many enjoyable fantasy tropes, like Mario and Luigi being at the center of an age-old prophecy, and a magical talking book that aids the plumbers on their quest through a mushroom-infested reality. By mid-1991 the project now seemed to be moving in a direction that satisfied Nintendo, the producers, and the directors. Despite a few false starts to the process, the many parties involved with the movie couldn't help but breathe a sigh of relief and give in to the notion that this just might work out after all. But little did they know that all this was soon to be sabotaged by the directors.

Morton and Jankel eventually deemed this latest script to be too blah, and had the writers amp it up to be more like *Ghostbusters*—larger than life, comedic, and centered around a snarky Bill Murray–type. With this in mind, Bennett and Runte introduced the concept of Dinohattan, a parallel urban universe where dinosaurs had never gone extinct and where King Koopa and his evil cronies ruled with an iron fist. As the writers were rushing to finish the draft, Bob Hoskins (and not a Bill Murray–esque actor like Tom Hanks) was now the top target to play Mario, and major revisions were needed. With the financiers pressing the filmmakers to get this movie into production, the directors fired the writers and replaced them with Dick Clement and Ian La Frenais, who delivered something more in the action-packed vein of *Die Hard* (and whose script even included a scene where Bruce Willis made a cameo, tunneling through the air ducts of King Koopa's castle). Nintendo and the producers saw the script and wanted something more grounded, which led Clement and La Frenais to write another draft that had the realism that Nintendo wanted, the grit that the producers desired, and the off-kilter dystopian feel that the directors craved. By appeasing all parties, this March 1992 draft was sent out to potential actors and proved sturdy enough to entice Bob Hoskins (Mario), John Leguizamo (Luigi), and Dennis Hopper (King Koopa) to sign on to the movie.

Production was slated for two months later in North Carolina, and things appeared to be back on track, but producers Jake Ebert and Roland Joffe began to worry that the script's current iteration was too far removed from the videogame's sensibilities and needed to be made lighter and more fun. Nintendo agreed that the tone was more mature, but with Sega beginning to attract older gamers, maybe more mature was a good thing. Ebert and Joffe understood where Nintendo was coming from, and they themselves did not want to make a syrupy children's movie, but they wanted something that was more accessible and less bizarre. This bizarre tone, of course, was the handiwork of Morton and Jankel, who had been hired to bring this exact sensibility. By this point, however, the producers realized that they had made a huge mistake with their choice of director, and with Nintendo's permission, they hoped to salvage any chance at a blockbuster by bringing in some script doctors. Nintendo consented, leading Eberts and Joffe to hire Ed Solomon (*Bill and Ted's Excellent Adventure*) and Ryan Rowe (*Tapeheads*) to punch things up before filming began in May. Because Eberts and Joffe believed that Morton and Jankel were the ones responsible for all of the problems, the producers forbade the directors from even speaking with the writers, which created a rift that lasted throughout production.

When the actors arrived on the set and received shooting scripts, they were shocked by how much the story had been changed, and they considered quitting the film. The producers then tried to rehire Dick Clement and Ian La Frenais to undoctor their already doctored script, but they were not available for the job. Instead, Parker Bennett and Terry Runte, who had written the draft that Clement and La Frenais had once upon a time redrafted, were hired and flown down to North Carolina. Upon arrival, Bennett and Runte worked closely with the producers, directors, and cast members to make script changes on the fly and play the unenviable role of creative peacemakers. Their already difficult job was made even more frustrating by the fact that the actors weren't talking to the directors, the directors weren't talking to the producers, and nobody was talking to Nintendo.

Over the next several weeks, things went from hell to the inferno's seventh circle. The directors lost any remaining allies when Morton poured hot coffee on an extra he didn't think looked dirty enough, but amazingly this obscene incident was soon trumped by another. Bob Hoskins and John Leguizamo disliked working on the film so much that they began drinking on the set, which may or may not have been the reason behind Leguizamo crashing a car and injuring his fictional brother during production, which resulted in Hoskins having to wear a cast that could be seen in various scenes throughout the film). With all the chaos, the production schedule ballooned from ten weeks to fifteen, forcing the producers to scrap the original climactic finale that would have featured Mario scaling the Brooklyn Bridge and saving the day by dropping an explosive Bob-omb down King Koopa's throat. Instead, they opted for the much cheaper alternative of Mario simply shooting King Koopa with a gun.

The entire production was an unmitigated disaster. Bill White knew that—he had seen it firsthand—but still, a part of him couldn't help but wonder if things might turn out okay after all. Drama and insubordination aside, they had at least shot enough footage to piece together a coherent ninety-minute film. And however bad it turned out to be, nothing could take away the fact that the movie would have stars, special effects, and the iconic Mario name. By this point, White was willing to admit that it would not be nominated for any Academy Awards, but there was still a good chance that it would make a ton of money. Take the game Super Mario Bros. 2, for instance, which was Nintendo of America's last-minute attempt to put a Band-Aid on Japan's sequel. That game was weird and kind of creepy, but still managed to sell ten million copies. So despite whatever it was that Kalinske thought he knew about the movie, White knew the truth, and he also knew the greater truths about hype-driven consumer culture.

"I don't know what you've heard," White said to Kalinske, "but I'm more than happy to bet you that it will wind up being the highest-grossing film of the year."

"Well, then," Kalinske replied, "I guess I need to get myself some better sources. I just figured that if the movie turned out to be as bad as a few folks have been saying, Nintendo is going to be looking for a fall guy."

"And you think I'll be the fall guy?"

"What happened to speaking in theoreticals?"

"Good point," White said. "I don't think we should be speaking at all. I appreciate the call, and honestly, I'm flattered by your insults, but it's time for me to go."

"Wait," Kalinske said with a dash of urgency. "I have just one more question."

"Ask fast," White said, "before I hang up."

"Fine, fine," Kalinske said. "I was just wondering if you might be able to tell me: what the hell is this Mode 7 thing?"

35.

SHOSHINKAI

Tony Harman stared out the window of a Tokaido Shinkansen bullet train, admiring the extraordinary way Japan's green-gray-gold countryside blurred together at 170 miles per hour. He was headed east, traveling in a private car with Mr. A., Mrs. A., and Mr. Y., all of them on their way to the 1992 Shoshinkai held in Tokyo. Shoshinkai was Japan's annual industry event similar to the Consumer Electronics Show, except that Shoshinkai only featured videogames and, even more specifically, it only featured videogames made by Nintendo and their licensees. It was an important occasion, headlined by a keynote speech from Nintendo's president, which explained why Mr. Y. was making the three-hour trip east and why Mr. A. had made a ten-hour plane trip to join him for this train ride.

Most of the commute had consisted of Mr. A. rolling his eyes as Mrs. A. translated Mr. Y.'s bombardment of questions for Harman. The inquiries were typically directed at Harman because the young American understood videogames better than most employees on either side of the ocean, and also because Mr. Y. loved challenging people with questions, and he suspected that his daughter and his son-in-law did not much appreciate jumping through hoops. Why do people enjoy playing videogames? he asked. Why should Nintendo make hardware when the real money is made on software? And why is it that only the Japanese know how to make good games? Harman answered all with savvy and sincerity until, at some point along the way, the other members of his party dozed off.

Harman found a strange joy in witnessing Mr. A. and Mr. Y. at rest. Both men were such unforgiving workaholics that at times it was truly hard to imagine that either ever slept. Mr. A. typically hung around NOA headquarters until after midnight, and Mr. Y. was rarely seen anywhere outside the office. Although both men possessed vastly different leadership styles (Mr. A. saw himself as one of the guys; Mr. Y. saw himself as the guy), each had earned the right to take it easy and enjoy life. But these were no ordinary businessmen; rather, they were artists, and nothing gave them greater joy than being authors of culture, purveyors of technology, and, most important, architects of emotion. In the same way that those responsible for envisioning, financing, and constructing this speedy train had indirectly awed Harman with a sense of power as the train rushed through the countryside, Mr. A. and Mr. Y. were responsible for creating similar feelings of possibility for an entire generation. With jobs like those, it was no wonder business was their pleasure. And it was no wonder that they would take a brief nap whenever their bodies and minds were willing to rest.

However, when Harman looked away from the window and found Mr. Y. examining him with those crisp, inquisitive metallic eyes, he began to wonder whether Nintendo's president had ever really fallen asleep. Maybe this was just another test. If it was, then from the looks of Mr. Y.'s clenched-teeth smile, Harman felt like he had passed.

Later that day, from inside the silver-domed Makuhari Messe convention center in Chiba City, thousands of adoring fans gathered to discover the latest and greatest from Nintendo. It was August 26, and between the addictive Super Mario Kart, the adorable Tiny Toon Adventures, and the assortment of indistinguishable Street Fighter II clones, these videogame zealots would ecstatically spend the next seventy-two hours waiting for the chance to try out the newest games. The lines were long and twisty, but the incredulous oh-my-God expressions on their faces told the whole story: it was as if they had died and gone to Mode 7 heaven. Nothing could snap them out of their wonderful stupor—nothing, that is, until their pious god arrived.

Hiroshi Yamauchi, the unparalleled Mr. Y., stepped to the podium.

Without prompting or prodding, members of the crowd uniformly dropped their controllers to listen to Nintendo's sixty-five-year-old president. Wearing, as he always did, a dark tie and a pair of caramel-tinted glasses, Yamauchi unveiled what was planned for the coming year. He began by cautioning the audience that the marvels of Mode 7 had only been the beginning. This best was yet to come, a joyous future indeed, because in early 1993 Nintendo would be launching something called the Super FX chip. Unlike anything ever made before, this custom-made RISC processing device would be used in select cartridges to render true, groundbreaking three-dimensional graphics.

The Super FX chip was Nintendo's answer to the Sega CD. Not directly, as neither had been developed with the other in mind, but metaphorically it highlighted the difference between each company's approach to technology. Sega, with its CD device and its recently rumored plans for a virtual-reality peripheral, planned to leap forward through innovative hardware. Nintendo, with its Mode 7 and this Super FX chip, was dedicated to evolving through upgrades in the software. Sega's approach was riskier and more costly, but if successful, it could shift the entire paradigm. If Sega CD caught on or virtual reality became the diversion du jour, then Sega could take the industry from 0 to 60 in half a heartbeat. Nintendo, however, ignored this sprint and chose to run a marathon through gradual innovations. Not only would consumers not have to routinely shell out small fortunes, but by transplanting technology into software, hardware's most vital organ, this would extend the lifespan of the Super Nintendo beyond that of its competitor.

This was part of the reason Nintendo never considered Sega to be a serious long-term threat, or at least it provided a valid philosophical justification whenever slumping sales met thumping denial. The other reason Nintendo didn't consider Sega to be the next coming was epitomized by the event currently taking place at the Makuhari Messe convention center. The name of Nintendo's annual trade show was derived from Japan's main association of toy industry retailers, not coincidentally called Shoshinkai. The Shoshinkai was a multitiered distribution network that, also not coincidentally, Yamauchi had come to control over the years. Although the organization had initially been formed to "promote friendly relations between industry participants," Nintendo's great products and its pinpoint attention to quality had provided the Shoshinkai something even more enticing: stability. Nobody got rich, but everyone made money and the Nintendo-controlled status quo became more important than promoting friendly relations for companies like Atari, NEC, and of course Sega.

Some might say that this was a monopoly, and if such a claim were lobbed at Yamauchi, he very likely might have said "thank you" and taken a bow. Unlike America, where monopolies and oligopolies are considered to be capitalism-crippling viruses, in Japan market control was par for the course. Look no further than the Japanese insurance industry, where not only have the top five firms (Toki, Taisho, Sumitomo, Nippon, and Yasuda) remained in the top five slots for the past forty years, but over that time the market shares of each have not increased or diminished beyond a few decimal points. That kind of stability is not so different from the kind of stability that appealed to the Shoshinkai, and further proved that in Japan, the notion of a monopoly was more like that in the board game Monopoly, where the goal was to build hotels on Park Place and wipe out the little guy who thought Baltic Avenue was a worthwhile investment. Time after time, this type of behavior was rewarded, with the only caveat being that if you're going to be the guy with the hotels on Park Place, then you must share some of that fortune with the thimble, the car, and the other pieces in the game. Nintendo did this by keeping the Shoshinkai content, sharing pieces of the pie with licensees like Capcom, Konami, and Namco, and by outsourcing the manufacturing to Japan's other electronic companies, like Sony, Sharp, and Ricoh.

This closely monitored system of carrots and sticks worked well for Nintendo and its network of partners, but—unsurprisingly—didn't sit well with its competitors. This dynamic was the reason, at least according to Sega of Japan, why they had failed to match Sega of America's success with Sonic, 16-bit systems, and a share of the cultural mind-set. From SOJ's perspective, they had been left out in the rain, faces pressed against the window of Nintendo's Park Place hotel, with no remaining choice but to beg for scraps from those leaving the royal feast. Sega of Japan's complaints were not unfounded, but neither were its woes all that different from Sega of America's early struggles. Perhaps there was no Shoshinkai in the United States, but there were retailers like Wal-Mart who didn't want to upset Nintendo or upend the wonderful new status quo. And while companies like Capcom, Konami, and Namco were not willing to work with Sega of Japan, it's not as though they were willing to work with Sega of America. In 1990, the third-party situation at SOA had not been any easier than it was for SOJ; if anything, it had probably been even worse in America, where there were fewer experienced programmers, no domestic manufacturing resources, and a software company bent on reverse-engineering the Genesis. So what was the difference between SOA and SOJ? Was Sega of Japan just a victim of bad luck, bad timing, and their own poorly crafted advertisements? Or was it that when Sega of America found themselves standing drenched outside the Park Place hotel, Tom Kalinske, Al Nilsen, Shinobu Toyoda, and the rest of those risk takers charged at the window together and broke through the glass?

Whatever the case, the fact remained that Nintendo dominated over 80 percent of the Japanese videogame market, and from the podium at Shoshinkai, Hiroshi Yamauchi hoped to increase his monopoly. After introducing the Super FX chip, he continued to speak about new games, innovations, and characters. What he went on to say was probably very interesting, at least judging by the collective oohs and aahs of the crowd, but Tony Harman had no idea what Nintendo's deity was talking about.

Harman sat with Nintendo's designers in the back of the auditorium and couldn't help but wonder if there was anything more boring than listening to someone deliver a passionate speech in a language you did not understand. Watching paint dry was not exactly exciting either, but at least that didn't make you feel oblivious. The only upside to Harman's boredom was that he got to be bored next to six of Nintendo's living legends. These were the guys responsible for the games, systems, and iconic characters that made Nintendo extraordinary. If Mr. A. and Mr. Y. were the architects of Nintendo, than these six were the designers, builders, and interior decorators:

1. **Gunpei Yokoi:** Nintendo's godfather of gaming, the inventor of Game & Watch, Game Boy, all of the early arcade games, and, essentially, the company's entry into videogames. Yokoi was the head of R&D1, a division that created both hardware and software as well as most of Nintendo's peripherals. Currently they were heavily focused on making games for Game Boy.
2. **Takehiro Izushi:** A key member of R&D1, where he had served as Yokoi's right-hand man since 1975. Izushi

spearheaded one of Nintendo's first commercial light guns, the Beam Gun Custom, and played an integral role in developing hardware for the Color TV–Game 6, Game & Watch, and Famicom.

3. **Yoshio Sakamoto:** A designer on R&D1's exclusive Team Shikamaru. This subgroup was responsible for designing characters and generating scripts for hit games like Metroid, Kid Icarus, and Wrecking Crew.
4. **Masayuki Uemura:** The designer of both the Famicom and the Super Famicom and the leader of R&D2, which focused almost exclusively on engineering and manufacturing oversight. Also, his wife was the one who had come up with the name Famicom.
5. **Genyo Takeda:** The head of R&D3, which was best known for technical software innovations like bank switching, MMC chips, and creating a battery backup that allowed players to save their games (designed specifically for The Legend of Zelda). Although they were the smallest group, they created a handful of notable games like Mike Tyson's Punch-Out!! and StarTropics.
6. **Shigeru Miyamoto:** Nintendo's greatest treasure, Miyamoto had been appointed the leader of R&D4, which was solely responsible for developing NES/Famicom games. In 1992, with the release of the Super Nintendo, this group was renamed the Entertainment Analysis Division (EAD) and was dedicated to developing 16-bit games. In addition to titles like F-Zero and Pilot Wings, EAD was responsible for making all of the Mario and Zelda games. Currently they were hard at work on a flying game that would launch the new Super FX chip.

Whenever Harman accompanied Mr. A. to Japan, he savored the time he got to spend with these men. Their minds just worked at a different level, and it was an inspiring thing to behold. Particularly Miyamoto, who so obviously saw the world differently and used his games to invite others into his memories. The castles that he used to explore as a boy became the enigmatic dungeons in The Legend of Zelda, the neighbor's vicious dog (held back only by a rusty chain) became the Chain-Chomp villains in Super Mario Bros. 3, and in a million other ways he managed to transform the invisible world that only he could see into the pixels of these games.

In the middle of Mr. Y.'s speech, Harman was watching Miyamoto and these other magicians, trying to see the world through their eyes, when suddenly they all simultaneously shot their heads up at once. In hushed tones and anxious murmurs they ricocheted whispers back and forth, debating, dissecting, and discussing a matter of contention until their conversation ceased and slowly they all looked up silently at Harman.

"What?" he asked sheepishly, having no idea what had spurred their concerns.

Miyamoto was the one who explained. Apparently, Mr. Y. had just announced to the world that not only had he bought the Seattle Mariners, but Nintendo was currently making the world's greatest baseball game, a game so perfectly wonderful that it would put all others to shame. The problem, however, was that he had never previously told any of the R&D heads that he wished to create such a game.

"Well," Harman said, taking this in, "one of us should tell him."

Harman's suggestion spurred another round of murmurs, and after this brief powwow the designers decided that nobody wanted to tell Nintendo's president that there was no baseball game under development. Instead they decided that one of them was going to have to produce the game, and to do it so quickly so that Mr. Y. would never suspect a thing.

"Okay," Harman said, skeptical of this plan. "Who's going to oversee this?"

"You," Miyamoto said, and the others chimed in with laughter.

"All right," Harman said, digesting this news. Those were some mighty big eggshells he'd have to walk on, but this was the opportunity Harman had been waiting for. Mr. Y. was so powerful that Nintendo's top magicians would rather make a multimillion-dollar game than have a quiet conversation to correct his mistake. Harman was going to make a baseball game for the Super Nintendo, but for next few moments he couldn't help but join in on their laughter.

36.

PROMETHEUS REVISITED

Luke Perry, *Beverly Hills 90210*'s sultan of sideburns, walked up to the podium at the eighth annual MTV Video Music Awards, airing live on September 9, 1992. Although each of the awards given out so far that night had been presented by celebrities in pairs (like Halle Berry and Jean-Claude Van Damme, or Marky Mark and Vanessa Williams), Perry, curiously, was alone. But the sly grin on his face revealed that he wouldn't remain alone for very long.

After quieting an assault of sexy shrieks, Perry got down to business. "Okay," he started, speaking with the MTV generation's requisite mixture of exhilaration and detachment. "I would like to introduce my copresenter, because nobody else had the *balls* to show up and do it!"

About twenty rows from the stage, Kalinske looked on as the sexy shrieks returned at full blast. He had come down to LA to watch the awards with Nilsen, Toyoda, Van Buskirk, and the rest of Sega of America's inner circle, minus Diane Adair, who was getting married in three days.

"So from a land far away and long, long ago," Perry continued, "it's a bird, it's a plane, it's a really bad smell . . ."

As the crowd swooned with anticipation, the Sega gang was all smiles. But their upbeat spirits had nothing to do with the incoming mystery guest and everything to do with the fact that tonight officially marked the beginning of Sega's elaborate plans to finally slay Nintendo.

"Ladies and gentlemen," Perry concluded, "it's Fartman!"

Amidst sounds of flatulence splatting from the speakers, radio shock jock Howard Stern descended to the stage as his alter ego Fartman. Dressed in a red cape, golden leotard, and protruding plastic ass, Stern showed off his alleged superpowers, causing a smoky explosion that nearly wiped out Perry. The crowd absolutely loved this. If Perry's cool appearance and casual cadence epitomized the look and feel of the MTV generation, then Stern's superheroics captured its sound and smell. It wasn't just about being gross—fart jokes had always been funny (or unfunny) ever since cavemen first discovered green beans. Rather, this was about subverting the system. It was about being outrageous at the most unlikely time, and because there was a level of self-awareness to the act, seemingly unsophisticated actions were upgraded to highly sophisticated comments on society. It was the new coda to the American dream: not only can you grow up to be anything you want, but you can do whatever you want as well, as long as you can look cool or clever doing it.

As Kalinske watched Perry and Stern announce the candidates for Best Metal Hard Rock Video, he didn't waste any neurons on whether this cultural shift signified the downfall of humanity. That was the obvious thing that someone his age was supposed to think, the same thing his father had thought once upon a time. When you turned forty, louder music and shorter attention spans were supposed to indicate the coming apocalypse. But Kalinske had young employees, younger kids, and a forever young marketing mind that was more interested in accepting and understanding than in criticizing and demanding. The world wasn't going to hell; it was just headed to the next level of cool, and Kalinske wanted Sega to be the ones welcoming everyone to that ethereal place.

"What do you think?" Nilsen asked, leaning over to Kalinske as Metallica rushed onto the stage to accept their award from Perry and Stern.

"Not exactly my cup of tea," Kalinske said. "But I get it, and I feel as though we have pretty good seats for whatever happens next."

Tom Kalinske was often right, but his comment at the show was terribly wrong. He and Nilsen had good seats when it came to enjoying host Dana

Carvey's finely tuned shenanigans, but there was one thing that being there prevented them from seeing: the official premiere of Sega's commercials for the "Welcome to the Next Level" campaign, which aired during the MTV Awards.

Jeff Goodby, however, got to see them all. After work, he, Silverstein, Sogard, and a few other guys from the agency went down to a dive bar on Union Street called the Bus Stop, where they got to experience the inaugural batch of ads with an unsuspecting focus group of drunken peers. For a crowd that loved music and loved even more what MTV had encouraged music to become, the bar blasted the Video Music Awards on a dozen televisions as if it were the Super Bowl.

"It's rather amazing, isn't it?" Goodby mused as the televisions depicted Howard Stern's persona flying up, up and away from the stage. "We spend our entire lives trying to find poetry or comedy in every place imaginable, but no matter what we accomplish, the truth is that nothing can beat a good fart joke."

"Shame on us," Silverstein commented, "trying to reinvent the wheel."

"All these years," Sogard added, "and it's been staring us right in the nose."

The riff session came to an abrupt end when a new Sega commercial popped up on the television sets. "Hey, man," Goodby shouted to the bartender, "turn it up." An hour ago the bartender might have laughed off such a request, but now he immediately complied. Although the patrons had started out arguing passionately about whether Van Halen's "Right Now" or Nirvana's "Smells Like Teen Spirit" deserved to win Video of the Year (with some side consideration of the supposed backstage feud between Axl Rose and Kurt Cobain), at some point the conversation had shifted to the Sega ads. Whether it was an ad for Taz-Mania, a teaser for Sonic 2, or a preview of the upcoming Sega CD, the spots were just electric; they didn't feel like commercials at all, but rather another part of the performance. Some of the commercials were only a heretical fifteen seconds long, but because there were so many quick cuts and because they spoke this audience's language so well, fifteen seconds was more than enough time.

When the commercial ended and the bartender lowered the TV to its regular volume, the guys from Goodby goofily high-fived each other. They were a little tipsy, and their collective coordination could use some work, but you know what? Fuck that. They had pulled it off, and they deserved to high-five however badly they pleased. It had been less than six weeks since the Crowne Plaza pitch, and since then they had won the account, shot a shitload of footage, and gotten five commercials ready to air. They now had proof that even without Andy Berlin, everything was going to be all right. And this revelation, in some random dive bar on a fateful Tuesday evening, meant the world to the guys from what would soon be called Goodby, Silverstein & Partners. Through thick and thin they hadn't lost their mojo, and that was a big deal because Kalinske wanted thirty more commercials by Christmas.

"Before we acknowledge how much work we have ahead of us," Goody said, "who's ready to start placing bets on how long it takes Nintendo to respond?"

"Quicker cuts?" Silverstein guessed. "More close-ups? Mario in a leather jacket?"

"The level after the Next Level?" Sogard threw out.

"Hey, as long as they don't resort to flatulence," Goodby said. "Although I do think we ought to address their newest spots. The ones with that Mode 7 thing, or whatever they're calling it." Ever since the release of Super Mario Kart in late August, Nintendo's ads, promotions, and PR had been touting this Mode 7. Obviously, no sane person fully understood what the hell it actually did, but the implication of Nintendo's technological superiority was obvious to anyone. "Sega needs to own that mind-share. When it comes to tech, we ought to be as synonymous with the future as *The Jetsons* is."

"So maybe we get 'his boy Elroy' to be a spokesman?" Silverstein joked.

"Or dare I say," Sogard ventured, "Jane, his wife?"

Goodby cackled. "Cut to some neon-lit crack den of the future where Judy Jetson is now suddenly feeling left out?"

"But back to reality," Silverstein said, "or at least our version of it. Does Sega have anything like Mode 7? Ideally, a Mode 8?"

"I have to believe that the CD is miles ahead of this," Sogard theorized.

For over a year now, both Sega and Nintendo had been making numerous announcements about their plans to soon release a CD-based videogame system. But as of September 1992, only Sega had actually unveiled concrete plans to move forward. In October, Sega would be cohosting a launch party in New York with new ally Sony Electronic Publishing, and then a month later, in mid-November, the Sega CD would hit stores just in time for the holiday season. Until that unveiling in New York, nobody knew exactly what to expect from the Sega CD, but Harold Sogard figured that whatever it could do, it would inevitably put this Mode 7 to shame.

"I'm inclined to believe the same," Goodby said. "But even so, the CD is a different kind of beast—more than just a console, I mean—and we still need a way to show everyone that our car is faster than theirs. Let's touch base with Sega's marketing and see what's there. If the Genesis has some gizmo that we haven't yet heard about, then we're golden; if not, then we'll just make something up."

The guys laughed, sipped their beers, and then returned their eyes to the 1992 MTV Video Music Awards, anxiously awaiting the next commercial break.

The scattershot commercials were unlike anything else on television, but that didn't necessarily make them effective. Since the MTV Awards, Kalinske had been receiving lots of congratulatory phone calls, which was promising, but it's not as though people were going to call and tell him how much the ads sucked. He was dying to get his hands on some new sales figures, sell-through numbers, or any other kind of unbiased consumer data, but that type of information was still a few weeks away. Until then, he'd have to keep relying on his gut instincts, Sega's incredible marketing team, and, to his surprise, a bunch of random eight-year-olds.

As taking Sega to the Next Level continued to take Kalinske further away from his family, he tried to make up for any lost moments by doing the little things whenever he had the time: caring for the baby, bringing home dinner, making it to his daughters' soccer matches. It wasn't much, not compared to the work Karen put into raising four kids, but it was critical to his well-being that he find small ways to show them he cared. It was in this spirit that one afternoon in early October, right before an upcoming business trip to New York, he broke away from the office to pick up his daughters from school.

He waited for them just outside the school building, where a sea of little faces swam past him on their way to meet their friends, greet their mothers, or hop on a bus. He was busily scanning the flood of students for Ashley and Nicole until he was distracted by something strange yet familiar.

"Sega!"

He looked around and found the voice. Some kid wearing a Dream Team T-shirt yelped the word again, this time louder. "Sega!"

His friends responded to his outburst with their own versions:

"Sega!"

"Se-gaaaa!"

"Sayyyyyyyyyyyyy-ga!"

Kalinske couldn't believe his ears; it was one of those rare moments in life where there is no choice but to actually do a double take. The kids were imitating Jimbo Mathison's Sega scream, and they were doing a pretty darn good job. This was way better than new sales figures and sell-through numbers. This was pop culture popping before his eyes, kernels thrown into the cultural microwave and coming out as fluffy popcorn. Even more remarkable was that these kids couldn't be more than ten years old, which was younger than Sega's intended audience. That was Mario's territory, and while Sega was more than happy to convert anyone ready to graduate to Sonic, this demographic was more of a long-term hope than a short-term goal. But evidently Sega's message had reached the teenagers loud and clear, and now it was trickling down to their younger Nintendo-aged siblings who wanted to be cool like their older brothers. If the ads were originally supposed to act as a dog whistle that signaled directly to the mutts of the world (teens, rebels, hard-core gamers), apparently there also was a large population of puppies out there whose ears could pick up this frequency—or, at the very least, puppies who didn't fully understand the noise but cared enough to fake it.

As Kalinske's daughters came into focus, a smile naturally came across his lips. He couldn't wait to smother each with a giant hug, but before setting aside the big ideas of corporate dad for the smaller moments in life, he allowed himself one last look at the herd of Sega screamers.

It was a beautiful thing, to unexpectedly hear his company's name shouted by kids no more than four feet tall, but that was nothing compared to the thrill of watching others do the same at a whopping forty feet tall. This was an unusual comparison to make, but one that Kalinske couldn't avoid while staring at Sony's gigantic Jumbotron in the heart of Times Square as it broadcasted Sega promos at larger than life size.

"I'm speechless," Nilsen admitted, standing beside his boss in a similar stupor. They were looking out the window of the Marriot Marquis's Broadway lounge, momentarily oblivious to the hundreds of bodies busily moving around them.

"Me too," Kalinske said with a single nod of agreement. And then for a few long seconds they continued to stare ahead, enjoying parallel moments of whatever it was they couldn't quite verbalize.

Kalinske and Nilsen had come to New York with a handful of colleagues to officially unveil the Sega CD at an event jointly hosted with Olaf Olafsson and the guys from Sony Electronic Publishing. It was October 15, 1992, and although the system and its games (six from Sega, five from Sony) wouldn't go on sale until next month, this was the public's first chance to see what Kalinske was hailing as the formal wedding between Hollywood and Silicon Valley. To back up this bold claim, he wished to create a spectacle worthy of the product itself, and called in a favor with Sony's Olaf Olafsson to borrow their Jumbotron for the day in order to showcase Sega's breakthrough CD-ROM technology. Or, to put it more poetically, as per the one-liner that Van Buskirk had given Kalinske to feed the journalists in attendance: "It takes a video screen the size of more than five hundred home televisions to indicate how big an idea we think Sega CD and interactive cinema will be to videogame players."

Using the Jumbotron to boast giant-sized game footage was a clever touch, but ultimately it wasn't all that different from how movie studios promoted big films, or how fast-food companies boasted about their latest burger creations. That was fine, but it wasn't groundbreaking enough for Sega. So instead of simply broadcasting footage, they rigged the Jumbotron so that it would become the (gigantic) screen for a Sega CD system set up in the Broadway lounge. This way guests at the launch party would leave with a once-in-a-lifetime experience, and those watching the big screen from below in Times Square would vicariously enjoy the experience of the Sega CD's videogames. This was Sega-worthy. This was Sonic-worthy. This was the Next Level.

For Kalinske and his employees at Sega of America, the Next Level was not just a marketing catchphrase but a full-fledged philosophy of how to tackle life. Personally and professionally, it was a challenge to work the hardest, think the fastest, and always find the fun in whatever needed to get done. It was a perpetual dare to dream like Walt Disney (who was dedicated to "plussing" everything that he and his Imagineers ever touched), innovate like Steve Jobs (who was always searching for new ways to put a dent in the universe), and take risks like the mythological trickster Prometheus (who stole fire from the gods and gave it to humanity—although if Prometheus had been a Sega employee, Kalinske would have expected not just fire but also a stylish fireworks ceremony too). That was the Next Level, or at least an approximation of the attitude, but the phrase and the values behind it were about more than just flowery words. In fact, those words would have been all but moot if not for the actions they inspired.

Coming off the heels of Boca, Sega's defensive mentality at CES had made sense. But Nintendo's very aggressive price drop signified that they were waking up, so to try to put his competitor back to sleep, Kalinske recommitted himself to staying on the offensive. And one of the best ways to take a bite out of Nintendo was to befriend those that Nintendo had bitten. In this manner, just as they had done with Sony and Tengen, Sega set out to form a cartel of enemies, headlined by alliances with the following:

1. Galoob: For nearly two years, Nintendo had been legally hammering Galoob, a California-based toy company, with plans to distribute a peripheral device called the Game Genie. This device, when attached to game cartridges, allowed players to enter codes that offered a multitude of perks, like unlimited health, an infinite number of lives, and even the ability to skip levels. Nintendo believed that this product ruined the integrity of their games and vigorously fought to keep the Game Genie out of stores. Kalinske and his troops, on the other hand, thought this was absurd. So instead of taking Galoob to court, they took them to dinner. And after the appeals and injunctions finally came to an end in December 1991, Sega became a full endorser of the product, giving it their official seal of approval and even throwing some marketing muscle behind the Game Genie.

2. Video Software Dealers Association (VSDA): Ever since it had sued Blockbuster Video in 1987 for renting out their videogames, it was no secret that Nintendo did not like the rental companies and that this feeling was mutual. While Nintendo believed that rentals prevented sales, Sega believed that this practice only served to whet a consumer's appetite. As a result, Sega forged a symbiotic relationship that began with Blockbuster (during the pivotal "Sixteen Weeks of Summer" campaign) and expanded to the entire rental community, highlighted by Nilsen's speech at the 1992 VSDA show, which ended not only with wild applause but promises from rental houses to bury Nintendo in any way possible. From big chains to tiny mom-and-pop outfits, Sega's relationship with the rental community proved to be a hidden competitive advantage, especially during an era when the rental industry flourished due to quicker releases of movies on VHS.

3. Disney: Although Nintendo liked to think of themselves as the Disney of videogames, their relationship with the actual House of Mouse was less than ideal. This appeared to change in early 1992 when Disney bought the distribution rights to the movie *Super Mario Bros.*, but the possibility of happily-ever-after vanished as the movie's hellish production got worse and worse. Nintendo's loss was Kalinske's gain, as Sega was able to obtain the videogame licensing rights to the 1992 hit *Aladdin*, as well as a collaboration deal with Disney's animation studios. This would mark the first time that Disney's hallowed animators would create a videogame, providing even more evidence that Hollywood and Silicon Valley were overlapping with Sega at the center.

While palling around with Nintendo's foes provided an immediate short-term benefit, Kalinske also initiated a series of long-term strategies designed to position Sega as a cutting-edge entertainment company and Nintendo as nothing more than "just a videogame company." In some cases, these plans were about identifying Nintendo's weaknesses and planting a flag where their competitor had not yet broken ground, while in other cases it was simply a matter of taking something that Nintendo already did well and doing it better. Regardless of the motivation, the goal was always to continually provide a Next Level experience, accomplished by some of the following:

1. Animation: Nintendo had a hit animated series, so Sega wanted to have two: one for Saturday mornings, and another for daily syndication. Pulling this off would take a miracle, so Toyoda recruited Michealene Christini, an outspoken, strong-willed miracle-maker with a background in producing cartoons (Marvel) and negotiating multimillion-dollar licensing deals (Mattel). Under her leadership, Sega set out to launch a pair of Sonic-based cartoons in 1993, one that aired Saturday mornings on ABC (*Sonic The Hedgehog*) and another that aired every day of the week after school (*The Adventures of Sonic The Hedgehog*). Both shows would feature the vocal talents of Jaleel White, who starred as Urkel on the hit sitcom *Family Matters*.

2. Videogames on demand: Once upon a time, Nintendo had tried and failed to turn the NES into an Internet service provider. Dial-up connections were too slow, and the idea of working with the Minnesota Lottery to bring digital gambling into living rooms hit too close to home. Sega, however, believed that there was something to this idea. Instead of using phone lines, however, they wanted to use cable lines and create so'mething called the Sega Channel, where Genesis owners could play dozens of videogames on demand. Although the concept felt slightly ahead of its time, Doug Glen believed that it could be done, and he proved it by leading Sega into serious discussions with Time Warner and TCI, who thought this could launch by late 1993.

3. Production: With the increasing convergence of movies and videogames, Kalinske pushed for Sega to build a multimedia studio in Redwood City. This state-of-the-art production facility, fully operational by late 1992, provided a one-stop shop to film movie-quality video, record studio-quality audio, and render jaw-dropping special effects, all in order to produce original content for the Sega CD's lifelike videogames. Ultimately, the goal would be to slowly grow this division to the point of one day producing theatrical films and releasing complete music albums, ideally to coincide with related game releases.

4. Virtual reality: Since 1991, Sega had been quietly developing a sci-fi-like set of virtual-reality goggles in order to provide the most immersive gaming experience imaginable. Nicknamed the Sega VR, this futuristic visor consisted of dual LCD screens, which merged together three separate 3-D technologies, and a series of tiny inertial sensors that tracked movements of the user's head. Completion of this project and proper safety testing were still more than a year away.

5. Edutainment: When it came to fun and entertaining videogames, Kalinske was willing to concede the youngest demographic to Nintendo. But he believed Nintendo had made a major blunder, both financially and sociologically, by not finding a better crossover into the realm of education. To fill this gap, Sega was developing a portable computer for kids that would combine the best elements of toys, videogames, and books. Of all the company's plans for expansion, this was the one that excited Kalinske the most, and the one he thought could make the biggest difference around the world.

These plans provide an incomplete snapshot of Sega's vision for the Next Level, but regardless of how neon the future dared to be, Kalinske knew that none of it would ever happen without videogames. That was Sega's heart and soul, the undeniable essence of the company, and they couldn't lose sight of that. It was true that Sega wanted to be more than just a videogame company, but to accomplish that, it was vitally important that they remain a premier videogame company. If ever Sega stopped providing the best hardware and software, then it could all end in the blink of an eye. Nope, not going to happen. Kalinske refused to ever let it come to that, which was the other reason he was in New York and why he needed to speak with Olaf Olafsson.

"Well," Kalinske said, scanning the busy room, "I think it's time that I find the man from Iceland."

"Good thinking," Nilsen said, still staring at the Jumbotron. "And I suppose it's time I provide a game demo or two. I think it's fair to say that New York's finest journalists may win Emmys and Pulitzers, but they sure can't play videogames."

With that, they went they separate ways, Nilsen to tutor the mainstream press on how to reach the next level of a videogame, and Kalinske through the crowd to find the president of Sony Electronic Publishing. On his way through the circuitry of smiling faces, Kalinske was targeted by Brenda Lynch and Ellen Beth Van Buskirk.

"There you are," Lynch said.

"There's someone from the *Times* we want you to meet," Van Buskirk said.

"Great," Kalinske said. "Can it wait a few minutes?"

"Of course," Van Buskirk said. "Go do your thing."

"All the better," Lynch confirmed. "More time for us to pump him full of propaganda before he meets the Great Oz."

"Thanks, ladies," Kalinske said. He was about to walk away when Van Buskirk darted a finger toward the window, leading his eyes to the Jumbotron.

"Come on," Van Buskirk said. "Are you seriously going to tell me that that isn't going to become an issue?" By now, Nilsen had taken over the Sega CD and was skillfully moving through Night Trap, one of the system's launch titles. Like most titles for the Sega CD, this wasn't a pixelated sprite-based videogame with standard levels, bosses, and power-ups, but rather something more akin to a choose-your-own-adventure movie. And in the case of Night Trap, that movie was a horror thriller. It told the story of five beautiful but naive college girls at a slumber party whose night of fun is ruined when vampiric Augers invade their secluded lake house. With outrageously absurd monsters and damsels in distress, Night Trap featured all the tropes of a cheesy horror movie. But now, for the very first time, when moviegoers would say things like, "No! Don't open the door!" the Sega CD provided the power to actually make that happen. It really was like being granted omnipotence while watching a horror film, but like most horror films, it included violence, sexual innuendo, and several gratuitous deaths, although the gore was never shown. These were the aspects that Van Buskirk meant would become an issue. This was the slaughtered elephant in the room. "There's going to be a backlash," Van Buskirk said. "I just can't see a world where that's not the case."

"I don't know," Lynch said. "It's not that bad. It's so goofy."

"Besides, we're shooting for an older demographic," Kalinske added. "This is no worse than any PG-13 movie in theaters these days. It's what the kids like."

"Who are you trying to convince?" Van Buskirk asked. "Me or you?"

"Look," Kalinske said, "I used to feel the same way. But I think your antennae might be up a little too high on this one. It's just one game. And it's not even a game we made; it's something we picked up to help out Sony.

This won't be anything like the caliber of product that we'll get from our studio."

"Slippery slippery goes that slope," Van Buskirk said, before shrugging and adding a smile. "But like you said, I'm not the one you need to convince. I just wanted to put it out there again."

"And please continue to do so," Kalinske urged. "I appreciate the foresight."

"Aye aye," Van Buskirk said.

"We'll keep our noses extra close to the ground," Lynch assured him.

"All right, then," Van Buskirk said, before glancing at her wrist. "It's been a few minutes. Are you ready for the reporter from the *Times*?"

"Come on, that's not fair," Kalinske said. "I've been with you this whole time. But don't worry, I'll be back soon. I just need to find our friend Mr. Olafsson."

"He should be over that way," Lynch said, nodding to her left. "He was looking for you a little bit ago."

"Excellent," Kalinske said. "Any last words?"

"Nope," Van Buskirk replied. "Just remember to stay on message."

Kalinske laughed and walked away. Moments later he couldn't help but laugh again, this time to himself, when he passed Doug Glen, who was staying precisely on message. "Sega's programmers coined the term 'Tru Video' to mean CD software that looks and sounds like a movie but plays like a great videogame," Kalinske overhead Glen explaining in his professorial voice. "Tru Video makes it easy to suspend disbelief and get drawn into the game fantasy. These new Sega games will have a powerful impact on the way we think about home entertainment."

Glen's comments were spot-on. Everything was, actually—the messaging, the marketing, even the little shrimp cocktails that were being brought around. Sega was firing on all cylinders and operating with a professionalism and playfulness that really did feel befitting of the Next Level. It was all going according to plan, except for a larger Sony issue that had to be figured out. As he took a second lap around the room, Kalinske continued to smile, nod, and act like he had all the answers. But there was one question that wouldn't stop nagging at him: where the hell was Olaf?

37.

THE SEGA-SONY-NINTENDO LOVE TRIANGLE

There he was, Tom Kalinske. Such a difficult man to pin down. Social butterfly, man of the people, and king of the chameleons. It seemed that he was always vanishing in plain sight, but there he was now, alive and in the flesh. "Tom," Olaf Olafsson said, tapping his target on the shoulder, "I've been looking for you."

"Funny," Kalinske said, "because I've been looking for you."

"Sure, sure," Olafsson said.

"Don't sure-sure me," Kalinske said. "I mean it! I've been around this room twice and am starting to suspect that the only way I could have missed you both times would be if you had been following right behind me."

Olafsson chuckled quietly. "Your sense of humor always kills me," he said. "Every single time. But enough of that for now. Fate has brought us together, and we have important matters that must be discussed."

The two of them casually gravitated toward an empty corner of the room. They'd spoken a bit earlier, while doing that whole smile-for-the-cameras thing, but this would be their first chance to actually speak with candor.

"What's on your mind?" Olafsson asked.

"Probably the same thing that's on yours," Kalinske said with a hint of frustration. "Newspaper articles hailing the wonderful relationship between Sony and Nintendo."

"It's complicated," Olafsson said, with the same level of annoyance.

After Nintendo had snubbed Sony for Philips back in 1991, it was generally assumed that this signified the end of any relationship between the two Japanese powerhouses. And for a time that was true, but four months later, at the Tokyo International Electronics Show in October 1991, Sony unveiled its PlayStation. Unlike the consoles that Sega and Nintendo were selling, the PlayStation aimed to deliver something more than just games: sophisticated educational products, like Compton's Encyclopedia, Microsoft Bookshelf, and National Geographic's Mammals of the World. To those previewing the console, these impressive interactive titles proved that the PlayStation was indeed capable of "more than just games." But because Sony didn't have a single game on display, this raised a question: where were all the games? Or, even if the games weren't ready to show off, where were they going to come from? The educational titles were impressive, but they had all clearly been licensed from other companies (Compton's, Microsoft, National Geographic, etc.), and without great original games, the PlayStation was just a computer in disguise.

This was Sony's fundamental problem. In the land of the console, content is king, and this realization led Sony back into Nintendo's arms. With fences somewhat mended, Sony returned to the negotiating table, but quickly found Nintendo to be as stubborn as ever. The only deal that Nintendo was willing to accept would allow them to control the CD system's software production with the same unforgiving authority they exerted over the cartridge business (limited quantities, strict licensing, lockout chips). Nintendo could afford to make enemies with such policies, but Sony was just getting its feet wet and didn't want to alienate the industry. Additionally, Sony would essentially be beholden to their supposed partner in every possible way, from the creative (Nintendo could control which games were made and when) to the financial (Nintendo would collect a majority of the licensing fees). If agreed upon, this relationship wouldn't be all that different from the master/slave dynamic that Olafsson had recounted earlier. Yet knowing all this, Sony's directors were still interested enough to consider moving forward, and the reason they felt this way had less to do with videogames and more to do with videotapes.

In October 1969, Sony had unveiled a prototype for the world's first videocassette recorder (VCR), which they dubbed the U-matic. Previously, video had been captured on film reels, which presented a number of challenges (like loading, playback, and duration), but the U-matic offered an all-in-one solution to make recording easy. And most important to a company like Sony, it presented a new consumer electronics category that would empower the masses with the power of recording video. Believing that this notion of videotapes would soon take the world by storm (and, in turn, spawn imitators), Sony approached JVC, Matsushita, and a handful of other top-tier electronic companies to sign a cross-licensing agreement and help establish unified technological standards. Although this agreement basically gave Sony's enemies a blueprint for how to build their own weapons, it ensured that Sony would at least receive a small royalty for every shot fired (and also temporarily deprived competitors of any incentive to create more powerful ammunition, as peace would be more profitable than war). Not long after that, Sony finished development of the device and launched the U-matic in September 1971. The new device was critically praised as being the unquestioned future of recording, but that future was currently too expensive for the present. With a price tag of $1,300, the U-matic failed to interest the consumer market, but it did quickly become the standard for television production and business communication. Although this was not the demographic that Sony initially planned to target, the business was highly lucrative, and they were more than willing to service a professional clientele until the time came when the technology became more affordable.

That moment arrived in September 1974, when Sony was preparing to manufacture a more consumer-friendly videotape format dubbed Betamax, which was basically a smaller, cheaper, and easier-to-use version of the U-matic system. Sony once again met with the top-tier electronics companies to negotiate another cross-licensing agreement, but this time their competitors were less receptive to working together. Although JVC, Matsushita, and RCA each cited a limited recording time (only one hour) as their reason for resistance, a more accurate explanation might be that they did not want to become further beholden to Sony. With the U-matic, Sony had approached its competitors with a humble let's-work-together mentality, but with the Betamax, Sony had adopted more of a take-it-or-leave-it attitude, which was made clear by higher licensing fees and little interest in working together to establish unified technological standards. Sony was surprised by the industry's reluctance to adopt the new format, but assumed it was only a matter of time before this became the worldwide standard, and the Betamax was introduced to market in April 1975. Initial sales were incredibly strong, as Sony enjoyed 100 percent of this market it had just created, but behind the scenes forces were at work to beat Sony at their own game.

Throughout 1975, JVC's engineers worked to develop a different consumer-friendly video format, which they named the Video Home System (VHS). In September of that year, a VHS prototype was completed and JVC quickly and quietly began to meet with other electronics companies and pitch their alternative to Sony's Betamax. The VHS resembled the Betamax in many ways, but there were two notable distinctions: the VHS recorded at a lower quality than the Betamax, which enabled longer recording times, and the VHS cost $100 less to license than the Betamax. By touting these differences and wisely reincarnating Sony's lost let's-work-together mentality, JVC persuaded Matsushita, Hitachi, and Mitsubishi to adopt their format, and the VHS was launched in October 1976. With strong allies, longer recording times, and a cheaper price point, the VHS quickly gave the Betamax a run for its money.

Although Sony was initially caught off guard by these developments, they responded in March 1977 with the release of a new and improved Betamax videotape. This new model could record for up to two hours, but in Sony's effort to rush this product to market, they made the mistake of not allowing for backward compatibility. Sony's original Betamax tapes didn't work on their new VCR, which alienated most of their initial customer base. This fatal flaw, along with a series of impressive maneuvers from JVC (like well-timed price drops and savvy European expansion), set the course for a slow, painful, and very expensive death for Sony's Betamax. By 1979, Betamax held only 40 percent of the market; by 1984 that number had fallen to 20 percent; and by 1988 Sony had officially thrown in the towel and started to produce VHS VCRs.

Several years had passed since the so-called videotape format war had reached a cease-fire, but Sony's scars could still be seen in every aspect of the company's play-it-safe demeanor. With videogames, Sony could not afford to make the same mistake it had made with VCRs, and the best way to avoid that fate was to join forces with a powerful company whose support of a certain format could stave off extinction. And there was only one company that could provide that insurance: Nintendo. Or, more accurately, that was the case when Sony initially considered entering the videogame industry, but then Sega began its ascent, and with each passing month the Nintendo-centric worldview blurred further. Through Olafsson's persistence, Sony now acknowledged Sega's growing presence and was willing to work with them on software, but when it came to hardware Sony still did not view Sega as a worthy suitor. Not yet, at least.

Until that perception changed, Sony's board of directors had adopted something of a Nintendo-or-bust attitude, and once again it was a Consumer Electronics Show that veered them toward bust. This time there was no public embarrassment, just a lot of private hand-wringing as negotiations reached a fever point. Both Sony and Nintendo had wanted to finalize a deal prior to the 1992 summer CES; not only was it the perfect venue for such an announcement, but there was a certain poetry to writing a happy ending at what had once been the scene of the crime. Days prior to the show it appeared the two companies were on the brink of an agreement (getting to the point where Nintendo's PR firm, Hill & Knowlton, had even prepared a celebratory press release), but eleventh-hour disagreements between the parties killed the deal. So instead of marrying Sony, Nintendo renewed its vows with Philips, and the Sony-Nintendo negotiations were indefinitely put on hold. Perhaps this was due to irreconcilable differences, but more likely it was a case of cold or even frozen feet on Sony's part. Despite all the back-and-forth, Sony still wasn't sure if the videogame industry was worth the gamble.

But that all changed at the pivotal management meeting on June 24, 1992.

After Sony's board of directors had made it clear that they had little interest in moving forward, Ken Kutaragi stunned them by revealing the console he had been secretly working on for months. In early 1992, he internally recruited the engineers behind Sony's System G (a special-effects engine that retailed for $250,000), and persuaded them to help him build a new and improved system capable of rendering 3-D graphics. By this point, he already had a basic design concept, although it was still in the architecture stage. Nevertheless, he believed it was vastly superior to anything out there, as well as to anything in Sega's and Nintendo's pipeline, and this was what he proudly presented to the executives on hand. This was the moment he had been waiting for.

Unfortunately, Sony's directors still weren't fully convinced. Like before, with the Super Famicom audio chip, they fixated on the defiance and not the brilliance. Luckily for Kutaragi, also like before, Ohga was there to protect him. During that meeting he quizzed Kutaragi on the capabilities of this proposed console and how such a product would fit into the overarching Sony brand. Kutaragi adeptly answered all of Ohga's questions and did so with a bravado that many in the room interpreted as arrogance. When the young engineer finished presenting his creation, Sony's chairman appeared torn. Kutaragi picked up on this and challenged his boss by asking if he was going to sit back and accept how Nintendo had treated Sony. This reminder of those fresh wounds filled Ohga with rage, and looking at what could be Sony's bandage emboldened him to trust his gut. "There is no hope of making further progress with a Nintendo-compatible 16-bit machine," he said. "Let's chart our own course."

Norio Ohga may have been the boss, but executing this order would not be an easy thing. His voice was loud, but not always louder than the collective scream of Sony's board of directors. If they wanted to cut this project off at the knees, or continue to play mix-and-match with Nintendo, they had enough leverage to do so. To protect Kutaragi from this type of torture, and to protect Sony from potentially serious internal conflict, Ohga did some corporate reshuffling. Kutaragi and nine members of his engineering team were moved to Sony Music, which had a separate facility in Tokyo's Aoyama district. At Sony Music, Kutaragi worked under one of his mentors, Shigeo Maruyama, and had the autonomy to beginning moving forward with his plans. It was still a small team, and taking this console from concept to completion was still a long shot, but at least they had a chance. And to give this project the best chance of going all the way, Kutaragi began to put together a team, which started with bringing in Phil Harrison in September 1992. Harrison knew games, having been the head of development for a software publisher named Mindscape, and his arrival marked a new level of seriousness for Sony's solo console efforts.

Still, those efforts couldn't be taken too seriously. As Kutaragi spearheaded hardware development and Olafsson pushed forward with the software, Sony's board of directors continued to play tag with Nintendo. And on October 12, 1992, Sony was finally "it." Exactly three days before Kalinske traveled to New York for the unveiling with Olafsson, there was a headline in the *Seattle Times* proclaiming "Nintendo, Sony Join Forces on CDs." Since the article in question didn't provide any specifics of what this supposed deal entailed, it was clearly meant to be a message from Sony's old guard that they were still a force to contend with.

"It's complicated," Olafsson reiterated, shaking his head. "I realize this is not quite the response you were hoping for, but I think this actually works to our advantage."

"How's that?" Kalinske asked, sincerely hoping his friend had a good answer.

"Because," Olafsson said with a wispy smile, "we can tilt the variables of the equation in our direction."

The equation was for a next-generation console, and the direction that both Kalinske and Olafsson had in mind entailed a joint relationship between Sega and Sony. Although the relationship between the men (and their companies) had perhaps begun with the intention of sticking it to Nintendo, it had evolved to something much greater than that. There was still the feeling that by working together they could be stronger than Nintendo, but what made the relationship possible was a shared vision. Videogames were no longer just videogames, but rather a Trojan horse to get inside the living room and be at the forefront of the entertainment revolution. This was the level after the Next Level, and it was time for Sega and Sony to go there together.

"This is my thinking," Olafsson explained, narrowing his eyes slightly. "This news will no doubt be greeted by those at Sony Music as it has been by both you and me—further undue nonsense with Nintendo. But beyond the frustration, there is a significant lesson to glean: Sony is still quite open to videogames, as long as they are running this race with a giant by their side. Well," Olafsson said, his eyes popping open, "a giant is a giant is a giant, and I have a feeling that Mr. Kutaragi would be much more comfortable with someone of your stature."

Kalinske considered this. "Do you really think he'd be able to work with SOJ?"

"Ah," Olafsson said with a smile. "You've heard of Mr. Kutaragi's reputation."

"I think everyone in this room has heard of his reputation."

"It's true, he is no picnic to work with. But it's also true that he's brilliant, and he cares deeply about his work. So I would suspect that he's smart enough to travel down the road most likely to take his work toward fruition."

"Good," Kalinske said, a weight lifting off his shoulders. "That all makes sense."

"For the most part," Olafsson said. "But what about on your end? There is the matter of Mr. Nakayama and his reputation."

"Gee, Olaf," Kalinske joked, "who spilled that secret?"

Olafsson tapped the side of his head. "Firsthand intelligence."

"Nakayama-san can also be a difficult man to work with," Kalinske acknowledged, "but when push comes to shove, he's always stood by my side."

"Good," Olafsson said. "Good, good."

With business out of the way, Kalinske and Olafsson discussed their families and plans for the holiday season, and then walked through the lounge together, enjoying what would be the first step in a long-term, symbiotic, and game-changing relationship.

"Will you have time to take in some sights before leaving?" Olafsson asked.

"Just one sight, unfortunately." Kalinske replied. "And it's in Long Island."

Olafsson raised an eyebrow, until he realized what could possibly be worth seeing out there. "Please say hello to Greg for me," Olafsson requested. "And ask him to tell you about his recent visit to Germany."

"Why, what happened in Germany?" Kalinske asked.

"I'd tell you myself, but how dare I deprive you the joy of curiosity?"

38.

SOMETHING WICKED THIS WAY COMES

"Is all that blood really necessary?" Kalinske asked, as he hovered over the arcade cabinet for a brand-new fighting game. Beside him were Greg Fischbach and Jim Scoroposki, the men whom he had trekked out to Long Island to see.

"It's part of the ambience," Fischbach cheerfully explained. He was a slender man with sandy blond hair and a refined fashion sense that somehow seemed to match the perpetual smirk on his face. "Besides, the kids are gonna love it!"

"Yep," added Scoroposki, a bulky fellow who was generally short on words.

"I'll bet," Kalinske reluctantly agreed as one of the characters on screen ripped out the spine of another. The game in question had been dreamed up by a pair of game designers named Ed Boon and John Tobias. Instead of using traditional animation, they used digitized graphics to create a fighting game they had originally hoped would star either Akido master Steven Segal or European star Jean-Claude Van Damme. After resistance from Segal and Van Damme, Boon and Tobias decided to create their own characters, their own rules, and their own mythology. The result was an intricately detailed and very, very bloody story that followed complex characters fighting against each other for the chance to represent earth in a battle against an evil monster in a cosmic tournament that would decide the fate of humanity. They spent ten months creating and perfecting their vision and weaved in all sorts of special moves and hidden secrets. They called their masterpiece Mortal Kombat. It was like Street Fighter II on speed.

"I've seen a lot of hits in my time," Fischbach explained, "but this one, trust me, it's the one."

Although Fischbach tended to speak about any subject as if he knew it inside and out, when it came to videogames he really did. He and Scoroposki had met in 1983 when they were both working for Activision, the software company made famous by being the first third-party licensee for the Atari 2600 (and then made wealthy by releasing hit games like Pitfall! and River Raid). Four years later, the two began to think about forming their own software company and following in Activision's footsteps—except in this case they sought to become the first American licensee for Nintendo.

At the time of their mutual brainstorm, Nintendo's success wasn't a foregone conclusion. The NES had just enjoyed a very successful 1986 Christmas season, but there were still plenty of naysayers claiming fad and predicting imminent doom. Skepticism was in the air, and as firsthand witnesses to the crash of '83, Fischbach and Scoroposki knew the hazards of such an endeavor. But they were young and optimistic enough to shrug off the risks in favor of the potential rewards. After deciding to take the plunge, the only thing left to decide upon was a name for their endeavor. Fischbach said that he would be okay with anything as long as it began with an A or a Z. Scoroposki did him one better and came up with Acclaim, which would trump their alma mater, Activision, in the phone book.

They contacted Howard Lincoln in the spring of 1987 and became the first American company to join Nintendo's third-party licensing program. Fischbach and Scoroposki quickly realized that their skill set wasn't necessarily in creating games from scratch, but rather in publishing titles made by others. They were businessmen at heart, blessed with the hustle genes in their DNA, and so they set to work obtaining the rights to games that they could then publish for Nintendo. They released their first title for the NES in September 1987, an action-adventure game called Star Voyager. As per their strategy, this wasn't a groundbreaking title, but actually an updated version of a game with the same name that had been released for the Atari 2600 in 1982. They made a deal with the game's developer, ASCII Entertainment, to release Star Voyager for the NES, and then followed it up with four more titles to fulfill their quota: Tiger Heli (originally a Japanese arcade game made by Toaplan), 3-D WorldRunner (developed by Square), Wizards & Warriors (developed by Rare), and Winter Games (a game made by Epyx that was originally released in 1985 for the Commodore 64). Acclaim got all of these titles out by Christmas 1987, and by the first quarter of 1988 the company had made more than $1 million in profits. Sensing the opportunity of a lifetime, Fischbach and Scoroposki took their young company public and received a windfall of cash, which they used to acquire other, smaller publishers and to expand their licensing endeavors. Now, instead of simply publishing content developed by others, they could purchase the rights to make games based on hit movies, television shows, and sports. This led to hit NES titles like Rambo (1988), Airwolf (1989), and WWF WrestleMania (1989). By 1989, Acclaim had reached $109.3 million in revenue. Fischbach and Scoroposki were pleased, of course, but their hustler DNA whispered for more. They wanted to expand, to build an empire, but that was difficult to do with Nintendo's stringent licensing agreement permitting only five games per year. They asked Lincoln to increase their quota, but Nintendo was unwilling to set a precedent of exception. There was, however, a loophole that Nintendo was willing to allow: if Acclaim purchased another licensee, then they could double their output. Fischbach and Scoroposki got the message, and in April 1990 Acclaim paid $13.75 million to acquire LJN Toys, an entertainment company known for making great toys (like Oodles and Thundercats) and lousy games (like Back to the Future and Friday the 13th). Between the two entities, they could now make ten games per year, and in 1990 Acclaim reached $140.7 million in revenue.

If the money was rolling in for Acclaim, then it had to be rolling in even faster for Nintendo, which physically manufactured the cartridges and also got a percentage of every game sold. Through mutual benefit and reciprocal affection, a deep friendship blossomed between Acclaim's Fischbach and Scoroposki and Nintendo's Lincoln and Arakawa. They spoke frequently by phone about matters both personal and professional, and even though Acclaim was based in New York and Nintendo in Washington, the four of them made an effort to get together for dinner every four to six weeks, sometimes on the East Coast, sometimes on the West Coast, and sometimes in the cities they remembered from the history books they'd studied as boys.

"Hey Greg," Kalinske said, as gory highlights from Mortal Kombat continued to scroll before his eyes. "I was with Olaf the other day and he suggested that I ask you about Germany. Any idea what that's supposed to mean?"

Fischbach's eyes lit up at the memory. Or maybe it was just the bloody red reflection on his face. "Right, that fateful day in Frankfurt," Fischbach said, shaking his head. "No shouting from Howard, but it was still a pretty wild time."

In the summer of 1991, Fischbach and Scoroposki had traveled to Germany for a dinner with Minoru Arakawa; his wife, Yoko; and Howard Lincoln. It had been set up like all their other around-the-world dinners, but this one would be different. After years of lucrative, symbiotic work with Nintendo, Acclaim would begin publishing games for the Sega Genesis. They had no idea how Arakawa and Lincoln would respond to this news and were not particularly eager to find out.

It would have been easy to say Acclaim's decision to work with Sega was about business, not friendship, but that wouldn't be completely accurate. As close as Fischbach was with Arakawa and Lincoln, he was equally close to Hayao Nakayama and considered Sega's leader to be a good friend. They had met a few years ago on one of Fischbach's increasingly frequent trips to Japan. Nakayama had known that he had no chance of luring Acclaim away from Nintendo's NES and to Sega's Master System, but he asked anyway. Fischbach, of course, had declined, but they did find a way to do some business together. Nintendo of America may have had a famously strict licensing agreement, but their contract applied only to America. In Japan, however, Acclaim had no exclusive relationship with Nintendo, and Nakayama was able to license games like WWF WrestleMania from Fischbach's company. As their business relationship blossomed, so did their friendship. When Fischbach traveled to Japan every other month, he could always count on two things: dinner with Nakayama, and that at some point during that dinner Nakayama would make a plea for Acclaim to publish games for Sega. At first it was cute and flattering, like a naive middle school girl asking out the high school quarterback, but now that girl was all grown up and a knockout. Fischbach didn't believe that anyone could effectively cut into

Nintendo's market, but Sega of America was making major moves and could not be ignored. Kalinske had changed the culture over there, Sonic was speeding through stores across America, and Sega's grassroots marketing efforts were so damn aggressive that they were impossible not to admire. For these reasons, the last time Fischbach had seen Nakayama and once again been asked to publish games for Sega, he'd responded that he would at least consider the idea. Acclaim was a public company, after all, and at the end of the day he answered to the shareholders. If releasing games for Nintendo and Sega would increase revenues, then that was what Acclaim ought to do. The question was whether, after Acclaim began working with Sega, Arakawa and Lincoln would continue to allow them to release games for Nintendo. If they didn't, then Acclaim would have to make a choice, and back then those kinds of choices always ended in favor of Nintendo.

Fischbach and Scoroposki reached Germany in the afternoon and, after checking in, had thought about bringing all of this up at the hotel where all of them were staying. But, feeling anxious, they decided to postpone the conversation until dinner, for which they had reservations at a gorgeous place about twenty miles outside the city. The five of them were unable to all fit into one cab, so Fischbach and Scoroposki hopped into the first taxi and then Mr. and Mrs. Arakawa and Howard Lincoln jumped into the next, following them to their destination. As such, Fischbach and Scoroposki found themselves in the backseat of a cream-colored Mercedes taxicab, coming up with potential segues to bring up the issue at hand. In the middle of the last-minute brainstorming session, however, their cab got into a horrible accident.

The Arakawas and Howard Lincoln saw it all happen before their eyes. The taxi in front of them, carrying Fischbach and Scoroposki, was making a left-hand turn, but the taxi driver must not have noticed the oncoming car, which broadsided the taxi, crushing it and sending it into a nearby ditch. The occupants of the taxi had to be dead—there was no doubt in the minds of anyone who had seen the accident.

Except that moments later, Fischbach, Scoroposki, and the driver stumbled out of the taxi. They were dinged up, with cuts and bruises, but had not suffered anything near the damage that had been done to the car, which was totaled. The folks from Nintendo hugged their friends tightly, partly out of joy and partly to confirm that they weren't seeing ghosts. Somehow Fischbach and Scoroposki had survived.

After that, the five of them crammed into one taxi and continued on to the restaurant. With top shelf Scotch in their hands, they celebrated life, friendship, and the pleasure of business, which was occasionally interrupted by Scoroposki finding small shards of glass in his hair.

At some point, even without a very good way to segue into it, Fischbach revealed that they would be publishing games for Sega. Lincoln and Arakawa instinctively looked to each other, and in that moment of hollow silence Fischbach braced for the blowback.

"Who cares?" Lincoln shrugged. "You guys are alive. That's all that matters!"

Arakawa nodded sincerely. "This is much more important."

The conversation then resumed with the same joy as before, all five of them partially aware that things would never be the same again. But still they drank, appreciating the incredible here and now.

"That's insane!" Kalinske declared, when Fischbach finished the story. "I wonder, do you really think the conversation would have gone differently if not for the incident?"

"I don't know," Fischbach said. "But I like to believe that some awful, awful German cabdriver unknowingly changed the history of videogames forever."

"That's a pretty bold claim," Kalinske declared.

"Yes, but have you been paying attention to this game?" Fischbach replied. "Trust me when I say that Mortal Kombat is going to change everything."

Kalinske agreed, and if Sega really stood any chance of knocking down Nintendo's door they needed to license the game. The copious amounts of blood and violence were certainly not ideal, but Kaliske had already made peace with this issue. "I want the exclusive rights to this. What's it going to take?"

"Actually," Fischbach began, "we were thinking about doing this one differently."

"Changing it up," Scoroposki added, but it didn't alleviate Kalinske's confusion.

"We're going to license the game to both you and Nintendo," Fischbach explained, "to release on both system the same day. That is, if you still want it."

The situation wasn't quite what he wanted, but as he thought about it more it seemed almost for the best. For years, Sega and Nintendo had been going back and forth, arguing over numbers, technical specifications and sell-in versus sell-through. Finally, though, there would be something to settle the debate once and for all. One game released on both systems at the same time on the same day. Whoever won the battle would not only get bragging rights, but likely also sow up a majority of the market share. Just like the fighters in the Mortal Kombat, one company would be left standing when it was all said and done. "Of course we still want it," Kalinske said, as the adrenaline rushed through his veins. "May the best man win, right?"

39.

AND AWAY WE GO!

On November 19, 1992, Tom Kalinske addressed Sega of America's nearly three hundred employees from the loading dock of the company's distribution warehouse in Hayward, California. There, with an unquenchable grin on his face, Kalinske delivered a passionate speech that Ellen Beth Van Buskirk had prepared for this momentous occasion.

TOM KALINSKE

Today we celebrate a tremendous Sega effort!

The applause cut off Kalinske right away; there was just no way to repress it any longer. Plus, it looked good for the cameras. Although this event had been designed as a thank-you to Sega's employees, it also cleverly doubled as a PR opportunity to officially kick Sonic 2sday into motion. It was the videogame world's equivalent of Santa loading his sleigh.

TOM KALINSKE

After fourteen months of development, marketing, and operational achievement, we are here to commemorate the first shipment of *Sonic The Hedgehog 2*.

This wasn't just fourteen months of work, but years of effort from hundreds of employees. Some were right here in the crowd, some were across the ocean, and others had planted the seeds once upon a time but were long gone by now. It was the ultimate team effort, and that's why everyone in attendance was wearing the same thing: a hooded sweatshirt with the words "We did it!" scrolling up the sleeve.

TOM KALINSKE

We have all grown to love this feisty little hedgehog . . .

At the heart of everything was Sonic The Hedgehog, Sega's little engine that could, did, and was ready to do again. According to a recent national survey, Sonic was now a more recognizable American icon than Mario, MC Hammer, and even Mickey Mouse.

TOM KALINSKE

And with this sequel, we can be proud of the team accomplishment, which has produced the very best videogame of all time.

For months, Sega of America had been guarding this new game as if it were enriched uranium, which in terms of the war with Nintendo was exactly what it was. Development was kept strictly under wraps, minus the four screenshots each month that Nilsen would select to share with the public.

TOM KALINSKE

Let me share with you what the national news media are saying about our blue spiked friend.

But that Hollywood-esque supersecret teaser mentality changed on October 8, with the beginning of Sega's Sonic 2 Store Tour. Since then, 345 retail locations (and three air force bases) in sixteen markets gave customers an exclusive opportunity to play the game.

TOM KALINSKE

"To watch Sonic kick into overdrive, or be hurled through a pneumatic tube at what seems like the speed of sound, is really amazing." *—Associated Press*

"What's faster than a speeding bullet, stronger than a locomotive and able to leap tall buildings with a single bound? Well, him too, but we're talking about Sonic The Hedgehog." *—GamePro*

"Not enough adjectives describe this game. . . . Faster, bigger, wilder and even more fun. Welcome back Sonic! Missed ya!" *—Richmond Times-Dispatch*

"Sonic 2 is, without a doubt, going to be one of the greatest action carts for Genesis." *—Electronic Gaming Monthly*

During the Sonic 2 Store Tour, 27,386 people played the game, 54 percent of them kids, 36 percent teens, and 8 percent adults. An additional 24,027 people stopped by to ask questions and watch others play. At the stores, three of the most frequent comments were "The new game is awesome," "I want this for Christmas," and "Sega is better than Nintendo."

TOM KALINSKE

Sonic has captured the hearts and minds of gamers and fans everywhere. I thank all of you for believing in Sonic and giving him better than the best you had to offer!

Once again, Kalinske was happy to be cut off by an explosion of applause. Normally, when giving speeches, it was his job to inspire the crowd, but today that dynamic felt reversed. And the best part was Sega's success, and that which would follow, was about more than just Sonic. As Peter Main had pointed out years earlier, the name of the game was the game, and Sega was not only making great ones, but games that other companies wouldn't have dared to make—and they were becoming bestsellers. There was no better example than Ecco the Dolphin, which had been greenlit at an Italian restaurant years earlier and now was roping in a female audience of gamers.

TOM KALINSKE

Each of you are responsible for making Sonic 2 the biggest selling videogame cartridge of all time. We are going to sell 2 million games here in the U.S. and another 2.5 million worldwide. In fact, we have already presold 20 percent of that volume—a phenomenon never seen before in the industry.

This second sentence was staggering, but it was the first that truly told the story. Without every single one of Sega's employees acting in exactly the way that they had, Sega's grand experiment would not be where it was today. Every little action mattered; every big idea had to be made real by the entire team.

TOM KALINSKE

What makes this cart even more special is that Sonic 2 sales will contribute more than $500,000 toward the Sega Charitable Youth Education and Health Foundation. I'm real proud of that achievement too.

And somewhere out there so was Anique, whose unique smile was no doubt lighting up the heavens.

TOM KALINSKE

I now want to recognize a few of the key people who went to the proverbial Mobius and back to create Sonic 2. Please raise your hands as I call out your name.

Sega's success was such an A-to-Z team effort that it felt almost cruel to single out only a handful of individuals, but it would have been crueler not to shine a momentary spotlight on those whose every action personified the Next Level.

TOM KALINSKE

Shinobu Toyoda.

To the man with legs long enough to straddle Japan and the United States.

TOM KALINSKE

Yuji Naka
Masaharu Yoshii
Yasushi Yamaguchi
Yutaka Sugano
All the STI artists and programmers

To those who dared to dream up this game.

TOM KALINSKE

Al Nilsen
Bob Harris
France Tantiado
Ellen Beth Van Buskirk

To those who made this more than just a game.

TOM KALINSKE

Richard Burns
Len Ciciretto and the sales group

To those who went door-to-door, selling Sega's revolution.

TOM KALINSKE

Glen Weisman
Laila Atassi
Joe Walkington
Roger Rambeau
Sandy Tallerico and the warehouse bunch

To those who turned operations into an art form.

TOM KALINSKE

Steve Apour, Ben Szymkowiak, and our valiant testers

To those producers, developers, and testers who could see the world in pixels.

TOM KALINSKE

Thank you again for your vision and hard work!

Amongst the raucous applause and the humble bows, there was one name notably missing from the list: Madeline Schroeder. In July, the mother of Sonic had left Sega, along with Jude Lange, to start their own software developer, Crystal Dynamics. This unexpected move signified an irreversible maturation for Sega, which had now earned enough clout to be viewed as a launching pad for bold career moves. This changing of the guard signaled that the company would likely experience an exit of many of those who had made Sega special, and an influx of new employees hoping to make a dent.

TOM KALINSKE

Speaking of hard work, we have a little more left to do before Sonic makes his way to the thirteen thousand retailer doors nationwide.

It had been two years since Kalinske took over, and in that time so much had changed. The Genesis was now in over thirteen thousand stores (up from four thousand), the Sega of America team now had nearly three hundred employees (up from fifty), and in 1992 the company's sales quadrupled to over $500 million.

TOM KALINSKE

We now need to ship all this great product. And we need to ship it to arrive all on November 24, Sonic 2sday, to each and every retailer.

There were two major events scheduled for Sonic 2sday. One would be held at Toys "R" Us in New York City's Herald Square, and the other would be at the store's location in Burbank, California. Teen celebrities had been confirmed for both events.

TOM KALINSKE

We're calling it the "Great Sonic Airlift," and we've enlisted the support of Emery worldwide to undertake this epic task. Now, I'd like to ask Mr. James Schutzenhoffer, the vice president of Emery's West Coast operations, and Mr. Al Nilsen, Sega's group marketing director, the man who created and directed the awesome marketing campaign behind Sonic 2's launch, to join me up here to get out the first shipment of Sonic The Hedgehog 2 cartridges.

Schutzenhoffer, blushing from the attention, stumbled over to Kalinske, follow by Nilsen, who needed a second to accept that this surreal-feeling moment was actually very real.

TOM KALINSKE

Yes, Sonic . . . you and Tails can join too.

As Sonic and Tails danced up to the loading dock, Kalinske held up a sealed copy of Sonic The Hedgehog 2. It was amazing, he thought, that Sonic had been around for less than two years but already felt as real as Mickey Mouse. Tails, of course, wasn't at that level yet, but he was on his way. And so would the next character, and the character after that, as long as Sega kept their eyes on the prize. It was even more amazing that his son, Brandon, would grow up in a world where Sonic would be as timeless as any other two-dimensional character. Even after Kalinske one day left this world, Sonic would be there, in some strange way, to watch over his children. No matter what else he did in life, there would always be Barbie and Sonic. But since Tom had no further influence over Barbie, it was all up to Sonic now.

TOM KALINSKE

This box commemorates the beginning of the Great Sonic Airlift. Congratulations, Sonic, and congratulations, Sega!

Photographs were taken, T-shirts were given out, and congratulatory hugs took place left and right. It was a wonderful occasion that would be remembered as even more golden when, just one week later, the American home console market would be split evenly between Sega and Nintendo.

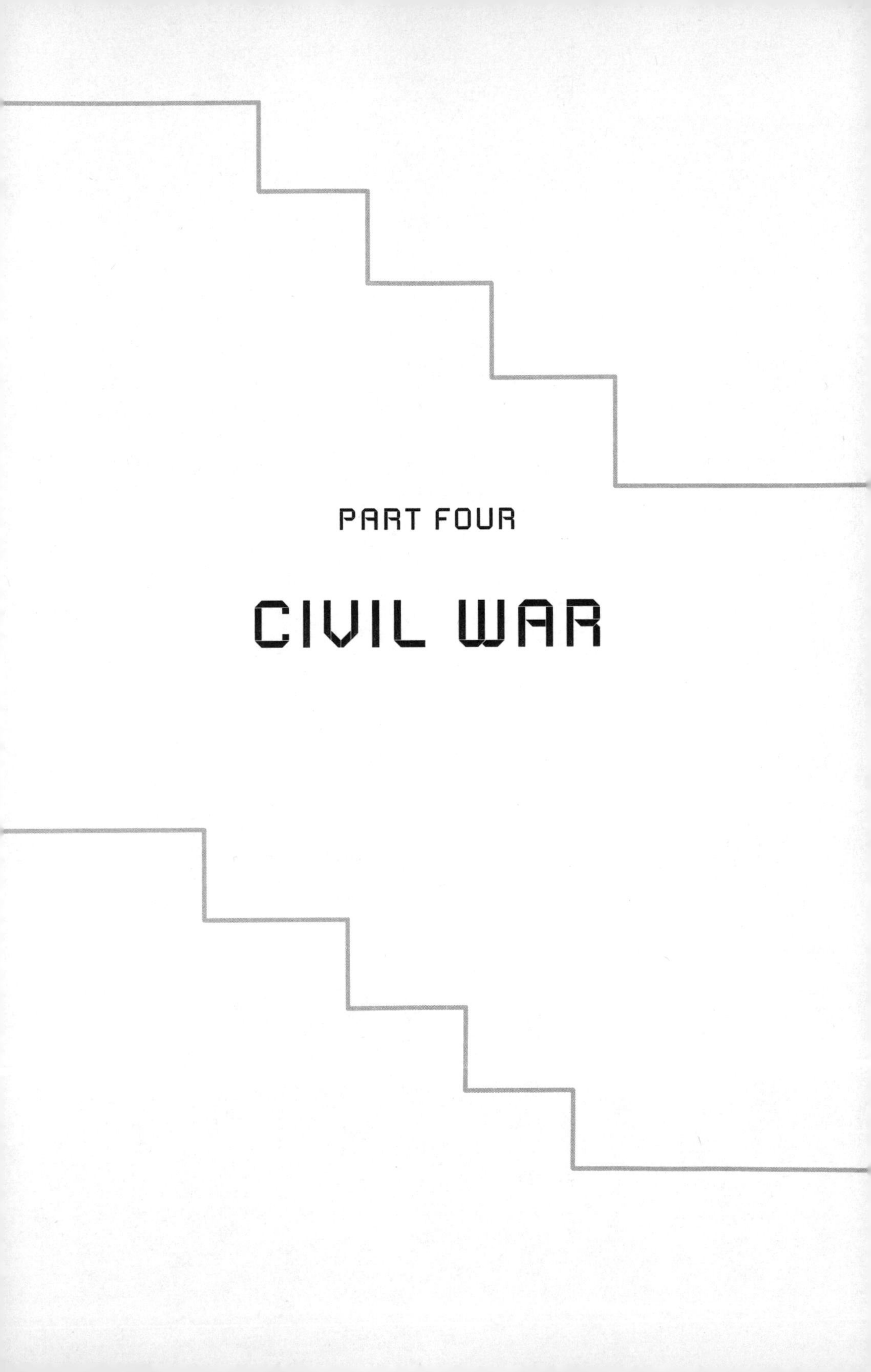

PART FOUR

CIVIL WAR

40.

HOW THE GRINCH STOLE CHRISTMAS

What the heck was wrong with Sega of Japan?

Seriously, what was it about those guys? Was it that they compulsively marched to the beat of their own drum? Or, like a petulant child, did they just like banging on drums and then smiling at the cacophony? There was something going on there, something ominously bizarre, but Tom Kalinske couldn't put his finger on it, and right now he didn't have the time to sleuth around and solve the mystery. It was November 24, 1992, Sonic 2sday had finally arrived, and like Santa Claus on Christmas, he needed to stay jolly in order to bring good around the world. And there was no better place to spread this message of joy than from the place that most resembled Santa's workshop south of the North Pole: the mammoth Toys "R" Us in the heart of New York City's bustling Herald Square.

Sega had taken over the back corner of Toys "R" Us to host what looked and felt like a political rally for Sonic The Hedgehog. Sonic himself was there to greet incoming fans, friends, and journalists, as was his huggable new sidekick Tails, on hand to complement but never overshadow the hedgehog. Both characters patrolled the perimeter of this rally, ushering guests toward a small stage where MTV's Adam Curry emceed the event.

Standing behind a podium with a Hollywood-worthy poster for Sonic 2, and in front of a thirty-foot "2 Fast! 2 Cool! 2 Day!" banner, Curry revved up the crowd for this groundbreaking occasion. After building up momentum for Generation X's first unofficial holiday, he welcomed Sega's nonhedgehog man of the hour. "And now to tell you more about all the exciting details is Sega of America's president and CEO, Tom Kalinske!"

As Kalinske walked onto the stage, the applause finally drove out of his mind any further musings about Sega of Japan. "Thanks, Adam," Kalinske said, taking his place. "This is a very exciting day for all of us at Sega. Today we celebrate the official launch of Sonic The Hedgehog 2."

Shinobu Toyoda tried to pay attention to Tom's every word, but he was too overcome with pride. So when his boss said, "In the last three years the videogame business has grown by 60 percent to become a $4-billion-a-year industry," what Toyoda heard was, We did it! And when his boss said, "The reasons for this growth are largely due to Sega developing new technologies and games that make playing more exciting and more fun" what Toyoda heard was, Can you believe that we did it?

A little over one year ago, Yuji Naka had quit working for Sega, but Toyoda had found a way to keep Sonic's creator happy, and that created a feeling of pride that he did not think would ever fully go away. To the outside world, convincing Naka that Sega of America was completely different from Sega of Japan may have appeared like nothing more than linguistic camouflage, but it really was much more than that. For many years, SOA was treated like little more than SOJ's errand boy. Although this dynamic may have spewed some unnecessary condescension, it wasn't all that far from the truth. SOJ created the concepts, the characters, and the games, and all SOA had to do was market and sell them. Okay, maybe marketing and selling to the 250 million people in the United States was not such a small thing, but it was more of a task than a tactical decision. This balance of power began to shift, however, when Tom Kalinske took over. It wasn't solely because of Sega of America's new CEO, although Nakayama's trust in his prestige went a long way; it also had to do with a confluence of people, plans, and pop culture. But most important, what Kalinske and his team did actually worked, and it worked in a way that it didn't in Japan.

Those at Sega of Japan watching this play out were likely quick to call their American counterparts "power-hungry." Although that may have been true to some extent, as humans tend not to prefer subservience, a better term to describe those at Sega of America would be "success-hungry." Tom Kalinske did not take pride in lording it over SOJ's executives, nor did Al Nilsen earn his self-esteem by telling Japan's marketing team what they were doing wrong. In fact, both quite literally went the extra mile to avoid those things; Kalinske traveled to Japan every month or two (unlike his predecessor, who only once went) and Nilsen often traveled with him, as well as constantly faxing over extensive marketing strategies (a kindness that was rarely, if ever, reciprocated).

Although one could identify a small spectrum of possible motives for their behavior (kindness, respect, or just covering their asses), what made both men, as well as the rest of their SOA colleagues, always go that extra mile with SOJ was their hunger for success. In 1990 they had had a taste, in 1991 they started a food fight, and by November 1992 they were addicted. What had once seemed like a long shot was now actually working, and that only made them want to work harder because they couldn't bear the thought of losing this feeling. That's why Sega of America had fought so hard over seemingly petty details regarding Sonic (1990), that's why they had been so meticulous in their attacks on the Super Nintendo (1991), and that's why they had started developing their own games in the United States (1992). SOA was SOJ's errand boy no more, and that's why Shinobu Toyoda had been able to persuade Yuji Naka to come to San Francisco and work with Mark Cerny and his team at the Sega Technical Institute.

The cross-cultural development of Sonic 2 left a collaborative imprint on the sequel that would help position this title as the videogame world's first true blockbuster. If creators from such different nations were both pleased with this game, then surely that universal appeal would translate to many of the countries between America and Asia. From the very first demo, Toyoda had no doubt about the quality of this joint effort, but he did harbor concerns about the pace of Naka's quest for perfection. If Sega of America was serious about declaring a Sonic holiday in November, then Toyoda needed to ensure that the game's development didn't spill into December, January, or February. This concern was part of the reason that SOA's executives had been so willing to green-light Nilsen's latest so-crazy-it-might-actually-work plan; by selecting a Tuesday in November and then creating a media circus around that date, Naka would realize this was more than just an air-quote "deadline." And if Naka hadn't fully understood what was at stake, Toyoda hammered home the point in June when he personally delivered airline tickets to everyone on Naka's Sonic Team. They would all be traveling to New York for Sonic 2sday, so if they didn't want to endure the greatest embarrassment of their lives, then they had better make sure the game was done on time. And it was, which made Toyoda smirk with pride while respectfully nodding as Kalinske spoke at Toys "R" Us.

"Today we have received word," Kalinske continued, "that Sonic 2 is already off to an incredibly fast start. The game has been on store shelves in the United Kingdom less than a day and already it's sold eight hundred thousand units."

The process of coordinating this global launch had brought Sega of America and Sega of Europe (SOE) closer than ever. SOE embraced SOA's win-or-go-home attitude and set up several launch events in major cities across Europe, highlighted by a pair of media-pleasers in the United Kingdom. First, a fleet of London's famous double-decker buses were outfitted with Mega Drive systems and spent the day traveling to schools around the city. Next, Sonic hot-air balloons were released throughout the day. These events, like those that Sega of America had arranged over the years, were designed as low-cost ways to attract an audience of more than just gamers, and each was also cleverly choreographed with an eye toward presenting media stories that would write themselves (i.e., "Sonic Balloons Invade Britain!"). Although SOA tended to emphasize speed, technology, and alternative thinking, and SOE's marketing typically focused on positioning Sega as Europe's luxury videogame brand (the Aston Martin to Nintendo's Vauxhall), Sonic 2sday proved that they had more in common than they originally thought, and by working together they could be more than the sum of their parts.

"And I have just spoken to our folks in Japan," Kalinske said next, conscious not to grit his teeth as he cheered Sonic 2sday's success in Asia. But it was a little hard to sport a smile, because for some reason Sega of Japan had decided that Sonic 2sday should not be on that Tuesday, as it was in every other country in the world. "Sonic is causing an equal frenzy in that country," Kalinske explained, "where the game has been available since Saturday."

Kalinske may have been forced to smile, but Nilsen couldn't help but roll his eyes. A few weeks earlier, Sega of Japan had decided that they would prefer to release Sonic 2 on Saturday, November 21. For someone trying to coordinate the world's first global launch, this sudden change of heart was frustrating, but more frustrating was the reason, or rather the lack thereof. After three years of working at SOA, Nilsen had come to expect a certain level of veiled capriciousness from SOJ (he often found out about their real marketing and product development plans by poking through the shipment of Japanese videogames that Shinobu Toyoda received each week), but to do this with Sonic 2sday, and to do it so close to the big day . . . that felt almost malicious. Someone driven by power might have threatened to dismantle Sonic 2sday everywhere if Japan did not conform, but since Nilsen and his SOA colleagues were instead driven by success, they ignored SOJ's odd decision and focused on finding a way to make it fit into the larger narrative. Returning to the blockbuster analogy, they chose to spin Japan's Saturday launch as an exclusive sneak preview.

For Kalinske, peddling this version of the truth to the media was no problem, but selling it to his employees was another story. That was the disheartening part, to have to look them in the eye and optimistically explain how this was actually a good thing. Sure, it would have been a lot easier to share in his employees' collective groan, but the problem with managing that way is that although the bad news is easier to swallow, it also inherently lingers longer. That's why Kalinske always walked around the office with a smile, saving the frowns for his own time.

"We developed Sonic 2 to be the fastest videogame in the world as well as the coolest," Kalinske continued. "Sonic has a new attitude, a new friend, and a whole set of new maneuvers. But rather than me tell you how gnarly they are, I've invited some pretty famous Sonic experts with us today who are anxious to give you their review of Sonic 2 and to describe the game features in greater detail. First up, from the NBC-TV sitcom *Saved by the Bell*, is Dustin Diamond. Dustin, come on up here."

Dustin Diamond ran onstage energetically, and when the moment was right, Kalinske stepped aside and let the geeky television star enjoy the limelight.

In slow motion, Dustin Diamond's smile disappears, he stops waving to the crowd, and then he runs backward off the stage as Tom Kalinske moves back to the podium.

"Keep rewinding for a few seconds," Ellen Van Buskirk said as she, Brenda Lynch, and a film editor reviewed footage of the presentation from a small news van parked in Herald Square. "I want to go back to when Tom starts talking about Japan."

"Less than a day and already it's sold eight hundred thousand units," Kalinske said on the video when the editor cued up the spot. "And I have just spoken to our folks in Japan. Sonic is causing an equal frenzy in that country."

"All right, stop it there," Van Buskirk said, pausing the tape. "Do we want to use that as the lead-in for Dai's speech?"

As requested, the film crew hired in Japan had given Lynch and Van Buskirk a lot of B-roll featuring excited kids buying Sonic 2. In addition to capturing that frenzy, they had also wanted some sound bites, which they received in the form of an interview with SOJ's Dai Sakurai at a special in-store videogame competition.

"It's a good idea," Lynch said, "but I don't think we need it. As soon as Dai opens his mouth, it's clear that we're not in Kansas anymore, and, all things being equal, I'd rather come in on the Sonic ring sound playing over Japanese teens shouting."

"Good point," Van Buskirk said. "All right, scratch that. Let's go back to Dustin Diamond playing the game and then get some stuff with the rest of the celebrities."

Lynch and Van Buskirk were racing to finish what they believed to be the cornerstone of Sonic 2sday: the international video news release (VNR). After the rally ended, they had only thirty minutes to cut together the VNR before it would be beamed via two satellites to every television newsroom in the United States. Outside of the United States, it would also be sent to every television news outlet in Europe, North Africa, and the Middle East via the Brightstar satellite, as well as to all of Asia through the Pan Am satellite.

Because time was very much of the essence, they had crafted an overview of what the video should look like weeks in advance. This twenty-six-point outline included things like "Opening scene of banners and window displays outside Toys 'R' Us at Herald Square" (#3 on the list), "Sound bite in English and then Japanese of a senior Sega executive talking about the excitement" (#14), and "Zoom out from London scene setter of the double-decker bus crossing the Thames with Parliament in the background" (#22).

Throughout the day, Lynch and Van Buskirk had been filling in most of their blanks with the footage they'd been receiving from around the world, but the whole VNR hinged on Tom Kalinske's keynote speech at Toys "R" Us in New York, and it was almost time to send his words around the globe.

"How's it going?" Al Nilsen asked, poking his head into the van.

"Fast," Lynch said as they scrolled through footage of Diamond playing Sonic 2.

"That's an understatement," Van Buskirk added, pointing to the screen.

"Good," Nilsen said, squeezing into the van. It was barely big enough for three, and certainly not big enough for a fourth of Nilsen's size, but in times of great stress the Sega team worked as one, and under such circumstances things like time, space, and physics momentarily seemed not to matter very much. "What do you have left?"

"Number five," Lynch said.

"Fast and furious action of the teen stars playing games," Van Buskirk clarified.

Building off the relationships forged at the *Sega Star Kid Challenge* television event, Sega had continued to work closely with many of Hollywood's youngest stars. In addition to Dustin Diamond, the New York event featured a handful of other teen celebrities, including Joey Lawrence (*Blossom*), Jonathan Taylor Thomas (*Home Improvement*), and Michael Cade (*California Dreams*). Getting these stars to show up was easy—it was about maintaining a relationship and paying their appearance fee—but getting them to show up and not look foolish playing videogames . . . well, that was a different story. To teach them how to expertly play Sonic 2 (or at least figure out which ones were better off just watching from the sidelines), Sega had hosted what amounted to a videogame tutoring session the day before. Luckily, like most members of their generation, many of them were able to quickly learn how to play the game. A practice session may seem trivial, but it was important to Sega that these young celebrities give the impression that they really did love videogames and weren't just heartthrobs for hire.

"With Sega's Blast Processing capabilities," Diamond said while playing the game beside Kalinske onstage, "the galaxy's most famous blue dude with attitude has all-new worlds, zones, music, and maneuvers, like the corkscrew, spin-dash, Sonic shield, and an unreal blue tube roller coaster run for head-on, gold-ring-chasing fun."

"Make sure you start with the Blast Processing," Lynch instructed the editor.

"And let's mix that with some Joey Lawrence and call it a day!" Van Buskirk said. "At least until the West Coast feed."

In addition to the rally at Herald Square, Sega was also holding a similar event at the Toys "R" Us in Burbank, California. This West Coast event would be almost identical to the one in New York, except in Burbank the keynote speech would be delivered by Ed Volkwein, would be emceed by Power 106 FM's George McFly, and would include a different roster of teen celebrities headlined by *Saved by the Bell*'s Mario Lopez. But that wasn't for another few hours; all that mattered right now was New York.

As Nilsen watched, applauded, and then approved the final video, Van Buskirk's mind took a short respite from traveling at a million miles per hour. Wow, she thought. Just wow. And like everyone else at Sega of America, she enjoyed a brief but exhilarating we-did-it moment before getting back to work and preparing for the next Next Level.

All this talk of the Next Level was becoming a self-fulfilling prophecy. From a pure marketing standpoint, Bill White admired this very much, but from the standpoint of a vice president in Nintendo's marketing department, he was much less enthusiastic. For years Nintendo had been foolishly allowing Sega to steal slivers of the market, but now they'd done more than that; they'd managed to steal Christmas. Well, not Christmas exactly—that was still a few weeks away—but they'd stolen something much more important than the anniversary of an immaculate conception and miraculous birth. Sega had stolen the Christmas shopping season.

For the past five years, Nintendo had been marking its holiday territory with something called the North Pole Poll. What they did was hire the Gallup Organization, famous for their political polling data, and pay them to independently conduct a telephone survey of children around the country ages seven to sixteen and ask about what they wanted for Christmas. Every year the poll had been conducted, the most coveted item was always a videogame system. And 1992 was no different, with 63 percent of kids starting off their Christmas lists with a plea for the newest videogame system. What kind of videogame system? Ha—well, the survey didn't say, because in previous years the answer had been obvious, and this year Nintendo was too afraid to ask. So instead of having Gallup ask the kids whether they preferred Sega or Nintendo, a separate survey of the country's retailers (the ones Nintendo was closest with) was conducted, and lo and behold, it forecasted that the Super Nintendo Entertainment System would be this year's hottest seller. And what's this? Amazingly, those same retailers forecasted that the best-selling game would be Street Fighter II!

Sham or not, it didn't really matter. The purpose of the North Pole Poll (and this year's unsubstantiated survey of retailers) was to get the story repeated over and over on the air (along with the words "survey commissioned by Nintendo" and, ideally, some photos of the SNES) until it became a self-fulfilling prophecy. That had always been how it worked, but this year Sega had crashed the party.

On November 27, results of the North Pole Poll went out to every media outlet around the country and, like every year before, most ran the story. But this year, when the anchors talked about how every kid wanted Santa to bring them a new videogame system, they used clips from the video news release that Sega had made for the new Sonic game. "Sixty-three percent of responding children said they really want a videogame system for the holidays," they said over footage of Sonic hot-air balloons. "Fifty-four percent want a portable videogame system" was the statistic that went out over shots of teen celebrities playing Sega. And "Forty-three percent want videogames" was delivered over Tom Kalinske giving a speech at Toys "R" Us.

The timing of Sonic 2sday was no accident, and the way that Sega got that video played all around the globe . . . White would have called it brilliant if it weren't so damn infuriating. There was a time when Nintendo could have crushed the competition, but Arakawa had wanted to avoid stooping to that level. Well, now that both companies were virtually tied for market share, meeting Sega at their level would no longer entail stooping. This was certainly not what Bill White wanted for the holidays, but if this was what it took for Arakawa to wake up and start fighting back, then maybe the North Pole Poll debacle was a Christmas miracle after all.

A fairy-tale future seemed to lie ahead for Nintendo's competitor, but beneath Sega's many mattresses of success was a nagging pea-shaped question.

41.

FUGU

"What the heck is wrong with Sega of Japan?" Tom Kalinske asked with a slow but determined shake of the head. "Seriously, what is it with those guys?"

Kalinske was in Japan, having breakfast at Tokyo's Le Meridien Pacific hotel with his favorite inside source at SOJ: Mike Fischer, one of the few, and certainly the highest-ranking, of the Americans working at Sega of Japan. Although Fischer was only twenty-seven years old, he had impressed Nakayama with his sitcom-style wit, unpretentious intellectualism, and infectious enthusiasm. Personality-wise, he had a lot in common with a young Tom Kalinske, but where Kalinske used those gifts publicly to sway the masses, Fischer preferred to operate behind the scenes. That's part of the reason he loved these off-the-books meetings with Sega of America's president. Normally, the two of them swapped pieces of information with the excitement of kids trading baseball cards. And that was how it had started today, until Kalinske interrupted the typical haggling session by asking for Fischer's most valuable card.

"What's wrong with Sega of Japan?" Fischer repeated to himself. For a moment, he politely tried to avoid laughing at the inquiry, but ultimately couldn't stop himself from cracking up. "Jeez, Tom!" he replied, rogue giggles slipping out between his words. "How on earth were you able to sneak such a loaded question past those goons at airport security?"

"All right," Kalinske admitted, "I suppose I could have toned it down a notch or two. But then again, isn't the whole beauty of our breakfasts together supposed to be that we get to stop playing politics for fifteen minutes?"

"This is true," Fischer replied with a nod so earnest it made the cup of coffee in his hand bob. "Plus for me there's the added joy of, you know, actually getting to speak in my native tongue. Oh, English, how I've missed thee."

Fischer had been missing the idiosyncrasies of English since 1988. After graduating from a small engineering college, he yearned for something bigger, so he sold his Honda scooter and Cannondale mountain bike to buy a one-way ticket to Japan. For two years he bounced around the country teaching English before getting a job in Sega's expanding consumer department. The company was becoming serious about growing their business abroad, so Nakayama wanted to bring in a crop of young non-Japanese go-getters to help facilitate that growth. After proving himself to be the most capable of these new recruits, Fischer was promoted to manage communications between the parent company and Sega's subsidiary in America. Like many of his colleagues at SOJ, Fischer didn't have particularly high hopes for SOA, but his barometer was reset when Tom Kalinske took over a few months later.

Few outside the toy industry had ever heard of Kalinske, but by a strange stroke of coincidence Mike Fischer happened to be in this exclusive minority. In college, Fischer had written his thesis on the econometrics of the modern toy industry, which basically meant he spent his senior year studying the innovative strategies of Stephen Hassenfeld (Hasbro), Bruce L. Stein (Kenner), and of course Tom Kalinske (Mattel). Through this research, Fischer knew better than anyone what Sega was getting in Kalinske, and from that moment forward, made a commitment to ensure that Sega of America's new CEO was given every opportunity to succeed. Make no mistake, he was first and foremost loyal to Sega of Japan, but he sincerely believed that what was best for SOA was best for SOJ, and so he had no reservations about stacking the deck when necessary. Sometimes that meant dropping hints about games in development, sometimes that meant simply warning SOA when Nakayama was in a bad mood, and other times, like today, it meant meeting Kalinske for breakfast to make sure that he entered the apiary with the right kind of bee suit.

"Fine, fine," Fischer replied playfully. "Let's have an honest discussion.

Rid yourself of the weight of that golden crown and I'll set aside this pesky jester's hat. What is it, specifically, that you'd like to know?"

"Mike, I've been with Sega for over two years now," Kalinske said. "I honestly could not be happier with the work that our team has done—we've really started to turn this thing around. But I get the strange feeling that the more successful we are in America, the less pleased they are in Japan."

"That's ridiculous!" Fischer exclaimed.

"Is it?" Kalinske asked sincerely.

"No, it's not ridiculous," Fischer admitted, "but it should be ridiculous!"

"I know!" Kalinske said, shaking his head. A sad, strange silence followed, which both men chose to fill by picking at their omelets.

"So what's the problem here?" Kalinske said finally. "Is it jealousy? We have 50 percent of the market and they have less than 15 percent?"

"I'd like to pause for a moment to remind you that I am one of the 'they.'"

"Yes," Kalinske said, "but whatever this attitude is, I feel it from you too."

Fischer considered. "I don't think 'jealousy' is the right word."

"Then tell me what the right word is."

The younger man thought about this for a little while. "When I was a kid," he began, "there were a few times when I wanted to do something, something that seemed completely harmless, but for whatever reason my parents wouldn't let me do it. I would ask them why not—why couldn't I go play football with the neighbors, or why couldn't I sleep over at this friend's house—and the only explanation they provided was, 'Because I'm the parent and you're the child.'"

"My parents used to say the same thing," Kalinske added.

"So you understand, then."

"No, I still don't. Because even though my parents said that, they never arbitrarily changed the date of Sonic 2sday! Nor did they ignore faxes, fail to return phone calls, and turn up their nose at my marketing plans even though their own were simply not working."

"Your parents had marketing plans?" Fischer joked.

Kalinske sighed and glanced toward the lobby for any sign of Nilsen or Toyoda. They had flown with him to Japan, Nilsen as part of his new job responsibilities and Toyoda to help Kalinske pitch the joint hardware venture with Sony. They'd head over to SOJ together, but Nilsen and Toyoda must still have been upstairs getting ready.

"Who came with you?" Fischer asked. "Shinobu and Paul?"

Kalinske shook his head. "Shinobu and Al."

"Ah, Al. How is he enjoying the promotion?"

"Too early to say," Kalinske answered. Following the success of Sonic 2sday, Nilsen became SOA's group director of global marketing, a new position created for the dual purposes of allowing him to continue to think big (but now on a global scale) and encouraging him to serve as something of a marketing liaison between America, Europe, and Japan. Plus, in light of the growing fragility in relations between America and Japan, it seemed like a critical time to have someone Kalinske trusted carrying the world on his shoulders. With Nilsen now bouncing from country to country, Diane Adair was promoted to take over marketing for the Genesis. The timing of this shake-up also fit nicely with SOA's plans to double in size by the end of 1993. Although Nilsen had done an excellent job managing his team for Sonic 2sday, there was no denying the unique physics of Nilsen's approach to marketing, and Kalinske wanted him to keep thinking bigger and not sweating the small stuff. Which worked out nicely, because Adair's meticulous style and teacher-like sensibility were well suited to reorganize the marketing department and bring new blood into the fold. And while Kalinske knew that it would be weird not to have Nilsen around the office as much as before, the potential of his replicating the magic of Sonic 2sday was too great to ignore. It was time to take the genie out of the bottle. "We'll see," Kalinske said. "But I'm sincerely hoping that Al can teach Japan's marketing team a thing or two about what a goldmine Sonic can be for them."

"He better," Fischer replied. "We need the money to pay the sea people, don't we?"

"Don't even get me started," Kalinske said, with a roll of the eye. One of Goodby's latest ads, which satirized those grainy Jacques Costeau films, looked and felt like all those ocean explorer videos, except in the Sega commercial everyone on the crew was crammed in a room playing Ecco the Dolphin. It was a great spot: it totally captured the Sega spirit but also captured the attention of the Cousteau Society, which was suing Sega and Goodby for three million dollars. "I hope they use the money to buy a sense of humor," Kalinske said with a cadence indicating he was just about ready to leave.

"Hey," Fischer said, suddenly smiling at some kind of memory. "Did I ever tell you where the idea of the hedgehog came from?"

"I thought it was the mascot contest," Kalinske guessed.

"Well, yes, but where the idea for the contest came from."

"I don't believe so."

Fischer leaned in. "This is one of my favorite stories. Naka-san gets all the credit, because he designed the game and, you know, he's this powerful personality, but because of that I think Oshima-san gets lost in the shuffle. So one time I went up to him and asked Sonic's true creator where that spark of an idea had come from. He's really shy, this unassuming kid, and I expected him to say something like 'It was a team effort,' or 'It was just one of those things,' but he smiles really small and he says: 'I just put Felix the Cat on the body of Mickey Mouse.'"

"Oh, come on," Kalinske said, "that can't be true."

"It is, I swear," Fischer chortled. "And as shocked as you are to find this out, Oshima-san is more shocked that nobody ever called him out on this."

After the two men laughed about how their lives had changed because of a twice-plagiarized creation, Fischer looked at his watch and adopted a less frivolous tone. "Look, I don't think that consistently being wrong inherently makes someone right. But I'm curious why you are asking me this now. The examples you mentioned, Sonic 2sday aside, have been the case for some time, no?"

It was a good question, and Kalinske could choose to answer it with varying degrees of frankness. "I'm asking," Kalinske said, with some coyness, "because I want to ask SOJ to do something, and I need them to say yes."

Fischer nodded. "I assume that if you wanted to give me any more information you would have, so I won't ask, but it's hard for me to advise you on this without knowing what you have in mind."

Kalinske looked across the table and considered how much to reveal. It seemed only fair to fill him in, since Fischer had been so forthcoming over the years. After all, Fischer had been the one who would tell him what Nakayama had to say about SOA after Kalinske left the room, and also the one who told him which days Nakayama was angry and carrying a golf club around the office like Al Capone with a baseball bat. "I want us to work with Sony," Kalinske said quietly, "on a next-generation console."

Fischer was momentarily taken aback, but very quickly the idea sank in. It made a lot of sense. "How far along is this?"

"The guys who did the hardware for that failed Nintendo venture are working out of Sony Music now. With assistance from Olaf and Mickey, we seem to have their approval. Now it's our turn."

"What does Joe Miller have to say about the specs?"

"Nothing yet," Kalinske said with a wave of his hand. "We're keeping this one close to the vest until it becomes real."

Fischer nodded, his cheeks flush with enthusiasm. "This could be huge, Tom."

"Exactly."

"We split the R&D," Fischer said, speaking softly to himself, "use their manufacturing capabilities, combine our distribution networks, and—"

"Crush Nintendo," Kalinske finished in a whisper.

"That seems highly likely," Fischer agreed. "And I have to imagine that my colleagues over here will see this as our best chance of finally getting a bigger share of the Japanese market."

"Yes, well, I'd like to do more than imagine," Kalinske said. "I want to make sure that politics don't get in the way of this."

"Me too," Fischer agreed, racking his brain for suggestions. "Well, as you know, Sato's team has already made great strides on a 32-bit system. So whether it's true or not, make sure that Sato doesn't feel like he's wasted his time. You know, tell him that Sony wants to defer to Sega's expertise. It might also be worth appealing to the arcade guys, so they can keep R&D excited on the software front."

"Got it," Kalinske said, mentally revising his plans for the day. "But what about Nakayama-san? Do you think he understands everything that we're doing in America? I hope he doesn't think we're too crazy."

"Are you serious?" Fischer asked, practically doing a double take. Kalinske was so intuitive in so many ways, but for some reason he failed to see how Nakayama really felt about him. "Tom, Nakayama trusts you more than he trusts his own staff."

Kalinske shook his head. "That can't be right."

"Wait, you really didn't know that?"

"We have a good relationship, I suppose. But I never quite know where I stand."

"You should, you're at the top!" Fischer replied. "I mean, just last week we were in the boardroom and the producers were presenting games to Nakayama-san. After every presentation his first question was 'What does Tom think?' He had no interest unless you had some interest."

Kalinske didn't respond, but Fischer could tell that he was rather flabbergasted (and Fischer was flabbergasted that Kalinske didn't know this already). "Hey," Fischer said, "remember how earlier I told you that my folks used to say, 'We're the parents and you're the child'?"

"Yeah . . ."

"Well, now that I think about it a little more, I'm realizing that was the wrong parental analogy. The real problem is that Mom and Dad like you the best, and your siblings are a little sick of hearing about their favorite son."

Fragments from the conversation with Fischer rattled around Kalinske's mind as he, Nilsen, and Toyoda were given a tour of Sega's Japanese R&D lab. This was always one of the best parts of visiting SOJ, the part that felt a bit like rushing toward a just-smashed piñata, but on recent trips there seemed to be fewer pieces of candy inside.

"Is it just me," Kalinske whispered to Nilsen and Toyoda as their hosts stepped several paces ahead, "or are they showing us less and less these days?"

Toyoda shrugged, but Nilsen had an answer. "Both, actually."

Kalinske looked puzzled at this response. "How can it be both?"

Nilsen quickly glanced over his shoulder and then explained that what SOJ presented changed with each visitor. So when the CEO of SOA was in town, they only wanted to show off the very best, hiding anything that might get nixed in its developmental phase.

Kalinske lightly shook his head. "Well that's kind of dumb, don't you think?"

"Don't worry," Nilsen added. "They'll show me the real stuff later today, and I'll report back to you guys."

Before Kalinske could muster up what would surely be a sarcastic response, Nakayama arrived to greet the team from Sega of America. After trading formalities, they resumed the R&D team's tour. As they did, Nakayama lingered a few steps behind the group from America, more interested in observing their reactions to the products than in the products themselves.

The item that Sega's R&D team appeared to be most excited about landed somewhere on the spectrum between hardware and peripherals. It was a 32-bit upgrade for the Genesis, something that connected to the 16-bit console and made it twice as powerful. "Would it play the regular Genesis cartridges," Kalinske wondered, "or would this require a different type of software?"

The R&D team proudly explained that it would require new and much more powerful software to play in 32 bits. But most important, Sega would be the first company to reach the next frontier, as they'd done with 16 bits. Rarely did Sega of Japan's employees smile, but in this case they did, which made it that much more awkward when Kalinske, Nilsen, and Toyoda offered little more than a thanks-anyway smile. The idea of a supercharged add-on was fine—that's essentially what the Sega CD was, after all—but if they were going to make a 32-bit system, then they might as well just make a new console and not just use the old one as little more than a battery. Kalinske could feel the waves of subtle resentment emanating from the SOJ research team, but better to endure that than head down a multimillion-dollar rabbit hole that would lead nowhere.

But that negative sentiment immediately evaporated when Kalinske was shown the Pico, a lunchbox-sized, plastic-coated portable device that would be released in Japan during the summer of 1993. The Pico looked like a colorful cross between a storybook, a Game Gear, and an Etch-a-Sketch. It came equipped with a "magic pen" that allowed children to enter the world of their entertainment. Much the way he'd felt when he first saw the Genesis, Kalinske was absolutely blown away by the potential of what this could become. It was the edutainment machine that Sega had been working on, a portable computer for kids that would demonstrate the educational power of videogames, while stealing away Nintendo's youngest demographic in the process. It was a marvelous intersection of education, entertainment, technology, and storytelling, and for a moment Kalinske was tempted to say, Who cares about next-generation videogames? Let's just do this and do it right! But then he remembered that he didn't have to choose just one thing. He'd built a company with enough resources to let him do all these wonderful things at once, and so after congratulating Sega's R&D team he made a comment that he hoped would be prophetic.

"If you can make this cheap enough to sell for a hundred dollars," Kalinske declared, "then I promise you this will be the number one toy of the year." The creators of the device blushed at these words, and Nakayama smiled like a truffle hunter whose pig had just sniffed out a big find.

As they all excitedly discussed constructive tweaks to the Pico, members of SOJ's marketing team came to whisk Nilsen away. "Duty calls," Nilsen explained, but on his way out he made sure to pass Kalinske and whisper prophetic words of his own. "We cannot bring over that 32-bit thing," he cautioned. "It just can't happen."

"I know," Kalinske said, bidding farewell to his new global marketing guru.

After Nilsen left, the rest of the day was spent in a series of meetings, more friendly than functional, until Kalinske and Toyoda finally discussed the Sony possibility with Nakayama and a few of his top lieutenants. Among that group was Hideki Sato, who had been Sega's director and deputy general manager of R&D since 1989, when his team brilliantly refashioned Sega's System 16 arcade board into the Mega Drive. Although Nakayama would have final approval on how Sega proceeded, it would be Sato's reaction that would inform his decision. Knowing this, and following Fischer's advice, Kalinske carefully proposed the idea of working with Sony in a complementary manner and highlighted that this would not sacrifice any of the great work that Sato's team had already done on the next generation of hardware.

The reaction to the joint venture was mixed at best. Although those from Sega of Japan saw some business sense in a joint venture, they appeared to be fundamentally resistant to the idea. Whether this mentality was spurred by corporate pride (we don't need them), creative ownership (we want to do this our way), or competitive skepticism (Sony just wants to see what we've got and steal it), Kalinske was not sure, but this certainly wasn't shaping up to be the no-brainer that he'd been hoping it would be. But true to Fischer's words, Nakayama seemed to trust Kalinske more than his own staff. Unlike in the boardroom two years ago, Kalinske wasn't given carte blanche to run with this opportunity, but Nakayama wanted Sega to very seriously explore this option and form a review team to work with Sony over the next several months. If Kalinske was right and this was a golden opportunity, then that would become evident and a wonderful partnership would be formed. But if he was wrong, then Sega had no obligation to continue working with Sony and could continue forward on their own. Sato and the other members of

R&D seemed to appreciate this resolution and harbored no noticeable ill will about moving forward in this fashion. For Kalinske, it wasn't quite a win, but given the initial feedback, it was more than he'd been expecting. Plus he was rather certain that this really was a golden opportunity, so getting the win was just a matter of time.

To celebrate this and Sega's other successes, Kalinske went out with Nakayama to the same geisha club that they had visited a couple of years earlier. Amongst the girlish giggles and swoops of sake, it was hard not to appreciate how far they had come together. And there was still so much further they could go together, a prospect that Kalinske was now more confident and excited after that day's events. Between the Pico, the partnership with Sony, and Nilsen's acceptance of a role as global marketing ambassador, Kalinske couldn't help but feel a hint of guilt for ever doubting Nakayama and those at Sega of Japan. They had their differences and certainly their occasional doubts about each other, but at the end of the day they were all part of the same family, and mutual success would make that family a big happy one.

Kalinske had decided to stop wondering what was wrong with Sega of Japan, but now Nilsen couldn't get the question out of his mind. Unlike his boss, Nilsen's query wasn't inspired by frustration so much as it was by genuine curiosity about why Sega of America and Sega of Japan had experienced such differing levels of success. They both had the same products, the same games, and the same blue hedgehog, yet with a smaller budget SOA had 50 percent of the American market whereas SOJ only had 15 percent of the Japanese market.

The easiest conclusion to reach might be that SOJ's personnel simply lacked the talent to execute, but Nilsen had spent enough time with them to know that such an assumption was heinously incorrect; the producers were virtuosos, the R&D team was relentless, and the executives were as sharp as anyone at SOA. The second-easiest conclusion to reach might be that they were just lazy, that they had the right people and the right products but simply didn't care enough to go sweep a nation. But Nilsen knew this was also terribly false: everyone he knew at SOJ worked around the clock, paid attention to the tiniest of details, and bled Sega blue. So then what accounted for the large disparity between SOA and SOJ? Cultural tastes?

Vastly different market conditions? Nintendo's parent company in Japan was that much more formidable than Arakawa's group in America?

Nilsen didn't know the answer, but he was happy to stop thinking about it for a little while when he went out to dinner with some members from SOJ's marketing team. Nilsen and about five junior executives, who in Japan are called salarymen, left the office together, happily sharing stories as they hit the streets. They walked for about five minutes toward a bustling alley with a hypnotic mix of neon signs and blinking electronic stores before the salarymen starting pointing and laughing at a restaurant on the second floor. Usually they spoke in English or tried their best to communicate with gestures, but Nilsen had no idea what was going until a few minutes later, when he was sitting on the floor of a private room in a dimly lit restaurant.

Nilsen began looking through the menu, until one of the salarymen pulled it away. They came here often, it seemed, and they would take care of all the details for their friend from America. After a long flight and a taxing day, Nilsen was more than happy to defer to his hosts for the evening, until he detected a few subtle scuffs of laughter when the food order was placed. Something was going on, and given the unusual delicacies of Japan, it was likely something weird. Octopus? Starfish? Bottlenose dolphin? Although none of these species got Nilsen's mouth watering, he was an adventurous eater and a team player, so as long as whatever they ordered for him wasn't breathing, he figured that he'd find a way to get it down. But when the food arrived, his suspicions were confirmed—there was more at stake here than just stomaching the unknown.

Fugu. That's what they'd ordered for him, and only him. As the dish was placed in front of Nilsen, the salarymen explained that this was a Japanese pufferfish. But unlike other creatures in the sea, raw fugu was lethally poisonous due to its heightened levels of tetrodotoxin. For this reason, it could be prepared only by qualified chefs who had undergone rigorous training. Even so, it was not a perfect science, and eating fugu resulted in about fifty illnesses and ten deaths each year.

With thin, sashimi-style slices of this possibly toxic dish sitting in front of him, Nilsen joined in the titillation of laughter around the table. "How do you think my parents will feel," Nilsen asked, "when they find out their son was killed by a pufferfish?"

The laughter reached a crescendo, but all eyes remained fixed on Nilsen.

Did he have the courage to tempt fate and eat the fugu on his plate, or was he just another foolish *gaijan* passing through the land of the rising sun? The expression in their eyes told Nilsen that they were very much wishing for the latter, granting them license to laugh at him and not with him, but to their dismay he lifted a pair of chopsticks.

Then, without hesitation, Nilsen picked up a slice of the treacherous fish and took a bite. Not bad; not bad at all. Amidst gasps from the salarymen, he took another bite and then another one after that. "It's actually very bland," Nilsen said, but received no response because the men from Sega of Japan were speechless with shock.

After another bite, Nilsen pushed the plate forward and asked, "Who wants to try a piece?" He scanned the table, and as he made eye contact with each one they physically recoiled at the challenge, their laughter gone and replaced with sudden dread.

"Come on," Nilsen implored, "there must be someone willing to take the risk."

There were only ten deaths each year. Ten out of hundreds of thousands, and besides, there was a hospital down the street. Nothing bad was going to happen, so it was worth trying at least one bite, but the fear in their eyes said otherwise. And that's when Nilsen finally realized the fundamental difference between Sega of America and Sega of Japan. They weren't willing to take the risk, to race Sonic against Mario or welcome a generation to the Next Level. These people were highly talented and certainly not lazy, but deep down they weren't as interested in winning as they were in not losing.

Without risk, there is no reward, and so Nilsen lifted his chopsticks and pulled the plate toward him, proudly eating every last piece of the pufferfish all by himself.

42.

BARBARIANS AT THE GATE

Going into Christmas 1992, Nintendo was experiencing the same problem facing many a gawky, pimple-populated teenager: they just weren't cool. And like that dateless teenager, doing typically cool things (like trying out for the football team, or buying a baseball team) only seemed to make things worse. The primary reason for this was Sega, whose speedy, snarky "Welcome to the Next Level" campaign had, by contrast, branded Nintendo with the dreaded scarlet U (uncool, unhip, unfit for anyone older than ten). Given the prevalence of these ads and the pop-cultural chord they had struck, Nintendo's employees were well aware of the tectonic shift in perception. But if any of them had been holding out hope that Sega was merely a passing fancy, that notion went out the window following a presentation from Market Data Corporation (MDC).

"Slowing growth," the representative from MDC said, for what seemed like the hundredth time. "The recession, of course, accounts for some of that slowing growth, but our investigation reveals other factors at play." Earlier in the year, when the tide had really begun to turn, NOA had commissioned MDC to study the changing landscape of the videogame industry. In the course of their research, they spent time with nearly eight hundred families, studying the habits of videogame players in much the same way that Goodby, Berlin & Silverstein's account planners had done not too long ago.

"Kids who own a Nintendo console still play a good deal," the representative continued. "About 2.3 hours per day, actually, based on the individuals included in our sample, but among those we surveyed they described the interaction as 'less involved.' This increasing apathy can also be used to explain the—"

"Let me guess," Peter Main interrupted. "Slowing growth?"

"Yes," the representative answered over a chirping of chuckles, evidently unashamed of his apparent predictability. "And the response we kept hearing over and over, to explain the aforementioned sentiment of apathy, is that playing Nintendo is simply quote-unquote 'not cool' anymore."

"Hey, I have an idea," Bill White interjected. "Maybe we should throw in a pair of sunglasses with every Super Nintendo bundle."

After another flurry of giggles, the representative continued. "We asked each of our interview subjects to rate each console currently available at retail in a variety of areas. Sega ranked highest in the majority of categories, including image, value, and technology," he explained. "But Nintendo scored highest in terms of fun, excitement, and game selection."

Arakawa nodded at this, not because it was an acceptable silver lining but because, to him, it was the only lining that really mattered. Gameplay was everything; all the rest was just noise. Had he voiced this sentiment, nobody in the room would have objected to his logic, but where they likely would have disagreed was on the importance of the so-called noise. Years ago, Peter Main had coined the slogan "The name of the game is the game," and he still felt that to be true, but there was no denying that Sega's sizzle had changed the rules of the game. And the biggest change, at least from Nintendo's perspective, was that Sega had turned this into a war of style vs. substance.

To anyone vaguely familiar with the videogame industry, there appeared to be no reason that Nintendo couldn't match Sega's style punch for punch; in fact, it baffled many that Nintendo had yet to publicly respond to Sega's jabs. After all, they had the personnel, the products, and the financial resources to redefine, or at least refurbish, the Nintendo brand for a new generation. This didn't have to mean shooting edgy Sega-like commercials, advertising during prime time, or promising the Next Next Level, but Nintendo could have crafted a campaign that reminded the world of all the fun that this level had to offer. Or they could have developed a speedy Sonic-killer game, in the same way that Sega had once upon a time created a Mario-killer. Or if Nintendo planned to stick with Mario, they could have slightly modernized his look (back in 1990, they had given him a nose job to make him look less stereotypically Italian; why not now equip him with torn jeans or a flaming plunger)? The point was that Sega had painted Nintendo into a corner, but there were a million ways they could have tried to escape this fate. For another example, they could market Mario Paint, an innovative art game that came with a computer-like mouse, as a high-tech art program and not a family-friendly color-by-numbers package. As Nintendo's market share dwindled, it became easy to offer up criticisms like these, but there was one major reason Nintendo didn't directly respond to how Sega had chosen to portray them: Sega was pretty much spot-on.

Nintendo really was the videogame equivalent of Disney. Not just with cute characters and squeaky-clean family values, but also with that magical sense of nostalgia that even children can feel. And just like Disney and Mickey Mouse, Nintendo and Mario were uncool in exactly the same way: because they weren't supposed to be cool. They were supposed to be fun, timeless, and magical, and with those great strengths came the great weakness of appealing to teens and young adults. If Nintendo preferred to remain as is, they would continue to concede this growing portion of the market, but if they wanted to evolve, they risked losing their identity. As a result, Nintendo was faced with a simple proposition: acceptance or adaptation.

"Our research is by no means conclusive," said the Market Data Corporation's bearer of bad news, "but if nothing else, it indicates that Nintendo needs to implement new strategies to retain a dominant position in the market."

Although this comment reeked of obviousness, those present listened with a show of thoughtfulness. They did so because each had a slightly different perspective on how Nintendo should progress, and each was secretly hoping that this would be the moment when everyone else in the room realized they had been right all along.

Arakawa, naturally, wanted to stay the course. Tilden generally agreed but thought that more could be done to spruce up the brand and take advantage of Nintendo's iconic intellectual properties. Main agreed with Tilden in tone but was more concerned with size. He believed that the videogame industry was becoming more and more like the film industry, and he wanted to go bigger and bolder for the launches of Nintendo's blockbuster titles. The advertising guys from Leo Burnett wanted to reinvent the wheel because that's what advertising guys always want to do. And last but not least, White didn't want to respond to Sega per se; rather, he preferred to do what they were doing, but do it even better. He wanted Nintendo to aggressively evolve and become more like Sega, believing that there was no better proof of the market's changing tastes than a look at the company's best-selling games in 1992:

1.	Street Fighter II	1,300,000
2.	Legend of Zelda	1,000,000
3.	Super Mario Kart	550,000
4.	Mario Paint	550,000
5.	Turtles IV	350,000
5.	Madden Football	350,000
6.	NCAA Basketball	250,000
7.	Play Action Football	250,000
8.	Super Star Wars	150,000
9.	NHLPA Hockey	150,000

Nintendo's top title of 1992 was a fighting game, and although there was no blood or gore on it, there was nothing Disney-esque about Ryu's *haduken* or E. Honda's hundred-hand slap. And of those top ten games, only five were first-party titles (indicating that other companies were better at capitalizing on consumer tastes) and four of them were sports titles (which Nintendo had typically shied away from but this year lucked out on because they had finally been able to make a deal with Electronic Arts after Trip Hawkins abruptly left the company). And beyond the games, the effectiveness of Sega's advertising served as further proof that fresh marketing was critical to continued success (and watching any sporting event was a reminder that even market leaders like McDonald's and Coca-Cola were constantly reinventing themselves). Come on, White thought exasperatedly. Sega had just pulled even with Nintendo; if ever there was a moment to pull out the big guns, now was it.

From an urgency standpoint, White may have been correct, but unfortunately for him the stars just weren't aligned for a big change at this time. Had it been two years earlier (prelaunch) or two years later (toward the end of the console's life cycle), Nintendo would have been more likely to roll the dice, but as they headed into 1993, there was just too much at stake. Sega had not only swiped away a large portion of Nintendo's market share but also had proven that there was room for more than one in the console business. And now that the precedent had been set, the barbarians were lined up at the gate and five reputable hardware makers lurked on the horizon:

1. **3DO:** A sleek, sophisticated, 32-bit system that planned to be more than just a console. With a tagline proclaiming it to be "the most advanced home gaming system in the universe," this do-it-all dynamo would also play movies, music, and CD-ROM content. Although this sounded overly ambitious, if there was anyone who could pull it off, it was Trip Hawkins, the founder of Electronic Arts, who believed so much in this grand new vision that he was willing to abandon the software company he had started from scratch. Even more ambitiously, his 3DO company wouldn't manufacture the hardware itself (like Sega and Nintendo both did) but would instead license the technology, the incentive being that while these manufacturers would have to bear expensive production costs, they would only have to pay very low royalty rates. In 1983 Hawkins had founded Electronic Arts to disrupt the software industry, and now, ten years later, he was looking to do the same with hardware.

   Release date: fall 1993

2. **Atari:** After sitting on the sidelines for many years, Atari planned to return to the console business with a bang. So while everyone else was just starting to think about 32 bits, they were preparing for the launch of their 64-bit system, called the Jaguar. Nobody knew quite what to expect from the return of Atari, but the system's "Do the math" slogan indicated that Atari was counting on its pedigree and standing out when it came to bits and bytes.

   Release date: November 1993

3. **NEC:** In October 1989 Sega launched the Genesis and NEC launched the Turbo-Grafx. Of the two 16-bit consoles,

Nintendo was more worried about the latter because of NEC's enormous financial resources. A bad launch lineup, poor marketing, and the long shadow of Nintendo had quickly doomed the Turbo-Grafx, but NEC seemed to have learned from that experience (just as Sega had learned from the failure of their Master System) and was ready to get back in the game. By the middle of 1992, NEC had developed a 32-bit system architecture called Iron Man, which earned rave reviews behind the scenes at the Consumer Electronics Show.

Release date: early 1994

4. **Bandai:** Although Bandai had no experience in the videogame industry, they were the third-largest toy company in the world (behind Hasbro and Mattel), which automatically equipped them with three important advantages: brand recognition, top-tier intellectual properties, and strong distribution networks. Given these assets, Bandai began working on a console code-named BA-X, which they designed to target a young demographic through edutainment and anime content.

   Release date: late 1994

5. **Sony:** Following the failed romance with Sony, Nintendo had little idea what the company was up to. Rumors circulated that Sony was serious about moving forward on their own, and there appeared to be some truth to this, as evidenced by Norio Ohga's shielding Ken Kutaragi and moving him under the umbrella of Sony Music. Although this may have given the PlayStation a stay of execution, it was hard for Nintendo to imagine that Sony would actually take this project all the way. Unless they had the support of a stable partner who knew the videogame business, they risked another Betamax debacle. And now that Nintendo was out of the running, that left only Sega, but there had been no whispers of any such relationship, and an alliance like that was simply too big to keep quiet.

   Release date: TBD

Each of these potential competitors had a nuanced set of strengths and weaknesses, and of course not all of them would flourish, but if even just one made a serious dent in the market, then the pie that Nintendo once had feasted on all by itself would be split into three pieces instead of two. Given the lucrative nature of the videogame business, a more populated pie-eating contest seemed all but an inevitability. And it was for this reason that, in these changing times, Arakawa was even more committed to staying the course and maintaining Nintendo's identity. Tweaks could be made, but in a coming war between style and substance, it was important that Nintendo remain more substantial than ever.

"Can we at least discuss the magazine?" White asked, hoping to leave the room with some sort of concession, even something small. He had unsuccessfully broached this topic before, but figured that with the research results now would be a good time to try again. "*Nintendo Power* is great, and I know Gail has done a great job; we can all see that it's, you know, great. But we have to recognize that by hoarding all of the secrets to our games, it's really pissing off the other magazines and, with that, lots of people in the industry. And let's face it—that's one of the areas where Sega really torched us."

Although Arakawa often took long moments to juggle pros and cons (or at least to make it appear that way), he did not like to be challenged in this manner and his answer to White was brisk. "No." He shook his head. "There is nothing of this matter to discuss."

In lieu of a conversation, a stiff tension permeated the room. White wanted to shout something back, at least to spur a discussion, but he knew there was nothing he could possibly say that would open up Arakawa's mind, and that only made him want to shout even more. But it was pointless, and everyone there could feel the staggering lack of efficacy.

As if to break up the subtly uncomfortable mood, the representative from MDC offered some optimism. "It's not all bad news. We did find some signs of progress," he said, quickly glancing at the papers in his hand. "More girls between the ages of six and fourteen are becoming primary players. And of all the people we surveyed, 96 percent knew who Mario was, and 83 percent of people said that they liked him."

It wasn't enough to put minds at ease, but it was enough to unknot everyone's stomachs. And it served as a reminder that beneath the numbers, Nintendo always had the advantage of Mario and everything he represented: Miyamoto's genius, and the no-nonsense perseverance that had enabled NOA to resurrect an industry from the ashes. Defeating Sega would require a recommitment to many of the same values: focus, belief, and the confidence to follow a vision in the face of adversity.

This was what made Nintendo so wonderfully successful in the past, from playing cards to videogames, and this strategy to double down on substance over style likely would have worked, if not what for Sega did next.

43.

MAGIC CARPET RIDE

Being a grown-up was not supposed to be this much fun.

Being a kid? Probably. Being a college student? Definitely. But being a grown-up? No way, that was supposed to be nothing more than pushing papers, rolling eyes, and constantly struggling against a slew of unwanted responsibilities. Being a grown-up, especially one with a corporate job, was supposed to be the beginning of the end. But for Diane Adair, who was flying to the Mojave Desert in order to film Sega's newest commercial, being a grown-up felt like nothing less than the beginning of a wonderful new beginning.

Personally and professionally, there were changes everywhere, but none more emblematic than that of her name. It had all started about a year ago, just prior to Christmas 1991, with an occasion that included a teddy bear, a persistent doorbell, and, with a strange symbolic appropriateness, a fat lady singing. It was Saturday, December 14, and Diane Adair was into the final hours of planning a surprise party for her boyfriend, Don. The plan was that while he was at Candlestick Park, watching Steve Young dismantle Kansas City, she would decorate the house, welcome guests, and entertain family and friends until the fateful shout of "Surprise!" It was perfect, or would have been if Don hadn't gone ahead and ruined her plans for his birthday party.

"You're going to be late," she said, trying to hustle him out the door.

"I know, I know," he said, resisting the light pushes on his elbow. "But first I need you to go to the Christmas tree. Bow is over there, and it looks like he's got something for you. You know how impatient he can be."

Bow was a fluffy, tan-colored teddy bear that Don had given to Diane at a Christmas party two years earlier. It was the first formal event they'd attended together, and in a strange way, the tan stuffed bear came to represent the formalization of their affections. Bow held a special place in their hearts, an inanimate pet that only the two of them could fully appreciate. On Don's instructions, Diane walked toward the Christmas tree and saw that there was a small jewelry box resting right below the bear's spiffy black bow tie. An early Christmas gift? How thoughtful! But couldn't he have just waited eleven more days? Well, the quicker she opened this, the quicker he would leave and she could get on with planning his party. But all thoughts about the surprise party were trumped by the surprise she felt when she opened the box, saw the ring, and then looked up to see Don getting down on one knee and asking her if she would marry him. "So," he concluded, "what do you say?"

Tears welled up in her eyes, but before Diane could answer the question, the doorbell rang. It was Don's best friend, ready to take him to the game, so eager to get going that he quickly rang the doorbell twice more before poking his head in the doorway. "Don, are you ready?" he asked, before he noticed Diane crying and apologized profusely. "Sorry, I didn't realize you were fighting, I'll just come back in a little while."

"No," Diane said, trying to explain the situation, but before she could do that Don pressed her for an answer (especially since the only thing she had said since he had asked was no). "Yes, of course," she said, smiling and laughing, but before she could embrace her husband-to-be, the doorbell once again rang. This time Don answered it, intending to clarify the situation for his friend, but when he opened the door he found instead an obese woman dressed in a Viking costume. "Um . . ." he said, because there's really nothing else appropriate to say under these circumstances. Diane had hired this woman to sing at the party this evening, but apparently she had arrived early and knew how to embrace the moment. And as the two-hundred-plus-pound Viking woman launched into an over-the-top operatic version of "Happy Birthday," there was really nothing appropriate for Diane to say under these circumstances except for "Surprise."

They got married nine months later, on September 12, 1992, a wonderful day that began with Diane Adair and ended with Diane Fornasier. She had gladly taken his name (and then taken a honeymoon in France), which seemed to metaphorically represent her transition from giver to taker. All her life, Diane Adair had been a giver, the kind of person who instinctively shares credit, who sincerely believes in the Golden Rule, and who would actually give someone in need the shirt off her back (if such a thing were socially acceptable for girls). Some might argue that her altruism was inspiring, while others might argue that it was naive, but Diane didn't care all that much because she didn't want to argue at all. Conflict was okay, if it came to that, but shouting, shoving, and winning first place in the blame game didn't get her juices going. Besides, needless conflict got in the way of giving 100 percent toward the task at hand. But over the past couple of years, a change had taken place within her. It began with her taking the job at Sega, grew when she took on Nintendo (Game Gear vs. Game Boy), and was finalized by taking Don's name. This evolution of confidence and self-esteem was then confirmed once again when Diane Fornasier took over Al Nilsen's responsibilities and was put in charge of marketing the Genesis.

Although Sega's marketing department had already developed a scintillating reputation for shock and awe, Fornasier's tenure on the Genesis would begin with a shot-heard-round-the-world type of bang. Following the launch of the "Welcome to the Next Level" campaign, the team at SOA could not have been any happier with how they had successfully redefined the Sega brand and, simultaneously, positioned Nintendo. Amazingly, Nintendo had not yet fought back, which only further enabled Sega to paint them into a kiddie-sized corner. As a result, Sega felt like they were dominating Nintendo on every front in the advertising war except for one: technology. They didn't think they were losing this battle (Sega CD promised the future, and Sonic's speed exemplified the present), but Sega's Next Level thinking wasn't content with just winning; it was about winning big. And the only thing standing in the way of total domination was Mode 7.

Ever since the release of Super Mario Kart, Nintendo had made a noticeable effort to tout their 16-bit system's Mode 7 technology, which enabled the console to display some rudimentary 3-D graphics. Although technological jargon often has a way of eluding mainstream consumers, Nintendo's Mode 7 resonated with a wide enough audience to cause SOA's marketing department to ask: "Do we have that?" When the answer came back no, the marketing department scrambled to look for something, preferably something with an ad-friendly name, that the Genesis had and the SNES did not. To find a cool-sounding needle somewhere in Sega's haystack, France Tantiado, a member of Fornasier's group, met with a seasoned producer named Michael Latham. Although nothing immediately came to mind, Latham grabbed a hardware manual and began paging through it. Surely there must be something, Tantiado suggested, or how else could Sonic be so much faster than Mario? While looking through the manual, Latham found something that kind of, sort of, maybe fit the bill: Burst Mode, which in theory allowed the Genesis to process code faster than Nintendo's chip could. Although this sounded like exactly what the marketing team wanted, Latham explained that Burst Mode actually had very little to do with the graphics, velocity, and overall performance of Sega's games. To say that Burst Mode was the reason that Sonic could move so fast would be like saying that cheetahs were faster than elephants were because of their spots. But still, it was something that Sega had and Nintendo did not, which was exactly what the marketing team had wanted. Tantiado loved the concept but hated the name, which prompted Latham to brainstorm possible nicknames and eventually settle on Blast Processing.

Although Blast Processing was really Burst Mode, which was really nothing more than the spots on a cheetah, this phrase went up the flagpole and caught fire internally, especially after Hiroshi Yamauchi announced that Nintendo had been developing something called the Super FX chip. From Jeff Goodby to Tom Kalinske, it seemed nearly everyone loved the idea of thumbing their nose at Nintendo by flaunting this magical Blast Processing. The lone voice of dissent came from Al Nilsen, who felt this crossed a line; there were so many positive things to say about Sega and the company's rags-to-riches story, why resort to fiction? His concerns were noted and debated, but ultimately he was outvoted and Sega chose to start deploying the phrase on Sonic 2sday (by having Dustin Diamond gush about "Sega's Blast Processing capabilities"). Following a positive response from the buzzword-loving mainstream press, this new term would come to embody the difference between Sega and Nintendo and serve as the focal point of a series of commercials to kick off 1993, which is precisely why Diane Fornasier was headed out to the Mojave Desert.

The spot called for a "long open road," which is exactly what Fornasier found when she arrived on the set. The highway where they would be filming looked almost exactly like it had in the storyboards, except that in real life the road was surrounded by miles and miles of frosted desert terrain. Not ideal weather for those on set, but perfect for the ad they were shooting. Titled "Top Fuel," it would begin with a Sega Genesis and Super Nintendo sitting in the middle of the road. The camera would then pan back and forth between the two as a narrator explains that "the Sega Genesis has Blast Processing, the Super Nintendo doesn't" and then asks the question that was on everyone's mind (because Sega had just put it there): so what does Blast Processing do? Answering that was the fun part, and it's why they had to shoot this in the middle of nowhere.

"How fast will it go?" Fornasier asked, looking at the Formula One race car.

"About 150 miles per hour," one of the ad guys said, "160 if we're lucky."

Fornasier was feeling lucky, but she knew that advertisers and clients always have a slightly different definition of this concept. Although both sides wanted the same thing—critical, commercial, and watercooler success—the agency guys would have been perfectly happy to accomplish that by not showing a single second of game footage. That's where Fornasier came in. Typically, someone in her position would farm out production work (and avoid the desert's thirty-degree winter temperature), but she wanted to be on set to ensure there was a semblance of substance to the agency's style.

"Just 160?" another ad guy asked. "Nah, I heard that bad boy can sling it up to 170. Might not work in this weather, but it's definitely a possibility."

"I think 150 will be more than fine," Fornasier said. The centerpiece of the commercial, and the part that vaguely answered the question about the meaning of Blast Processing, was a television playing Sonic 2 attached to the back of the race car. And although it's fair to assume that the coolness of this aesthetic would increase with each additional mile per hour, Fornasier also knew that there would be an inverse relationship between the race car's velocity and a viewer's ability to figure out what the heck was playing on the TV. "On second thought, there are no shots of the speedometer, so 140 might even be okay too."

"Oh, come on!" the agency guys playfully exclaimed.

"Hey," Fornasier said, proudly taking charge, "back when I worked at

Del Monte, my boss used to take out a stopwatch and count how many seconds it took for the logo to appear on-screen. It's up to you guys how to film this, but if you want to try to break a speed record, then I'm bringing a stopwatch to the next shoot."

The guys from Goodby made it work, as they always managed to do, and they made sure everyone had fun in the process. For the next several hours they filmed the television-clad race car, and then they moved on to shoot the part that would piss off Nintendo. Not only had Sega chosen to fabricate something called Blast Processing, but they decided to end the commercial with one final insult. After the race car speeds offscreen, the narrator returns with another question: "And what if you don't have Blast Processing?" As he speaks, the camera pans to a wimpy, blah-blah-white milk truck on the side of the road. The clunky thing can hardly get started—it's a piece of junk from bumper to bumper—but when it finally manages to lumber down the road, a television is revealed on this vehicle as well. And it's playing Super Mario Kart, the poor little racing game that now seems so slow by comparison. Just another victim blasted by Sega's marketing process.

What Kalinske loved so much about Blast Processing was how perfectly it unified Sega's defenses for the coming year. Not only did it provide cover in the battle against Nintendo, but it also armed Sega with additional ammo to take down the slew of 32-bit consoles on the horizon (like Atari's Jaguar and Trip Hawkins's 3DO machine). Additionally, Kalinske was always hyperconscious about the "story" that Sega was trying to sell to the world, and now that his company's underdog tale was reaching its climax, it was time to shift the narrative from the little engine that could to the engine that powered the videogame industry. Blast Processing fed right into the new storyline, and the best part was that if anyone tried to dismiss it, they'd be the ones cast as villains. If Nintendo (or Atari, or Trip Hawkins) called out Sega and accused them of doing exactly what they had done, then they'd look like the jealous killjoy. Besides, they wouldn't be able to prove anything anyhow, because technically it was true: Genesis *did* have Blast Processing, which would go on to do so much for Sega, even though it actually did almost nothing at all.

From a tactical standpoint, Sega appeared to be invincible in the short run. That mentality, however, began to change when Kalinske received a call from a psychologist named Arthur Pober.

"Wait, wait," Kalinske said, speaking into his office phone. "Why don't you slow down and start over again?"

"Sorry, Tom," Pober apologized. "But it doesn't matter how slow I say it, because the fact remains the same: the United States government is looking to crack down on the videogame industry."

The words made Kalinske momentarily numb. "Please tell me this is a prank."

"Unfortunately not," Pober said. "I have this on good authority, and I suspected that you would appreciate this as a courtesy call from me to you."

Arthur Pober was a proudly blunt Brooklynite who was currently serving as director of the Children's Advertising Review Unit (CARU). Founded in 1974, CARU was a self-regulatory group that had been created to promote responsible children's advertising across various mediums; from commercials during Saturday-morning cartoons to advertisements in the Sunday circular. In effect, CARU was the advertising police, at least when it came to protecting children. But because this was a self-regulatory agency, its officers were typically members of the toy industry, which is why Tom Kalinske had been a member since Mattel asked him to join in 1978.

"Have you spoken to Howard?" Kalinske asked. Sega's leader wasn't the only member of CARU; so was Howard Lincoln, who had been a member of the group since Nintendo's heyday.

"Not yet," Pober said, "but he's my next call."

"Good," Kalinske said, hoping that Lincoln would be just as rattled as he was. "In your expert opinion, what are we looking at here? How bad could this be?"

"Best-case scenario," Pober mused, "the government gets bored and stops nosing around. Worst-case scenario? They hold congressional hearings and wind up regulating the gaming industry."

"That would effectively kill the entire business."

"Hey, you wanted to know how bad it could be."

Kalinske couldn't believe what he was hearing, but at the same time it felt like the other shoe had finally dropped. Ever since his conversation with Emil Heidkamp at CES about the dangers posed by the increasing realism of videogames, Kalinske had been feeling that this was only a matter of time. "Do you know what prompted all of this?"

"I really don't know," Pober explained. "But if I had to guess, it would be as a result of some recent commercials. I haven't seen them myself, but I keep hearing about subliminal messages and some kind of scream. Does that mean anything to you?"

Kalinske shook his head. Of course it did, and as he weighed the gravity of this situation, he couldn't help but think back on the story of Prometheus, the Greek mythological figure who had stolen fire from the gods and delivered it to all of humanity. He had so nobly risked his life, but had done so for the sake of moving civilization forward. "How much time, if you had to guess, do you think we've got before this hits the fan?"

"A lot," Pober said, "and that's if anything even comes of this at all. Like I said at the beginning, these are just rumblings. Could be nothing."

"Let's hope," Kalinske said, shaking his head.

Pober agreed and then bid him adieu, leaving Kalinske alone in his office with only his thoughts of Prometheus. Now he remembered the details from the second half of the story. After delivering the gift of fire and advancing humanity, Prometheus was punished by the deities for committing this terrible crime. How dare he bring the mortals one step closer to immortality! And how dare he think himself a hero to mankind when what he'd really done was to rob them of their beautiful ignorance? A price must be paid for this act of hubris, a heavy price indeed, and so Zeus, the god of gods, sentenced the thief to eternal torment for his irreversible transgression. Bound to a rock, the immortal Prometheus would be, and each day an eagle would descend from the heavens to feed on his liver, inflicting unspeakable pain. Prometheus would suffer like nothing ever before, but that physical toll would never compare to the pain of knowing that each night, without fail, his shredded liver would grow back, and each morning the eagle would return to repeat the cycle over and over, until forever never came.

The message was rather clear: the keepers of the world had a habit of punishing the champions of progress. And Kalinske vowed to do everything in his power to make sure that didn't happen.

• • •

Howard Lincoln, however, felt like he had already done everything in his power. "It's not us," Lincoln explained to Arthur Pober, "I can tell you that much. We strictly monitor all of the content on our systems. Every single second of it."

"Hey," Pober said, "I'm not here to point fingers."

"I mean it," Lincoln continued. What better example of this was there than the case of Mortal Kombat? Both Sega and Nintendo had tried to get exclusive rights to the violent arcade game, but Acclaim (who owned the console rights to Midway's games) wanted to release it on both systems. That was Acclaim's choice, and that was fine, but just because Sega was going to release this gory game as is didn't mean Nintendo would do the same. For the SNES version, Lincoln asked the developers to decrease the level of violence and substitute a bland gray sweat in place of the game's blazing red blood. This was not an easy decision, nor was it a business one; Lincoln knew that by watering down the violence, Nintendo's version would likely be outsold by the one Kalinske would be dealing. But just because it was more profitable, that didn't make it right. Lincoln knew that in the coming months there would be those who wondered who was right and who was wrong, but he didn't care at all about that, because he knew without question that for Nintendo this was the only way. "Goddamn," Lincoln continued. "If it weren't for Nintendo, this industry would just be a bunch of pornography."

"I know, I know," Pober said. "But I still wanted to call out of courtesy. I doubt that anything will come of this, but you have a right to know what's going on."

"Thank you for doing so. I do appreciate the call," Lincoln said. "I'm curious, have you relayed this to my good friend Tom Kalinske?"

"Not yet," Pober said, "but he's my next call."

"Good," Lincoln said. For the past two years, Kalinske had been trying to label Nintendo as the Disney of videogames, so Howard Lincoln couldn't help but take a little comfort in the fact that if anything came of this, he'd have the likes of Mickey Mouse, Donald Duck, and Goofy in his corner. "When you speak with Tom, you really ought to ask him how he's currently feeling about the notion of karma."

• • •

There was a lot of evidence to suggest that Nintendo really should be considered the "Disney of videogames," but in 1993, there was another developer who wanted to steal this moniker away from Nintendo, and that was the actual Walt Disney Company.

Since 1981, when Nintendo introduced the Mickey Mouse Game & Watch, Disney had been dipping its toe into the business of videogames. There had been a few notable efforts to develop games internally, but for the most part the extent of Disney's toe-dipping entailed licensing out characters to unaffiliated developers. The list of licensees included Nintendo (who released a few Game & Watch titles), Sega (who crafted hits like Quackshot and Castle of Illusion), and most notably Capcom (who picked up the rights to several pieces of intellectual property in the Magic Kingdom way back in 1987). This clever acquisition was the work of Joe Morici, Capcom's forward-thinking VP of sales and marketing. Throughout the 1980s, Morici's claim to fame was that he had come up with the name Mega Man (the Japanese-created character had originally been christened Rock Man), but by 1993 it would have been difficult for him to begin any bragging session without first mentioning Disney. Over the past five years, the combination of Capcom's talented developers and Disney's storied properties had resulted in hits like Mickey Mousecapade (1987), Ducktales (1989), and Who Framed Roger Rabbit? (1991). The relationship also led to The Little Mermaid (1992), which was notable for being one of the first games designed for and marketed toward girls. It was a match made in heaven, but as Capcom continued to capitalize on these intellectual properties, Disney started to believe that maybe they should be the ones making these games (and raking in the windfalls that followed).

For Disney, the time to delve deeper into the videogame industry couldn't have come at a better time. Although the company had an unparalleled reputation for making great animated films, it had been several years since Disney had produced a classic. With features like *The Black Cauldron* and *The Great Mouse Detective* the eighties had been an underwhelming decade, but that trend changed in the nineties. *The Little Mermaid* rang in the new decade, *Beauty and the Beast* followed after that, and then came *Aladdin*. To capitalize on these hot properties, Disney entered into a partnership with Sega and Virgin Interactive, a third-party developer that Shinobu Toyoda had recommended.

There were already a million reasons why Kalinske was thrilled to be in business with Disney, but the relationship became even more vital when he learned about the government's recent interest in videogames. If they did decide to get involved, Sega (and not Nintendo) would likely be the target. There was no denying that Sega aimed at an older audience, and there was no doubt that the folks in Washington would selectively overlook this fact. In the event that this really did come to pass, Kalinske wanted to have as many squeaky clean friends standing beside him as he could. So what better way to announce the beginning of this beautiful relationship than a press conference at the Consumer Electronics Show?

Jeffrey Katzenberg, the head of Walt Disney Studios, had a similar but much bigger idea. This was his first foray into videogame publishing, and he wanted to make a splash—a Disney-sized splash. Katzenberg saw the original plans for the press conference, decided they were not grand enough, and then made arrangements for a press conference that fit his vision. What he provided was nearly magical: Katzenberg and Disney re-created Agrabah in a hotel in Las Vegas, complete with live animals, staged performances, and real palm trees. With speeches by Katzenberg and Kalinske as well as a keynote by Virgin's founder Richard Branson, it was the event of the 1993 Winter CES.

After the event, Katzenberg was amped up and excited, but also worried about how the industry would react to his game. There was also a strange variable that raised the stakes of Katzenberg's and Sega's game: Capcom (as per their previous licensing deal) was also releasing a version of Aladdin, this one for the Super Nintendo. All of this paraded through Kalinske's mind as he accompanied Katzenberg toward Sega's booth for the very first demo of the game. With everything at stake, Kalinske was a little bit nervous. He would have been much more nervous, however, if he had any idea what was going on with the actual game chip at that very moment.

Video games at that time were written on EPROMS, erasable programmable read-only memory chips. During Katzenberg's press conference, Diane Fornasier and the rest of the marketing team were to arrange Sega's kiosk and prepare an enticing setup for the demonstration of the Aladdin EPROM. The problem was that as the press conference started, she realized that nobody had the EPROM. The game was missing. In this pre-cell-phone era, the team communicated on walkie-talkies, and after a mad, static-filled scramble they eventually realized who had the game chip. It was a woman on Sega's marketing team, but unfortunately she had not shown up that day. With time running out, Fornasier sent a team to her hotel room where they discovered her huddled over the toilet, red in the face and violenty ill. After providing requisite concern, they eventually learned that the disk was in the hotel's safe, and then prodded her into remembering the combination between the graphic tirades of vomit. It was not pretty, nor easy to forget, but they got the EPROM, and then rushed back to the show.

Meanwhile, Katzenberg was on his way to the booth. Kalinske got a message from someone on a walkie-talkie to stall Katzenberg from arriving at the kiosk. He orchestrated about ten good minutes of stalling, attempting to highlight the wonders of CES to a confused Katzenberg, who was trying to lead the press to his game. The Disney movie mogul tried to break away, but Nilsen made a last-ditch effort to slow him down, getting Katzenberg to try riding one of Sega's racing simulator games (which bought about five more minutes). Finally, though, Katzenberg had had enough and took everyone to the booth for the demonstration. While he was in front of the booth, talking about the challenges of creating a game that was true to the movie, the team arrived with the EPROM and slipped it to Fornasier, who inserted it casually into the Sega Genesis on Katzenberg's cue as if nothing had happened. Aladdin went on to be named Best Genesis Game of 1993 by Electronic Gaming Monthly.

44.

CRAZY LIKE A FOX

As "A Whole New World" melodically played from the speakers scattered around Disney's Vegas-based Agrabah, Nintendo's Tony Harman couldn't help but agree with the message behind the theme song from the movie *Aladdin*. It was a whole new world indeed, he thought as he watched footage from the game. So much of what Sega released was just well-packaged junk, but this game right here was an absolute gem.

What made Sega's Aladdin different from the Capcom/SNES version was that it had been built with a new style of graphics called digicel animation. This breakthrough, created by Virgin's David Perry and Neil Young, enabled hand-drawn film cells to be directly scanned into development software. As a result, they were able to work directly with Disney's animators, and had access to over 250,000 cells from the movie throughout the development process. Harman was awed by the final product, which looked and felt like it was controlling the protagonist from an animated film. Despite his mortal dislike of Sega, and his belief that with only twelve levels its lasting interest was limited, he had to admit that this was a game-changer. Just one man's opinion, of course, but Harman's ability to evaluate videogames was second to none.

Arakawa valued these opinions so much that he assembled something called the product launch committee, which was basically his way of getting the marketing guys (Main, White, Sakaley, and Tilden) to speak with NOA's product development guy (Harman) and discuss all aspects of a game's release—things like which games deserved their own commercials, and how best to promote these titles in *Nintendo Power*. Harman respected NOA's marketing staff, but it always floored him that not a single one of them was a gamer. Sure, these were businessmen and it was true that one need not adore widgets in order to sell them effectively, but it also couldn't hurt, right?

Like most people who love playing videogames, Harman secretly dreamed of one day creating a hit, and he finally got closer to that goal after Mr. Y.'s announcement at Shoshinkai of a Nintendo baseball game. Although most of Nintendo's games were made by Japanese developers (either by Nintendo themselves or by a third-party game maker in the region), for this baseball game Harman chose to look toward Europe. He'd been closely monitoring the technological advancements over there and was greatly impressed by some of the work being done. Harman's European fascination had already paid off handsomely once (by finding Argonaut Studios, which later developed the Super FX chip), and he expected it to do so again, this time with a respected British publisher named Software Creations.

Harman had high hopes for his baseball game, and so did Nintendo of America (who was negotiating with Seattle Mariners superstar Ken Griffey Jr. to make him the face of this game), but as great a sports game as he believed that this would be, that's all it would ever be—a great sports game. Hopefully it would make a lot of money and provide countless hours of fun, but it would never be a classic like Super Mario Bros. or The Legend of Zelda. The even more secret dream behind Harman's secret dream was to make a game starring one of Nintendo's iconic characters. It was an understandable desire, but Nintendo was so protective of those franchises and wouldn't let somebody outside Japan touch them. There was something admirable about this stance, but there was something almost racist about it as well. Nintendo didn't believe that anybody outside Japan could achieve greatness. Not only did Harman believe that this could be done, but he also believed that by not taking this chance, Nintendo risked missing out on some of these breakthroughs. And seeing the Aladdin game at CES served as proof to him of this fact. So he decided to write a paper, a manifesto of sorts, that discussed the criteria of what makes a great game, and use this to prove that one could be made outside Japan.

• • •

Bill White couldn't have cared less about where the games were made; all he cared about was that they were marketable, and that he be allowed to promote the hell out of them. He had been given the autonomy to do so during his early years at Nintendo (*The Wizard*, the Pepsi deal, the Nintendo World Championships), but as time wore on, he increasingly got the feeling that he was being handcuffed. Nowadays, there wasn't much more to do than smile, nod, and spread Nintendo optimism. So while Harman was wandering around the trade show, White was back at the company's massive booth trying to turn every piece of good news into headline-worthy greatness.

"Christmas was phenomenal," he said, speaking to a reporter from Bloomberg. "Videogames came back in unbelievable fashion. And it gives us tremendous momentum coming into 1993, where we expect hardware and software sales to rise by 19 percent."

The reporter nodded, jotting this down. "Very impressive, but what does Nintendo have to say about its flagship 8-bit system, which has fallen considerably? You've gone from nine million units sold in 1989 to less than three million this past year."

"I look at that as a positive," White said. "To continue sustaining sales of the original NES, when more people are buying the faster and more sophisticated Super Nintendo systems—to me, that speaks volumes about Nintendo's commitment to quality."

"And what about CD-ROM technology?" the reporter asked. "Sega already has a system out there, and the 3DO company has recently announced plans to launch a CD-based system by end of the year."

"That's a great question," White said. "Nintendo hopes to bring out a 32-bit CD-ROM accessory by the end of the year. But if we don't feel it meets our standards, it won't be released. In this environment, it's easy to get caught up in the arms race aspect of it all, but that's just not how we at Nintendo believe in doing business. That being said, our commitment to technology is second to none. Are you familiar with Nintendo's new Super FX chip."

White used this self-created segue to promote Star Fox, and hyped Nintendo's new franchise with comparisons to *Star Trek* and *Star Wars*. As the interview neared its end, the reporter asked White if he could pass along any materials to publish alongside the article. That would make sense, White thought, but alas, he could not. All exclusive content went first to *Nintendo Power*, so there was nothing to do but smile and nod. The handcuffs, they were feeling very tight these days.

One month later, while White was fantasizing about a Houdini-esque escape and Harman was coming up with ways to improve upon Aladdin, Fornasier was desperately trying to avoid going to the bathroom. Her bladder was not particularly thrilled by this decision, but she felt like there was just too much at stake for her to leave the room, even if only for a minute. That's just how it went during Sega's annual planning meetings, and this one would prove to be particularly pivotal, as it led to the company's best year yet.

"So there's a chance that Sonic 3 won't be ready for Christmas this year," Fornasier explained. In an ideal world, Sega would release a new Sonic title every holiday season, but it was only January 1993 and Naka had already cautioned that the newest game might be late. Since Christmas tended to make or break a videogame company's year, finding a potential replacement was a top priority. "I know I'm crazy for even asking this," Fornasier continued, "but is there by chance a consensus on what everyone thinks our best game will be for '93?"

Before she even finished asking the question, nearly every single person in the room was already shouting out a response. Aladdin! Jurassic Park! Eternal Champions! Not only were they just yelling the names of the games, but shipping quantities and suggested prices as well. It was a lot of noise, but Fornasier hadn't expected any less. The conference room was filled with all the top people from sales, marketing, and operations, as well as the producers for each of the upcoming year's games. Of course each of them shouted out their own games; their job performances (and likely their annual bonuses) heavily depended on the decisions made in this meeting. If large initial orders were placed for a game, and significant marketing dollars were put behind it, there was a strong chance it would be a hit. So Fornasier asking everyone in the room about the best game was kind of like Santa Claus visiting a first-grade classroom and saying "I have a finite number of gifts this year, so can you please tell me which kids have been the nicest?"

"Okay, okay, okay," Fornasier said, trying unsuccessfully to quiet the cacophony. This is exactly why she couldn't go to the bathroom: leave for just a second and alliances would be formed, favors called in, and everyone would channel their inner bully. "Come on, let's all settle down."

"Quiet!" Paul Rioux grunted, instantly shutting up the room.

"How about we try a different approach?" Fornasier suggested. "Let's talk about Star Fox for a moment." Nintendo's Star Fox would be coming out in March, and they were putting more behind this game than any they had ever released before. Not only would the folks in Redmond be shipping out a million units to more than 18,000 retailers, but they were also holding special citywide events in Atlanta, Cleveland, Dallas, Houston, Los Angeles, Salt Lake City, and Tampa. To disrupt the success of Nintendo's Star Fox, Sega had scheduled the release of several top titles for around this time. Games like X-Men, Toejam & Earl 2, and Ecco the Dolphin, the gorgeous end result of Ed Annunziata's dolphin fantasy from a couple years earlier. At this point, there was nothing more that Sega could do from a product standpoint, but Fornasier was curious if anything else could be done on the marketing front. "We all know Star Fox is going to be big, but there's still time for us to shrink the size."

"Thank God Nintendo didn't get this out for Christmas," someone from the marketing team said. "Could you imagine?"

"You know," Richard Burns said, interrupting everyone's imaginations, "I'm not so sure that's correct. There's a solid chance it would have gotten lost in Sonic 2sday and, to be perfectly honest, the retailers I've spoken with aren't too upset about having a mini-Christmas in March. It's actually something we ought to consider."

"It's too late," someone from the product development team replied. "We can't get anything else out by March, let alone anything big."

"Yes we can!" Fornasier declared, her mind noticeably whirring. "Well, not literally, of course, but Richard's got a great point. Of course the retailers are thrilled to get a bonus Christmas, so that's exactly what we should give them. A Christmas-like launch every month of the year!"

"Boom!" said Tom Abramson, slapping the table. "Like it, love it, done deal. Just like the greeting card industry: every month will have something big going on."

Okay, every month was a little too aggressive, but every six to eight weeks should do the trick. Why didn't we come up with this sooner, Fornasier wondered, especially since this resembled a similarly staggered promotional strategy from the consumer packaged goods industry? But it didn't really matter that the idea had never come up before; all that mattered was that it worked like gangbusters when she was at Del Monte, and it should be just as effective now at Sega.

And as the jockeying began for whose games should receive the full holiday treatment of promotion, marketing, and sales, Fornasier couldn't help but continue to ignore her bladder and be thankful that she hadn't gotten up to use the bathroom. If she had, she might have very well missed out on the gift of having Christmas all year round.

Not long after that meeting, a man named Bob Knapp was on his way from Osaka to Newark. During the long trip from Japan to New Jersey, he had a layover in San Francisco, where he was greeted with some disturbing news at Gate 81.

"I'm sorry, sir," the airline attendant explained. "But that's what my system says."

"That can't be accurate," Knapp said. "My inbound flight was first-class."

"Yup, it says that too," the attendant said. "But your final leg is economy."

"Hmmm, I see," Knapp replied. "Are there any upgrades available?"

"Yup, there's one," the lady beside him said. It was Ellen Beth Van Buskirk, headed east for Toy Fair, and with the millions of miles she'd accumulated, there was no way she wasn't getting that seat. "And it's about to be mine."

Knapp took a half step backward to fully observe this woman and then a half step forward to plead his case. "I think it makes more sense for me to take the seat. I've just flown in from Osaka. I'm sure you understand."

"What?" Van Buskirk asked. "Are you crazy?"

"Hardly," Knapp said, and he and Van Buskirk then proceeded to argue over who most deserved the seat in question. Not long after, the bickering turned to banter and then it evolved into flirtation. By the time the flight departed, Knapp had gotten the seat, but neither of them could wait to land in New Jersey, where they could continue to flirt with each other.

• • •

In the following weeks, while Van Buskirk was busy falling in love at first argument, Tony Harman finished writing his manifesto and came to the conclusion that a developer must have at least three things to make a great game. The first was a big budget, somewhere in the range of $3 million to $4 million, to ensure that corners were not cut. The second was coin-op experience, to truly understand the value of capturing a player's interest immediately. And the last was an iconic character, ideally one that has already been established (like Mario) or one steeped in its own captivating mythology (like Star Fox). Of course it took much more than just these three things, but this was the basis behind why company's like Capcom, Konami, and even Sega had a knack for making hits.

Not long after Harman finished the paper, Arakawa approached him with some unexpected travel plans. "Tomorrow," he said, "you will be going to Japan."

"Okay," Harman said, making it a point to never appear caught off guard in front of his boss. "Any particular reason?"

"I sent your paper to Mr. Yamauchi, and he would like to see you."

And so, after a silent gulp, Harman hopped on a flight with Arakawa and his wife, and twenty-four hours later the three of them wound up inside Yamauchi's office.

"He is pleased that you could fit in this visit on such short notice," Yoko Arakawa explained to Harman, translating for her father. Although Yamauchi's office was small, it was as intimidating as one would expect from Nintendo's quietly ruthless leader. It was really warm in there, somewhere around eighty degrees, and Yamauchi sat there in a white undershirt, his lower half hidden behind a large wooden desk. In front of him was a pristine coffee table, a small television, and a pair of couches on either side of the room. Some of the employees at NCL referred to his office as the "realm of the Mother Brain," making a reference to the giant, cranium-shaped, energy-sucking villain who appears at the end of Metroid. Yamauchi often had guests in and out of his office, and today was no different. In addition to hosting his daughter, his son-in-law, and Tony Harman, there as well, on either side of the desk, were Miyamoto, Yokoi, Takeda, and Sakamoto (each nobly standing upright and reluctantly sweating due to the room's toasty temperature).

"I have read your report and found it interesting enough to pass along to Nintendo's greatest experts," Yamauchi said, gesturing to the videogame legends surrounding Harman. "It is their opinion that you are wrong, and that only the Japanese can make a great game."

"With all due respect," Harman countered, with a noticeable amount of gall in his voice, "your experts, these men here, account for most of the best developers in the world. Most, but not all, and I truly believe that with the right resources a great game could be made outside Japan."

After the words were translated, peals of laughter permeated the room. Harman continued to make his case, highlighting key points from his paper and trying to appeal to Mr. Y.'s love of innovation, but ultimately it appeared to be no use. "Face it," Yoko Arakawa finally said, "you're not going to win."

Harman was prepared to leave with his tail between his legs (smiling, though, as his idea had made it all the way to the top), but he decided to try one more approach. "Let me just ask one more question," he said, taking a step toward Yamauchi. "How many bad television commercials do we make each year?"

This was not a particularly tactful inquiry, but Harman knew that and thought he knew Yamauchi well enough to believe this might make a dent. Everyone in the office tried to extrapolate the meaning of this question, but before any further clarification was requested, Yamauchi burst out in laughter. "The answer: many."

Harman nodded. "And how much does each one of these commercials cost you?"

Yamauchi quickly discussed this with the experts on his couches and then came back with an answer. "They say around $3 million."

Harman nodded once again. "Then why don't you give me $3 million and one year to make a great game? Maybe I'm wrong and won't succeed with this, but the worst-case scenario is that you'll just make one less bad commercial."

At this, Yamauchi smiled, the finest and most silvery smile Harman had ever seen, and then the legendary president of NCL stood up and accepted the deal, provided that this young American kept Miyamoto apprised of his progress.

• • •

"Progress?" Nilsen asked, over lunch with Kalinske. "I don't have any progress to report. That's the problem."

"Don't be so hard on yourself," Kalinske replied. He'd known Nilsen for several years, but never seen him quite like this. The hustle and bustle of his new job was wearing him down. Kalinske wanted to cheer up his old friend, remind him that all those frequent flier miles were for the good of this company they had built together, but he could tell from the haggard expression on Nilsen's face that he wasn't in the mood for remember-whens. "It'll get better before you know it. Just watch."

"But what if it doesn't?" Nilsen asked, more sincerely than rhetorically.

Kalinske assumed by this tone that he was fishing to see if getting back his old job was a possibility. Not that he necessarily wanted this, but Kalinske knew that Nilsen was the kind of guy who found power and pleasure in possibility. Kalinske also knew that if the conversation continued down this road, Nilsen probably wouldn't like where it went. The fact was that Fornasier was doing an excellent job, and whatever she lacked when it came to wild what-if ideas, she more than made up for with her strong management and communication skills. Make no mistake, Nilsen was a star, but with so many cooks now in Sega's kitchen, they needed a team player. That being said, Sega still needed Nilsen's star power, possibly now more than ever, but they needed it in different galaxies as the company expanded around the globe.

"Al," Kalinske said, trying to help his friend shine, "you're the guy who made Buster Douglas a bigger star after he lost the title. If anybody can make this work, it's you."

Nilsen nodded slowly—processing, not accepting. "But what if it doesn't?"

"But why?" Arakawa asked Tilden, looking at the April 1993 issue of *GamePro* magazine. On the cover, right there in front of them, was artwork from Nintendo's Star Fox. Not only had this artwork been intended for *Nintendo Power*, but White had specifically met with Arakawa, Tilden, and Harman to discuss sharing it with outside magazines and had explicitly been told not to do so.

"I don't know," Tilden said. "I'm just as surprised as you are." She was indeed incredibly surprised by this act of defiance, but her incredulity could not be compared to Arakawa's. To him, such a gesture was just unfathomable.

"But why?" Arakawa asked again, still stunned by the insubordination.

Following this discovery, Arakawa tasked Peter Main with firing Bill White, making him the first executive to be let go since the NES launched in America. Main, however, did not want to dismiss White. Not only did he consider White an invaluable asset, but he admitted to Arakawa that he had given his protégé permission to do what he had done. "You should fire me, if anything," Main offered, but Arakawa was not persuaded. This was not about chain of command, but about looking into someone's eye and telling them no, only to find out that a direct instruction had been ignored and trust violated. His mind was made up. It was time for Bill White to leave Nintendo, and Peter Main would have to play the role of executioner.

45.

UNACCEPTABLE

Hayao Nakayama looked closer at the final prototype of the Sega Pico. As he turned it over in his hands, inspecting it with a piercing look that was half glare and half stare, the R&D team members who had gathered in the conference room anxiously awaited his response. This portable edutainment device was scheduled to be released in Japan by early summer, so by this point in the process gaining Nakayama-san's approval was more formality than necessity. The parts had already been ordered and the molds had already been made; the Pico was happening whether the president of Sega liked it or not.

But oh, how much they wanted him to like it, to see Nakayama-san give them a smile. He was stingy with those smiles, that was just his way, but when they came, it was well worth the wait. And as he put down the prototype the room filled with a collective surge of hope.

Nakayama-san, however, was not yet ready to deliver either a smile or a frown. Clearly the R&D team had done well, for the device looked impressive. This was a very important thing, because Tom Kalinske believed this product would be a winner. But the product also looked expensive, and he wanted to know how much it would cost to make. Better yet, he wanted to know what the retail price would be. A sequence of stares around the table finally led to an answer. Someone suggested 15,000 yen, maybe 20,000—somewhere in that range (about $150 to $200).

As Nakayama-san processed this information, he picked up the Pico once again. Fifteen thousand yen? For this? But why? Tom Kalinske had said that it needed to sell for $100, for at that price he could make it into a sensation. Tom had said that very clearly, and these employees had been there when he said it. There was no excuse to be made, none.

Suddenly Nakayama-san smashed the device against the table. This was unacceptable. He lifted it up and smashed it again. Unacceptable. Another smash, louder this time. Unacceptable!

Over and over, until the silly blue thing was busted to pieces. And still he continued, smashing the machine until the message was made abundantly clear.

46.

BLOOD, SWEAT, AND TIERS

"Come on," Kalinske said with disbelief, leaning over his desk with the phone to his ear. "You must be exaggerating."

"If anything, I'm doing the opposite," Fischer claimed. "What's the opposite of exaggerate? Deexaggerate? Unexaggerate?"

"I don't think those are real words," Kalinske said.

"Fine," Fischer replied. "Then you'll just have to trust me. By the time he finished smashing the thing, there were pieces of it all over the room."

"I know it's the truth, but I just can't believe it," Kalinske said. For a brief moment he imagined what the reaction would be if he did something similar at SOA. Nilsen would probably start cracking up, and then everyone would scramble for the pieces as if it were a scavenger hunt. "I think the thing we need to do is get you a camcorder and somehow find a way for you to start secretly recording these things."

Fischer had a good laugh over that. "It's a good idea, but I worry it's not an original one. I'm pretty sure SOJ is already doing that to you guys."

"You better be kidding."

"I am, don't worry. But would you honestly be surprised?"

"At this point," Kalinske said, "I'm not sure there's anything that Sega of Japan could do that would truly take me by surprise. And yet I remain hopeful."

"Such is the nature of an optimist," Fischer suggested.

"Optimist on the outside," Kalinske revised, "pragmatist on the inside."

"And not a trace of pessimism to be found?"

"I wish that were the case," Kalinske said with a chuckle. "But you just told me that the head of SOJ, the man who signs my paychecks, had another temper tantrum. And this outburst was directed at the very same people who I'm hoping would follow our advice and work with Sony. So it's probably time to start looking for a backup plan. Speaking of which, there's someone I need to see about that."

Although Kalinske was the undisputed king of the Rolodex, when it came to contacts in the technology and advertising industries, then Doug Glen had to be crowned some kind of prince at least. His contacts had been an asset to Sega in a variety of areas over the years (Sega CD and Goodby, Berlin & Silverstein were good examples), and most recently with Sega Channel, as Glen had been instrumental in striking deals with TCI and Time Warner Cable to implement this revolutionary on-demand videogame service. These deals and the company's groundbreaking plans for the Sega Channel would be announced to the industry at this summer's Consumer Electronics Show, and then begin national testing in June. Ideally, if all went according to plan, it would be available to millions of subscribers nationwide in time for the 1993 holiday season.

"Did you want to discuss the Sega Channel?" Glen asked, entering Kalinske's office. "Or is this about the ratings? Either is fine, but knowing in advance helps me switch gears."

By ratings, Glen was referring to the videogame ratings system that Sega was currently trying to create. Following the courtesy call that Kalinske received from CARU's Arthur Pober, Sega had quickly become very interested in establishing a council that rated videogames in much the same way that the Motion Picture Association of America (MPAA) rated movies. In fact, one of the first things Sega did was approach MPAA president Jack Valenti, but he and other movie industry folks looked down on videogames and didn't want them to sully the organization's reputation. With the MPAA no longer an option, Sega set out to create a new entity. It was a noble idea, and one that would certainly cover Sega's ass (and cleanse Kalinske's conscience), but it presented a host of challenges, like figuring out who should rate the games, what criteria should be used, and, most important, how to spearhead this endeavor without giving the impression that the ratings council just did whatever Sega said.

"Actually, I wanted to talk to you about something else entirely," Kalinske said. "I was wondering: do you know anyone in the hardware business?"

"Many people," Glen answered. "We can venture to Silicon Valley if you'd like."

"What about when it comes to hardware for videogames?"

Glen scratched his nose in thought, his big Brass Rat MIT ring shining in the sun. "Are you asking if there's anybody at Sega of Japan that I deem trustworthy?"

"That wasn't my intention," Kalinske said. "Although I now realize I probably should have asked you that a few months ago."

"Then what is it you are curious about? Contacts at Nintendo?"

"Contacts who can help us build a next-generation console to defeat Nintendo."

Glen nodded as he quickly tried to figure out what scenario might have prompted this inquiry. "Well, it was my understanding that Sega of Japan was nearing completion of a 32-bit prototype. Is that not accurate?"

"No, no," Kalinske said. "That's accurate. But my worry is what happens if that prototype isn't everything that we want it to be."

"Gotcha," Glen replied. "Nobody comes to mind, but I'll put my ear to the ground. Is there any company or region where I should direct my efforts? And, for that matter, any that I should avoid?"

"You don't need to waste your time looking at Sony," Kalinske stated. "But otherwise the entire board is open. However, I would suggest that, in the spirit of team building, it would be helpful to find someone that SOJ might like working with."

"Message received," Glen replied, already sifting through his mental Rolodex. "Given the current trajectory of our conversation, a question comes to mind."

"What's that?"

"Regardless of whether or not my efforts prove fruitful, have you considered what would happen in the so-called next generation if Sega were to develop hardware inferior to Nintendo's?"

"I'll tell you, Doug, I've given that question a *lot* of thought."

"And is there a scenario in which that turn of events doesn't reverse everything that we've done to Nintendo over the past two years?"

"Only one," Kalinske replied. "We crush Nintendo's spirit so much during this 16-bit generation that by the time we fight that next war, they've fallen too far behind to ever catch up."

Glen blinked several times while considering this. "That doesn't seem likely."

"No, it does not," Kalinske echoed. "But a man can dream, can't he?"

In March 1993, Kalinske tried once again to turn that perpetual dream into reality, this time by inviting Seattle Mariners slugger Ken Griffey Jr. to come visit Sega and discuss making a game together. Griffey accepted the invitation and, just prior to the start of the baseball season, visited Redwood Shores with his wife, Melissa, and agent, Brian Goldberg.

After greeting his guests, but before giving them the deluxe tour, Kalinske delivered the disclaimer that everyone in the room appreciated, even though they knew it to be almost entirely false. "I just wanted to start off by saying that Sega's interest in developing a Ken Griffey Jr. baseball game has absolutely nothing to do with Nintendo's ownership of the Mariners."

"Oh, yeah, of course," Griffey said, acquiescing in the lie. His wife nodded.

"That never even crossed our minds," Goldberg added, making it unanimous.

The truth was that in this case the truth didn't matter. At least not when it came to intentions. Griffey and his agent were using Kalinske for leverage with Nintendo just as much as Kalinske was using them to try to embarrass his competitor. Best-case scenario: Sega and Griffey make a game together. Worst case: they don't, but Nintendo is forced to spend more money to sign Griffey.

"Speaking of which," Kalinske began, "how has everything been working out with that new ownership group? A smooth transition, I hope?"

Griffey and his wife spoke positively about the change and Nintendo's involvement with the team. Everything had been good so far, no complaints, and it was great that the team had been able to avoid being uprooted.

"And from a negotiation standpoint," Goldberg added, "I can confirm that they've been nothing short of a class act."

Goldberg was referring to the new four-year, $25 million contract that Griffey had signed during this past off-season, making him one of the highest-paid players in Major League Baseball (not too far behind the six-year, $43.75 million deal that Barry Bonds had just signed with the San Francisco Giants). As thought bubbles full of dollar signs filled the room, Kalinske couldn't help but wonder how much it would tick off Main, Lincoln, and Arakawa if this guy they were paying $25 million signed with Sega and not Nintendo.

After the false formalities were out of the way, they spent the next few hours playing all sorts of videogames (Griffey was particularly impressed by the Sega CD), while Kalinske bragged about his company and how his marketing team could be an asset to Griffey. "Just look at what we've done with Joe Montana," he explained. "Even though he's nearing the final stages of his career, the guy has never been hotter."

Although this may have sounded like nothing more than a throwaway sentence, just another chance to brag about what Sega could do, it also conveyed the other big reason that Kalinske wanted to sign Griffey. Since 1990, Joe Montana had been the primary face of Sega's sports games. Montana was perfect for the job—the man was a winner (with four Super Bowl rings) and also a class act (he never complained about doing commercials, promotions, or signings)—but there was no getting around the fact that his days of playing football would soon be coming to an end. At the end of the 1990 season, a year in which he had won MVP, Montana injured his elbow in the NFC Championship game and missed the next two seasons. While he was sidelined, his backup, a young lefty named Steve Young, miraculously transformed into one of the best quarterbacks in the league (and even won the 1992 MVP award), which all but ensured that either Montana's career was over or he'd soon be traded to play out his final years with some desperate franchise. Either scenario wasn't particularly appealing to Sega, which made Kalinske and his colleagues eager to find a new face for its sports games. In addition to this, there was one other big reason why the company was making a push to find a new, younger face for its sports games: a new subbrand called Sega Sports.

As part of the shakeup that moved Nilsen to global marketing and gave

Diane Fornasier the Genesis, she was also put in charge of brand and product management. At the time, Sega was expanding so quickly that she and other executives wanted to take advantage of this by building a variety of powerful subbrands. This would eventually lead to categories specifically for women, and another one dedicated to children, but the first place they decided to strike was the digital playing field. It was a logical place to begin, both because Sega had a strong roster of sports titles and because of the runaway success of the Electronic Arts subbrand EA Sports (whose John Madden Football and NHLPA Hockey '93 were among the top-selling Genesis games). After the decision was made to move forward with this, a study was commissioned to decide what to name the subbrand. Consultants came in, pitched ideas, spoke with ordinary people, and ultimately decided that the new Sega sports line should be called Sega Sports. So with Sega Sports poised to launch in late 1993 and Joe Montana not getting any younger, what would be more perfect than bringing on Ken Griffey Jr. right now?

With all this in play, Sega of America's CEO tried his hardest to woo the coveted superstar during his visit to Redwood Shores. After they finished with the tour, Griffey, his wife, and his agent spent a few hours back in Kalinske's office discussing baseball, videogames, and everything in between. It was a fun day for everyone, which left Kalinske even more hopeful that Griffey might spurn Nintendo and come over to the dark side.

Days later, news of the meeting leaked, and the *Seattle Times* ran a story by Bob Sherwin with the headline "Genesis of a Conflict: Griffey Talks Video to Sega, Not Nintendo." The article (abridged below) details how both parties instantly benefited, and further illustrates the philosophical differences between Sega and Nintendo:

SUNDAY, MARCH 21, 1993

PEORIA, Ariz. – Seattle outfielder Ken Griffey Jr. has become involved in some video-game intrigue that may not have Nintendo officials reaching for the joystick.

"In all fairness to Nintendo and the Mariners, we have let them know about it," Griffey agent Brian Goldberg said. "We've given Nintendo a chance to respond, in what other way they want to respond."

> But privately, the word around Nintendo is that if Griffey signs with Sega it would be considered a disloyal act.
>
> "They (Nintendo) had the first chance," Griffey said. "It wasn't like I came to them and said, 'Let's make a game.' They called me and asked me. But they did nothing. If you want to call that disloyal, you can."

As Kalinske read the article, there wasn't a thing in the world that could have wiped the smile off his face.

Weeks later, Kalinske's face was filled with another tough-to-remove smile. This one, however, had nothing to do with Griffey, but rather a man named Garske—Chris Garske, Sega of America's VP of licensing and acquisitions.

"I take it you like the idea?" Garske asked.

"Yes," Kalinske said, still smiling. "But I have to admit that I like the news almost as much as the idea." The news was that Nintendo would be censoring their version of Mortal Kombat by changing certain elements, like the color of the game's graphic red blood to a bland gray. Although Howard Lincoln had made this decision a few months ago, this was the first that Kalinske had heard of it. This could be huge! If Nintendo was going to release a dulled-down version of the game, then Sega needed to find a way to capitalize on that. But figuring out how exactly to do that was now a much trickier question after what Sega had been working on over these past few months.

After the MPAA had turned down the opportunity to rate videogames, Sega forged forward on their own. They created the Videogame Ratings Council (VRC) as a new governing body to rate and review all game content. Initially, it would only be games published by Sega, but the goal was for the VRC to rate games by all companies and for all systems. Impartiality was important, as was quickly establishing a reputation for integrity, which is why Kalinske made it a point to bring on Arthur Pober as the first head of the VRC. Building this organization out of thin air was a large part of the challenge, but so was actually creating the ratings themselves. How many tiers should this new system have, and where should the age cutoffs be? What was the difference between violence, cartoon violence, and graphic violence? And should more realistic games (like those for the Sega CD) be held to a higher standard than those for 16-bit systems?

After working with Pober, speaking with psychologists, and conducting weeks of internal discussions, Sega came up with a three-tiered system:

GA (General Audiences): Appropriate for all family members. No blood or graphic violence. No profanity, no sexual themes, and no use of drugs or alcohol.

MA-13 (Mature Audiences): Parental discretion advised. Game situations and characters are appropriate for teenage judgment. These games may contain some levels of blood and graphic violence.

MA-17 (Adults Only): Not appropriate for minors. Games include complex situations that call for mature judgment. Could have lots of blood, graphic violence, mature sexual themes, profanity, or drug or alcohol usage.

With the VRC created and these criteria now in place, Sega was ready to move full steam ahead with its rating system. But then there was the issue of Mortal Kombat. What kind of message would it send if one of the most anticipated games in years earned the harshest rating? And what would that mean for game sales? If only there was a way to make Sega's version more violent than Nintendo's, avoid the MA-17 rating, and maintain the integrity of the VRC, all at the same time. And that's where Garske's idea saved the day.

"Some sort of blood code," Garske explained to Kalinske. "So when you buy the game, it comes without any of that over-the-top gore and violence. But then all you have to do in order to get the game to look just like it does in the arcades is enter a code. A combination of buttons and then boom—blood everywhere."

"And it would be called the blood code?" Kalinske asked.

"Yeah," Garske replied. "Or we can change it to something else."

"No," Kalinske said. "I like it."

The blood code—a way for Sega to have its cake, eat it too, and also have some left over to throw in Nintendo's face.

47.

THE MAN WHO CAME FROM PEPSI

Beauty may be in the eye of the beholder, but as George Harrison sat through the final cut of the *Super Mario Bros.* movie, he felt rather certain that there were very few beholders who would find any beauty in this monstrosity.

"Why are Mario and Luigi at a nightclub?" he whispered aloud, prompting the Japanese man beside him to glance his way and politely smile.

Harrison was here, at a small movie theater in Pasadena, because Arakawa had asked him to fly down to Los Angeles, pick up the legendary Shigeru Miyamoto, and escort the creator of Mario to a special screening of the movie based on his videogame. Up to this point, Miyamoto had not yet seen any cuts of the film, so it was important that Harrison keep him in good spirits for the viewing. Amazingly, given his reputation, Miyamoto appeared to be nothing other than jovial from the moment Harrison picked him up. He remained upbeat as they inched through the LA traffic, didn't lose steam as the film butchered his iconic characters, and politely smiled whenever Harrison would sigh and whisper something rhetorical aloud.

"And why on earth would Mario dance like that?" Harrison muttered to himself, once again earning an affectionate smile from Miyamoto.

George Harrison was Nintendo of America's new director of marketing and corporation communications, taking over for the recently departed Bill White. This promotion, making him Peter Main's right-hand man, was an incredible opportunity, but it left no traces of enthusiasm on Harrison's face. There were a couple of reasons why this was the case. One was that Harrison had an uncanny talent for staying even-keeled. Whether he was hearing good news or bad, he always had the look of someone who'd just been informed what time it is: thankful, thoughtful, and calculating. The other reason he wasn't jumping for joy was that just nine months earlier, Harrison had been under the impression that he had already been given this job.

George Harrison was initially recruited by Nintendo in late 1991, just a couple of months after the launch of the Super Nintendo. At the time, he was director of new ventures at Quaker Oats, in charge of investing in external projects or acquiring small companies that would fold nicely into the brand. Although his work there was undoubtedly impressive, what really appealed to Peter Main was the line on Harrison's résumé right below that. Prior to joining Quaker Oats, Harrison had been the director of U.S. marketing for PepsiCo from 1981 to 1987, fighting on the front lines for the second place soda-maker during the so-called cola wars that dominated the decade.

Much like the current battle between Sega and Nintendo, the cola wars represented a heated corporate rivalry between Pepsi, the scrappy upstart, and Coca-Cola, the established market leader. Although the companies had been battling each other since the end of the nineteenth century (Coke was founded in 1886, Pepsi in 1898), Coke paid little attention to Pepsi, the perpetual follower, until 1975, when the underdog drastically amped up their marketing efforts and issued the now-famous "Pepsi Challenge." Originally, the challenge entailed blind taste tastes at malls around America, but it eventually expanded to be more than just an amateur science experiment; it came to embody the revitalized spirit of the company. Pepsi wasn't just challenging Coke, but challenging Americans to break the status quo and drink something different from their parents. This youth-oriented approach soon evolved into a full-fledged slogan "Pepsi: The choice of a New Generation." By 1983, Pepsi had begun to outsell Coke in supermarkets nationwide, and they were looking for something big to put them over the top. That something turned out to be Michael Jackson, who signed a record-breaking $5 million endorsement deal which made him the face of the Pepsi Generation and, quite literally, the King of Pop. Jackson was a huge part of Pepsi's rise, but the underdog's biggest coup didn't come until Coke made a blunder that almost ruined the company.

While Pepsi was conducting taste tests in public, Coke was conducting similar tests in private. The company's senior executives commissioned a top-secret internal research program, called Project Kansas, whose objective was to create, test, and brand a new, better-tasting cola. After years of tweaking chemical formulas and testing diverse focus groups, a sweeter soda was born. This reformulated beverage, imagined to be the future of the Coca-Cola brand, was named New Coke and launched in 1985. It didn't take long for the backlash to begin, and when it did, it was unlike anything a Fortune 500 company had ever experienced before. Coke's customer support received complaint calls around the clock, and within weeks they had received over four hundred thousand angry letters. The situation got so bad so fast that a consortium of Coca-Cola bottlers even sued their supplier for changing the product.

On July 10, 1985, less than three months after New Coke was launched, Peter Jennings, the anchor of *ABC News*, interrupted *General Hospital* to announce that Coca-Cola's executives had admitted that they were wrong and announced that Coke would be returning to its original formula. As quickly as the backlash had come, so did the celebration for the return of the iconic beverage. Within forty-eight hours of the announcement, over thirty thousand people called Coke's hotline to commend the decision, and Arkansas senator David Pryor famously called this "a meaningful moment in U.S. history." By the end of the year, Coke was back to the original formula everywhere, and the cola wars resumed pretty much where they had left off. Back and forth, back and forth it went, with both companies striving to make soda a metaphor to define lifestyle.

Living through these battles, George Harrison learned a great deal about marketing, brand equity, and the hidden psychological advantages of being an underdog. So much of what Pepsi did was about bloodying Coke's nose and provoking a reaction, and so much of it worked because Coke had been caught off guard with Pepsi and had to do twice as much to make up for earlier mistakes. As a result, Harrison's biggest takeaway from his experience at Pepsi was the importance of reaction: how to respond, when to respond, and, most important, when it was better to just stay on the sidelines.

In a world where most marketing folks want to come up with the next big thing (e.g., Sonic 2sday, a head-to-head mall tour—basically what Al Nilsen did so well), Nintendo was pleased to find someone who appreciated the importance of picking one's battles. This mentality fit perfectly into the Arakawa-inspired, Main-executed NOA devotion to long-term strategy over short-term shakeup. Plus, Harrison came with the added bonus of having a glimpse into the heart of their competitor, having worked for what could be considered the Sega of the cola wars. So in March 1992, Harrison accepted the job of Nintendo of America's director of advertising and promotion, making him the company's number two marketing man behind the incomparable Peter Main.

Or so he thought. But when he arrived in Redmond, he discovered that Bill White (the man he thought he'd been hired to replace) had been shifted to a newly created position running corporate communications. This surprise led him to believe that either White had been on thin ice for a long time or Nintendo was actually a terribly disorganized company. And based on everything he had witnessed thus far in his career at Nintendo, it certainly didn't appear to be the latter. Naturally, Harrison was frustrated by this turn of events (although it was hard to tell because of his perpetually even keel), but any ill will quickly evaporated when he fell in love with Nintendo's unique corporate culture and the fast-paced environment of the videogame industry. Compared to the food industry, this was rapid-fire. Pepsi was a blast, but the products didn't change. There was Pepsi, there was Diet Pepsi, and then you do it all over again. But at Nintendo there were new games each month and, similar to the Hollywood business model, it was a hit-driven industry, where opening weekend would typically make or break a title. There also turned out to be an unexpected fringe benefit to working at NOA: father-son bonding. Harrison didn't know the first thing about videogames, but his six-year-old son did, and the boy would stay up with his father, testing out titles and giving him insights on which games to advertise and which features to highlight. After years of playing corporate Chutes and Ladders, Harrison felt like he had finally found the perfect job. Although some outsiders were perplexed by why he would join Nintendo during a period of decline, this never bothered Harrison. From his perspective, Tom Kalinske was nothing more than the videogame industry version of the Man of La Mancha: getting everyone excited with all sorts of hype until everyone realized that Sega was nothing more than smoke and mirrors. Nintendo's slow and steady pace suited Harrison well, and then in early 1993 it suited him even better, when Bill White was suddenly gone and he got the job he'd once thought he already had.

"Fantastic!" someone shouted when the credits for *Super Mario Bros.* rolled down the screen. There was laughter, there was clapping, and when Harrison turned to Miyamoto there was still that smile. Harrison was nervous to hear what he had to say and braced for the worst; even soft-spoken creative geniuses must have a mean streak, and there seemed no better time to strike than when your masterpiece has been butchered.

"So, what did you think?" Harrison asked. The movie was obviously horrible; all that mattered now was how annoyed or upset Miyamoto would be.

Miyamoto cocked his head in what appeared to be a sincere period of thought. The movie was scheduled to be released next month, May 28, 1993, to be precise, but a furious reaction from the game's creator could potentially change all that. Finally he spoke, and when he did there was no yelling or screaming. Nothing that even, for a moment, betrayed his gentle stuffed-animal-come-to-life personality.

He was a good sport about it all, and appeared confident that the people who watched the film would be able to separate what they saw from what he had spent years creating. Like most people at Nintendo, it appeared that he possessed an instinctive talent for knowing which battles were worth fighting.

After the screening, Harrison flew back to Redmond and met with the president of NOA to make good on the second part of Arakawa's request.

"George," Arakawa greeted, ushering the young executive into his office. "Tell me, have you seen the film?"

"Yes," Harrison said with a gentlemanly nod as he took a seat opposite his boss's boss. "I watched it from start to finish with Mr. Miyamoto."

"Good," Arakawa said, curious but never quite showing it. "What did you think?"

"It's sort of a bad-news-and-good-news situation," Harrison said. In addition to looking out for Miyamoto, he had been tasked to watch the film and determine if it had turned out so badly that Nintendo ought to pay however many millions for the distribution rights and then never let it see the light of day. "Well, the bad news is that the movie is really bad. It's just terrible."

Arakawa nodded; this was hardly a surprise.

"But the good news is that it'll be in and out of theaters so quick, no one will notice it. So we're better off to just let it go and die then to pay a king's ransom merely to stick it on a shelf somewhere."

"Do you feel certain of this?"

Harrison considered the question. It was definitely one of the worst movies he had ever seen, and there was no way it could help Nintendo take back any market share from Sega. There was a temptation to just make it go away (after all, Nintendo had the money to make that happen without issue), and also an urge to publicly make light of the situation, but those were the kinds of reactions meant for the Pepsis of the world. The Cokes, the market leaders, needed to shrug off these kinds of things and just keep moving forward. "Yes," Harrison said. "I'm certain."

"Very good. Then we move on," Arakawa said. And so they did.

Meanwhile, as Nintendo was formidably playing the role of a market leader like Coke, Sega was plucking a page right out of Pepsi's playbook. Ten years earlier, Michael Jackson had helped boost Pepsi to the top (albeit momentarily), and that's who Sega was counting on to now do the same for them. It was a lot to ask from Jackson (especially after his earlier game, Moonwalker, had only been a mild success), but this time he'd have help from another celebrity: Sonic The Hedgehog. And so, as per his deal with Sega, Jackson was hired to create the sound track for Sonic 3. Kalinske not only believed that this would be Sega's biggest game yet, but that this would be the first in a new genre of videogame sound tracks that would be consumed like a pop music album. It all seemed so perfect, except there was one small hitch to the plan.

As suspected, Sonic 3 would not be ready in time for the 1993 holiday season, but with Fornasier's any-day-can-be-Christmas strategy, the delay was almost moot. Timing, at least initially, was not the problem with Sonic 3, but the size of the game was starting to become one. Since this would be

Yuji Naka's last Sonic title for the Genesis (after this, he would begin developing for Sega's next generation 32-bit console), he wanted it to be the best and biggest of the franchise, and for him to accomplish that to his liking the game would need to be 24 megs, which was 50 percent larger than normal. Although the size issue certainly presented challenges, it didn't really become a problem until Naka realized that making a quality 24-megabyte game would take longer than he had initially thought. On second thought, Sonic 3 probably wouldn't be ready until summer 1994. Delaying a game was never ideal, but it usually wasn't crippling because marketing plans could still be moved. With Sonic 3, however, things were a little more complicated because Sega's promotional wizard Tom Abramson (whose prowess would soon earn him *Advertising Age*'s Promotions and Event Person of the Year award for 1994) had recently attained marketing's Holy Grail: a Happy Meal at McDonald's. This was a major achievement for Sega, financially as well as symbolically, but the problem was that Sonic 3 would need to be released some time during the first quarter of 1994. Based on Naka's current prognostication, that was absolutely impossible. This left Sega with two options: either cut down Naka's game and release it earlier, or let Sonic's creator just do what he did best and lose the McDonald's promotion. When laid out like that, the decision was simple, until Paul Rioux complicated everything with a Hail Mary idea that could potentially solve everything.

Years earlier, Rioux had remembered seeing a contraption that, when attached to a videogame cartridge, could add new characters, levels, and additional content to the original game. It was almost like the Game Genie, except instead of granting powers, it revealed all new worlds. What if Sega could harness that technology? The big feature of Sonic 3 was a new playable character named Knuckles, and that had to be accounting for a good part of the delay, right? So what if Sega released the first half of Sonic 3 without this Knuckles character and then, a few months later, they sold a contraption that would essentially "unlock" the rest of the game? Would that work? Amazingly, it would. Sega could then release the equivalent of Sonic 3: Part 1 in February 1994 (with nothing but a small cameo from Knuckles), and then when Naka was done they could release the second half. The concept might be a little confusing to sell, but that could be solved with clever marketing, so the real question was whether two Sonic games in 1994 (plus a Sonic-themed pinball game in late 1993) would cause consumers to become sick of Sonic's hero. This was certainly a very realistic possibility, but Kalinske and company were never ones to back down from a challenge, especially one in which they had Michael Jackson on their side.

As Kalinske sat in his office reading about the "lock on" technology that would make Rioux's Sonic plan possible, the stars were aligning to knock Nintendo out of the solar system. Everything was coming together, until suddenly it was not.

"What do you mean?" Kalinske asked, speaking into the phone with a frustrated desperation. "You're kidding, right? Please, tell me you're kidding."

It was Olaf Olafsson on the other end. "I'm afraid not," he said with a sigh. "They couldn't make it work. 'Creative differences,' that was the party line."

"I can't believe this," Kalinske said, digesting the fact that Sega and Sony had abandoned the plans to jointly release a next generation console together. "I don't really know what to say, except that I'm sorry."

"Oh, please," Olafsson replied. "You tried your best."

"That's what scares me," Kalinske said. "I did, I really did, and they still couldn't find a way to make it work."

"Don't beat yourself up. It's just business. These things happen."

"Am I correct to assume that Sony will be moving forward without us, and that you will be launching a console on your own?"

"Hopefully," Olafsson answered.

"Well, if you do," Kalinske said, "then I wish you the best of luck."

"To you as well, my friend," Olafsson replied.

After the men hung up, each allowed himself a final moment to theorize about what could have been before moving on to the much more pressing theoretical question: is the videogame industry big enough to support having three horses in this race? And if not, which horse is headed to the glue factory?

48.

MARCH OF THE LEMMINGS

The world is full of misconceptions, but perhaps none more fatally fantastical than those involving the lemming. As legend has it, these feisty creatures are prone to combating periods of overpopulation by blindly marching one by one off tall cliffs and unceremoniously plummeting to their deaths. It's unclear where this global rumor began, but evidence suggests that its popularity spread from Disney's 1958 Academy Award–winning documentary *White Wilderness*, which highlighted this unusual and unnatural behavior. Although it was later discovered that the filmmakers had flown in the featured lemmings from Canada and had actually tossed them off the cliffs by hand, it was too late to reverse this morbid misconception. The false legacy was further perpetuated in the 1970s with an outrageous off-Broadway show (*National Lampoon's Lemmings*, which launched the careers of John Belushi and Chevy Chase), in the 1980s with a famous Super Bowl commercial (Apple Computer's 1985 ad featuring a flock of blindfolded businessmen following each other off a cliff), and then again in the 1990s with the release of an extremely popular computer game (Lemmings, where players must stop these pixelated creatures from marching to their doom). Although Sony's Olaf Olafsson was in the enlightened minority of individuals who knew the truth about lemmings, he also knew the metaphorical value behind this urban legend. And that's why in May 1993 he traveled to

Liverpool, England, for a pivotal meeting with Psygnosis, the publisher of the addictive, misconception-perpetuating computer game.

"On just the first day, we sold fifty-five thousand copies on the Amiga," a Psygnosis employee explained while giving Olafsson a tour of the company's headquarters in South Harrington. Although every game publisher's offices are naturally tech-heavy, the piled-high workstations and futuristic 3-D renderings at Psygnosis gave Olafsson the impression that he had died and gone to high-tech heaven. "And several reviewers actually gave Lemmings an unprecedented perfect score!"

"Very impressive," Olafsson commented, pleased with everything that he had been shown thus far. "Well worth the plane ride to Liverpool."

After things had fallen apart with Sega, it became clear that the only way Sony could generate the developmental talent and technological resources to support their own console would be to acquire a publisher or partner with someone else. And since Sega and Nintendo were the only viable hardware makers (sincere apologies to Atari, SNK, and NEC), purchasing a game publisher was now a top priority. Since the top Japanese game makers would likely be too expensive (and also tip off Nintendo to Sony's big plans), the most likely candidates for acquisition were the big U.S. publishers like Acclaim, Activision, or maybe even Electronic Arts. Although each of these companies had impressive track records, Olafsson was less concerned about what a potential acquisition had already accomplished, and more interested in what they were capable of doing next. It was this forward-thinking logic that led him into deep discussions with John Ellis and Ian Hetherington, the managing directors of Psygnosis.

Ellis, Hetherington, and a gentleman named David Lawson had founded Psygnosis in 1984 with the goal of fusing their devout interests in art, rock music, and videogames. In the following years, the company's artistic ideals and strategy for selling games had much in common with the early years of Electronic Arts; both focused almost exclusively on computer games, both flaunted a bohemian fascination with graphics and technology, and both gorgeously marketed their software like music albums (Psygnosis, in fact, often had the packaging designed by Roger Dean, the artist famed for his album covers for the rock band Yes). But come the early 1990s, Electronic Arts transitioned to making games for consoles, while Psygnosis continued to remain dedicated to higher-powered computers. Why would two companies that started around the same time, with similar creative ambitions, take such divergent directions? Some of the difference can be attributed to EA being a publicly traded company (and beholden to bottom-line-focused shareholders), but most of the difference can be attributed to the console's lack of popularity in Europe. Consoles were significantly less popular in Europe than they were in Japan, the United States, and even South America (although in the last of these this demand was mostly supplied through black and gray markets).

To better understand the state of the British videogame industry, imagine the 1980s as a pop-cultural experiment in which the United States was the test subject and the United Kingdom was the control variable. For both countries, the decade began with a pandemic of Pac-Man fever that infected the masses until the videogame crash of 1983. In the aftermath of this disaster, names like Atari, Arcadia, and Coleco were tossed aside in favor of Apple, Amiga, and Commodore, as both countries simultaneously swore off videogames and prepared to get their interactive entertainment fix from the burgeoning personal computer industry. As the science fiction novels had been prognosticating for years, computers were finally here to take over our lives (and our living rooms), giving rise to ambitious new software companies like Electronic Arts and Psygnosis. Given the high cost of computers, neither company enjoyed any sort of overnight success, but by 1987 both had developed a reputation for publishing highbrow, high-tech, caviar-quality products. And as the end of the decade approached and the prevalence of personal computers increased, the future looked bright for both companies until something unexpected happen in the United States: Nintendo.

In the process of miraculously resurrecting the videogame industry in America, Nintendo's triumph had the strange effect of killing the computer game industry. Not directly, nor completely, but the seemingly inevitable personal computer revolution was thrown off course by Nintendo's unexpected videogame evolution. At first, companies like Electronic Arts resisted, refusing to be stymied by the 8-bit console's less sophisticated parameters, but soon that resistance became futile. By 1990, only 15 percent of households owned a personal computer, while nearly 30 percent owned an NES. But that was all just statistical static when compared to the frequency of game companies striking it rich by making games for Nintendo. And as more software developers migrated to consoles, the quality of computer games stagnated, which in turn hurt the computer industry, which indirectly helped the console industry. The circle continued viciously on and on until the only games that really mattered came in 8 or 16 bits.

This, however, was not the case in England, where Nintendo-mania had never happened in the 1980s and, therefore, the computer industry had never taken a backseat to consoles. And because England had not been the petri dish for Nintendo's experiment to globalize videogames, by the time the NES finally invaded Britain (Nintendo of Europe was established in June 1990) the personal computer revolution was far enough along to temper the console's initial reception. It sold well, very well around the holiday season, but it never set the country ablaze as it had in Japan and America. And as a result of all this, the kids in England tended not to argue about Sega vs. Nintendo, but rather debate the merits of consoles vs. computers.

Although this was a hot-button issue in classrooms around Great Britain, consoles vs. computers was never much of a battle amongst the country's top game publishers. With the exception of Rare (who created NES classics like Battletoads and R.C. Pro-Am), most British developers had missed the 8-bit wave and, by default, remained committed to making computer games. Where EA had been forced to abandon making heady games like M.U.L.E. and The Bard's Tale and do things like reverse-engineer the Genesis and develop a subbrand called EA Sports, companies like Psygnosis had never faced this type of identity crisis. As a result, while Electronic Arts was making games like Lakers Versus Celtics, Psygnosis continued publishing titles like Shadow of the Beast, a groundbreaking side-scrolling action game renowned for its cutting-edge graphics, parallax scrolling backdrops, and an incredible score composed by David Whittaker. Creatively, this type of work was likely more fulfilling, but financially the downside was that Psygnosis made significantly less money than Electronic Arts. Still, beneath this lost opportunity, there was a hidden financial upside, one that wouldn't be realized until a few years from now. And that was precisely what had brought Olaf Olafsson to Liverpool.

"Thank you for inviting me out here," Olafsson said, sitting down with Ellis and Hetherington to further discuss the reason for his visit. "I'm rather impressed by what I've seen today. It confirms my greatest hopes and expectations."

"We appreciate your saying so," Ellis replied.

"We do try our best," Hetherington added.

"It shows," Olafsson said, his mind quickly flashing back to everything he had seen: Lemmings, Shadow of the Beast, and the company's latest creation, Microcosm, a satirical action game centered around a bloody, futuristic corporate rivalry between Cybertech and Axiom, the galaxy's two largest conglomerates. Like the Sony game Sewer Shark and the Sony-turned-Sega game Night Trap, Microcosm was built with FMV animation, but it blew away those other titles because Psygnosis had rendered the graphics on workstations similar to those Hollywood used for special effects. When Microcosm was finished, Psygnosis was hoping that it would become one of the first games to bridge the gap between computers and consoles, with releases scheduled for both MS-DOS and the Amiga CD32, as well as Sega-CD and the 3DO system. This convergence between computer games and videogames was what really appealed to Olafsson. Now that the console world was catching up on the capabilities offered by computers, companies like Psygnosis were suddenly at the forefront of this next generation of gaming, particularly the kinds of games that Sony imagined for its PlayStation. "I think that Psygnosis would make a wonderful addition to the Sony brand," Olafsson said, "and I very much hope you feel the same."

They did, and on May 23, 1993, Sony Electronic Publishing acquired Psygnosis. To do so, they paid $48 million, which was an exorbitant sum that sounded even more astronomical when those hearing the news instinctively replied: Who the hell is Psygnosis? Why would Sony pay that much? Nearly fifty million bucks for a bunch of dopey little lemmings? Derision and mockery ensued, but Olafsson was deaf to these, because this deal had never been about the past, but exclusively about the future. And whereas in the past, the so-called future of videogames had looked at Europe as an afterthought, Sony wished to plant a flag in this market, which was primed for the convergence. Olafsson hinted at this in a press release issued that day, saying, "Psygnosis and its management will play an integral role in the development of industry-leading interactive entertainment, as well as our expansion into Europe." But that explanation didn't seem like nearly enough to adequately answer the lingering question of who the hell Psygnosis was.

"But why, oh why, would Sony pay that much?" Olafsson mockingly asked over lunch with Schulhof. Both smiled and laughed before moving on to the next order of business. It had little to do with the game Lemmings, and much more to do with the popular misconception. There was no truth to the myth that lemmings commit mass suicide, but like most great lies, this one was based on a figment of truth. Although they don't march off cliffs together, they do have a rare talent for marching in unison and obediently following the leader during times of migration. And with the next generation of videogames right around the corner, Psygnosis had an important asset that Sony considered much more valuable than just games. Psygnosis had something that Sony believed would induce lemming-like behavior among other game developers and steer them to the Sony PlayStation when that great migration finally took place.

49.

SWITCHING SIDES

"Come on, Bill, tell us already," a retailer pleaded between sips of beer. His request was followed by a chorus of boozy agreement from his industry brethren.

"You owe us," another shouted. "And we want to hear the story."

It was a cool June evening in Chicago, and several veteran retailers had stopped by the Sheraton Hotel for a couple of drinks at the bar with Bill White and a handful of employees from Sega. It had been a couple of months since White suddenly exited Nintendo for Sega, and the industry still had no idea what had happened. The lack of information eventually bred gossip, and the guys who had watched Nintendo grow (and profited from that growth) wanted to know which rumor was true. Had he really just snapped? Punched a coworker in the face? Slept with someone's wife? Or was the truth even more fascinating than all these presumptive fictions?

"There's no story," White claimed. "It was just time to move on."

"Bullshit!"

"Double bullshit!"

"Hey," White said, cutting off the cacophony, "every day kids go into your stores and choose Sega over Nintendo. So I just figured that it was time I did the same."

This comment earned a premium smile from Kalinske. The kid had a point, and it showed that he certainly knew how to think on his feet. Seeing him in action for the first time made Kalinske even more pleased that he'd been able to recruit White to Sega. Plus there was that wonderful added bonus of sticking it to Nintendo. And the timing couldn't have been better. A few weeks earlier, Ken Griffey Jr. had been in the Bay Area to play the Oakland Athletics. Kalinske was supposed to meet with him and his agent again and hopefully finalize a deal, but days before their tentative meeting Nintendo had managed to sign the slugger to a videogame deal of their own. It appeared Nintendo was finally waking up and acting like the market leader. This worried Kalinske, but not as much after he was able to steal White over to Sega.

As the night wore on, the retailers looked for any possible opening to try to get White to come clean. Eventually, when Sega's Richard Burns forced White into committing a so-called rite of passage (i.e. hazing of new employees), White was ready to give them a story. But it was not the one they had been requesting.

"We deserve to know!" one of the retailers exclaimed. "The guy was at Nintendo for six years. He ate, slept, and shat Mario. Then all of the sudden he's out of there? How do we know he isn't just some spy for Redmond? That this isn't all part of some ruse?"

Kalinske looked to White. The guy kind of had a point.

It generally takes a lot of creativity to disprove the idea you're a double agent, but White was up to the challenge. "You want to know how you can trust me?" White asked, climbing on top of the table. "Here's how," he said, unbuttoning his pants.

"Um, Bill," Kalinske asked, "what are you doing there?"

White smirked at Kalinske and the others around the table, then dropped his pants and shorts and pointed his pale rump in the direction of Nintendo's office in Redmond, Washington. Hey, they wanted a story, didn't they?

Nilsen was less amused with White's "story" than others were, but he knew that likely had less to do with the moon over Chicago and more to do with the crippling fatigue that he'd been feeling as of late. This new job as global marketing director was crushing him, so much so that he actually looked forward to the typically stressful Consumer Electronics Show as a chance to relax. Or, at the least, as a chance to stay in one place for more than forty-eight hours. The travel was killing him, all that time spent up in the air, and the worst part was that he was never able to sleep on planes. So his new job had basically induced a case of self-inflicted insomnia.

"I cannot express how nice it is to be having dinner here with you," Nilsen said to the licensing and media folks from Viacom. Between Nickelodeon and MTV, these folks had been great partners with Sega over the years, and the relationship had grown beyond the superficial. It was genuinely nice to see them.

"We feel the same way," said the president of Viacom's new media group. "We really enjoy working with you." The words put a smile on Nilsen's face, but it wasn't until midway through the meal that he realized how true the statement really was. "In fact," he continued, "so much so, that we want you to come work for us."

The question felt surreal, or maybe it was the concept of actually leaving Sega. Had he been asked six months ago, he would have laughed at the suggestion, but now he couldn't help but entertain it. The job was in New York, which was less than ideal, but at least it was in one place. And it was with fun people and fun properties. But still, could he really, truly leave Sega? The company he had helped build, the family there that sometimes felt closer than blood relatives?

"I'm flattered," Nilsen finally said. "I'll need to think about it."

"Of course, of course."

As Nilsen finished his meal he still couldn't believe that he was considering the offer, but the more he thought about it, the more he realized that, Viacom or not, he couldn't keep doing what he'd been doing these past six months.

50.

THE TIPPING POINT

The main attraction of Summer CES was, by far, Acclaim's Mortal Kombat. In just a few months, the game would be released for both the Genesis and SNES on a day the software publisher was calling "Mortal Monday," September 13, 1993.

"Have you taken a close look at this game?" asked Takuya Kozuki, Konami of America's president, as he and Emil Heidkamp neared Acclaim's booth at the show.

"No, but I have already seen enough," Heidkamp replied, ignoring the game footage playing on the nearby assortment of large, larger, and largest televisions.

"Come, let us see," Kozuki requested.

Heidkamp followed his boss to one of the large televisions, where a demo was playing Sega's bloody version of Mortal Kombat. A yellow ninja callously harpoons an opponent. A blue ninja shoots ice and then decks his frozen rival with an uppercut. A mercenary with a metal plate on his face throws daggers and then does something called a "psycho kick."

"Oh, wow," Kozuki remarked. "Did you see that one?"

Was this game really so bad? These were cartoon characters, after all. Heidkamp took a step forward, close enough to see the actual pixels that created the violence. Was this the tipping point, or just another foot or two further down on the slippery slope? Heidkamp stared at the pixels—the blues, the greens, and all of those many reds. For a moment it was all a blur, and then instantly it wasn't. On the screen, the yellow ninja spewed fire and charred his opponent to a crisp, the blue ninja ripped out a woman's spine, and the man with the half-metal face tore out someone's heart. There were more moves like these, and each earned the player extra points.

"Emil," Kozuki said, attempting not to sound rehearsed, "we must do a game like this. Don't you think?"

"Remember our deal?" Heidkamp asked, shaking his head. "Besides, Kozuki-san, we are already doing great."

"Yes, but one can always do better."

"No, not like this."

"Now, now," Kozuki said, turning to Heidkamp. "Think of them as fairy tales."

"We have had a great run together," Heidkamp replied. "What a great ride this has been. But if we're going to do this type of game, then it is time for me to leave."

Takuya Kozuki looked upward in thought, but there was a sense that the thoughts filtering through his mind had already been circling around in there for days. Kozuki and Heidkamp walked the floor a final time together before cordially shaking hands and going their separate ways.

When the Consumer Electronics Show ended, Nilsen flew out to London. Or was it France? Or was it actually Brazil? He couldn't remember; they were all blurring together. And then after traveling to London (or France, or Brazil), the president of a third-party developer in Japan urgently requested his presence for a meeting. The company was thinking about expanding into the U.S. marketplace and wanted to sit down with Nilsen and learn the business to see if this opportunity was worth the risk, and what kind of product and marketing support could be provided by Sega and its subsidiaries. So later that day Nilsen hopped on another plane (another chance to push the limits of insomnia), and he arrived in Tokyo the next morning.

Baggage claim. Restroom. Cab. Location change.

The this-and-then-that wasn't usually the hard part; the real devil of it all was doing this-and-then-that with a smile. It was staying energetic, motivated, and with the desire to conquer the world. If the job had led to big changes, then maybe it would have been easier. But after six months in this new role Nilsen wasn't sure that he was making any difference at all. In the room, everyone loved his ideas, the potential, the possibility, and the risk, but then he'd fly off and excitement would give way to bureaucratic inertia and business as usual.

Meanwhile, back in Redwood Shores, SOA continued to move at a million miles per hour. But from afar, Nilsen wasn't so sure that everyone was moving in the same direction. Sega was emerging as the leader of the videogame industry, but what kind of leader did the company plan to be? What would be its hallmark? The gimmicky Blast Processor? The bloody game Mortal Kombat? The athletes behind Sega Sports? And what happened to being the good old-fashioned blue dudes with attitude? Or maybe those things all meshed together nicely, Nilsen thought; maybe he was just upset not to be at the center of it all. He wanted to be there in Redwood Shores, but instead he was walking into an unmemorable silver building for the meeting that had brought him to Japan.

Show ID. Speak broken Japanese. Sign in. Tiny elevator upward.

Even though Nilsen had some concerns about what was happening back at Sega of America, he had no doubt that Kalinske would find a way to make everything magically work out. That's what he did, over and over. Tom was a magician, always pulling Barbies and He-Men and hedgehogs out of his hat, but what if someone was tampering with his magic wand? It wasn't really Redwood Shores that Nilsen was worried about, but quiet, unspoken friction between SOA and SOJ. Having spent more time in Japan recently, he saw it firsthand: the glances, the subtle comments, the references to "those guys over there." Was this kind of thing normal, or was it a cause for concern? More important, if it did turn out to be the latter, what could actually be done to remedy the situation? How could peace be found during an invisible war?

Elevator dings. Arrive on floor. Speak broken Japanese. Wait.

And wait.

And wait some more.

Then someone came over to let him know the meeting had been canceled. Apologies. Smiles. Let's reschedule. Then more waiting, more flying, and more hours spent not sleeping through the night on his way back to London. Or France. Or Brazil.

"Is this a joke?" Kalinske asked as Nilsen paced around his office.

"No, Tom, I'm very serious," Nilsen replied. "I wasn't actively looking for something else, but the more I think about Viacom's offer, the more sense it's beginning to make." Viacom's offer, as it turned out, was actually three offers: he could choose between vice president of Viacom's new media division, vice president of consumer products marketing for Nickelodeon, or a hybrid job to oscillate between the two (with some MTV work thrown in from time to time). The main takeaway was that Viacom was flexible; they didn't care exactly what Nilsen did as long as he was doing it for them.

"Are you feeling unappreciated?" Kalinske asked. "Is that it?"

"Me? No, not at all. I don't need a pat on the back whenever I come up with a good idea. It's not like that."

"Then tell me what it's really like."

"I'm trying, Tom," Nilsen replied, taking a deep breath. "It's a combination of things, really, but it starts with all the traveling."

"But you love to travel!"

"I thought so too. But not like this. Nobody could like to travel like this."

"You're right," Kalinske said. "And I imagine it can't be good for your health."

"Exactly!" Nilsen replied. "And you know I can't sleep on planes."

"That's right," Kalinske said. "Well, then, why don't we take a closer look at your schedule and see where we can cut?"

"It's not just the travel," Nilsen said, shaking his head. "It's also the job itself."

Kalinske knew that Nilsen wasn't thrilled with the job, but he hadn't quite realized the level of his dissatisfaction. "What is it about the job that you don't like? Is it something specific that maybe we can fix, or just a general feeling?"

Nilsen stopped pacing and took a seat. This was a good question, one he hadn't asked himself nearly enough. Everything had blurred together to such an extent that he'd stopped taking notice of the difference between specifics and generalizations. "Well, a lot of times," Nilsen began, thinking back on his least favorite meetings, "most of the time, I just feel like, 'I'll go in,' and then what happens is . . ." he said, then trailing off. At that moment he realized that this wasn't about specifics or generalizations, but actually about influence. And when he played it all over in his head, he realized that he had none. "I'm like a diplomat without a country."

"What do you mean?" Kalinske asked.

"I'm heading this group and traveling around the world, but nobody I speak with actually has to listen to me," Nilsen said, figuring this all out as he spoke. "I'm supposed to be the head of global marketing, but when I go to Sega of Europe, I'm just an employee from SOA, and when I go to Sega of Japan, I'm just an employee of SOA. And then the worst part is that when I finally do come back here, they look at me like I'm no longer a part of SOA."

Kalinske didn't know what to say. This wasn't the first conversation they'd had about Nilsen's frustrations, but it was the first that made him think it was more than just Nilsen venting about jet lag. Still, as bad as it all sounded, he knew that Nilsen could never leave Sega. He was Sega; how could he go somewhere else? And especially now, when the company was poised to finally surpass Nintendo. "I hear what you're saying, and I don't disagree with any of it," Kalinske said. "These are all extremely valid concerns, and hearing what you're going through gives me an even deeper appreciation for everything that you've accomplished this past year."

"What have I accomplished?" Nilsen asked.

Kalinske laughed, thinking this was a joke. Turned out it wasn't. "Oh, Al, don't get down on yourself like that. Sega is on the precipice of something big, and you're the one holding everyone together."

Nilsen sighed. "Maybe, but I don't know how much more of this I can take."

"Can you just give it a little bit more time?" Kalinske asked.

"I don't think so," Nilsen said, surprising both of them with that response.

"So what are you saying?" Kalinske asked.

"I don't know," Nilsen replied. "But it has to mean something that I haven't told Viacom no yet, right?"

"It does," Kalinske said. "It means you need to take a step back and figure out what's really important to you. So while you're doing that, why don't I see what can be done to ease your burden?"

"All right," Nilsen said, standing up. "Thanks for listening."

"Always, Al," Kalinske replied. "I mean it—always."

Nilsen walked toward the door, but before leaving he turned around with one more question. "I just realized that in all this talk about me, I forgot to ask how you were doing. So: how are you doing?"

"Don't worry about me," Kalinske advised. "Things are going great."

Although this was likely the same answer that Tom Kalinske would have given even if the sky were falling, things at Sega really were going great. The Genesis and SNES were neck and neck. Game Gear was about to catch up to Nintendo's Game Boy. And Sega CD, Sega Channel, and development on Sonic 3 were all great examples of the company's continued commitment to reaching the Next Level. For the time being, Sega was firing on all cylinders, but it was the future that had Kalinske concerned.

Particularly when it came to a next-generation console. After Sega and Sony had spent nearly six months trying to jointly develop 32-bit hardware, it appeared that they couldn't come to an agreement on the system's architecture, and the entire thing fell apart. What it boiled down to was that Sony's Ken Kutaragi wanted to create a machine that was 100 percent dedicated to 3-D graphics, whereas Sega's Hideki Sato wanted to build a machine that could also accommodate the typical 2-D sprite-based gaming. This didn't make any sense to Kalinske; weren't three dimensions undoubtedly better than two? Hadn't Sega CD proven that players craved lifelike graphics? What was the issue here? But when Kalinske pressed for an answer, he was told that this was better because developers would have a very difficult time making games in 3-D. And when Kalinske requested more information beyond that, he was told, in a variety of ways, that he wasn't an engineer and simply could not understand. Just like that, the future that Kalinske had envisioned was now nothing more than a what-if buried in the sands of time.

And the worst part was that Sony planned to continue with its 3-D system and enter the console market on its own, creating a whole other foe for Sega to contend with besides Nintendo. No, forget it, that wasn't even the worst part. Not only would Sony be entering the console market on their own, but they now had the advantage of knowing exactly what Sega had up its sleeve. Wait, scratch that as well, because there was something even worse: what Sega had up its sleeve was not very good at all.

"In a word," Joe Miller said, looking at a prototype of the system, "it's lousy."

"How lousy?" Kalinske asked, feeling his stomach sink.

SOJ had recently sent Shinobu Toyoda a prototype of this new 32-bit system, which they were calling the Saturn, and when it arrived he and Kalinske naturally brought it to Joe Miller, SOA's resident tech expert.

"I can't say for sure at this point," Miller explained. "I mean all I'm looking at is hardware here, chips and processors, but it's much less sophisticated than I would have expected. Whether or not it will run smoother than what Sony has planned, who knows? That will depend on a lot of factors, and good software has a proclivity for making tech problems disappear, but . . ." Instead of trying to finish the sentence, Miller just slowly shook his head.

Thank you, SOJ, Kalinske sarcastically thought over and over, and he would continue to sporadically shake his head in frustration until he realized there was still time to fix this. At the earliest, the Saturn wouldn't launch until 1995, which meant that there was still an opportunity to salvage the situation. SOJ didn't want to work with Sony? Fine. Then Kalinske would find someone they would be willing to work with.

As it turned out, it was Doug Glen who actually found the right match: Silicon Graphics (SGI). They were one of Silicon Valley's top manufacturers of high-performance hardware and software, most famous for inventing those magical computer systems that Hollywood used to make its most elaborate special effects (like in *Jurassic Park* and *Terminator 2*). Apparently they had developed a revolutionary new chip that would propel the videogame industry forward, and they were looking for a partner.

"This is fantastic!" Kalinske exclaimed, giving Glen a giant mental hug. SGI had a reputation that had to be taken seriously, but not the consumer electronics track record (or the perceived arrogance) to rub Sega of Japan the wrong way.

To see if this possible match was really as good as it sounded, Kalinske went with Miller and Glen to SGI's headquarters in Mountain View, California. There they had an incredible meeting with Jim Clark, the founder of Silicon Graphics, and Ed McCracken, the company's president. The chipset they had developed was apparently quite something. "It's going to be far more powerful than anything out there in the market today," Clark proclaimed. "I guarantee it!"

"Wow," Kalinske said. "I have to say, this all just sounds incredible."

"The next logical step in gaming," Miller added.

"Why don't you show them the demo?" Glen suggested.

"You've already got something to show us?" Kalinske asked. "You guys sure do know how to impress a guest."

"Just wait until you see what it can do," Clark said as the demo started. And he was right. Kalinske was completely and utterly blown away. He hadn't felt this way since seeing the Genesis for the first time. And boy oh boy, was it good to feel this way again.

Kalinske wasn't the only one checking out new technologies. Tony Harman was traveling around Europe in his search for the right company to make his great game. There was so much impressive work being done at a variety of places, but Harman didn't find what he was looking for until arriving in Leicestershire, England, where he met with a company called Rare.

Although each software company is fundamentally unique, Rare Ltd. had managed to rise above the pack and live up to the inherent boast behind their name. From corporate history to creative vision, they really were a rarity. That's why, in the eighties, they were the only British developer chosen to ride Nintendo's wave, and it's also why, come the nineties, Rare had seemingly fallen off the face of the earth.

It all started with the Stamper brothers, Tim and Chris. In 1982, after years of programming games for various arcade companies, the Stampers opened their own shop and started making games for the Sinclair ZX Spectrum. From a creative perspective, developing for the Spectrum made complete sense (at the time, it was the most sophisticated, fastest-processing personal computer available in the UK), but from a business perspective it was a little bit dicey (less than a million people actually owned the computer). Nevertheless, the Stampers were committed to making the best games for the best system available, so that's exactly what they did. And in 1983 they released their first game, Jetpac, which amazingly went on to sell more than 300,000 copies.

In a medium where players rarely gave a second thought to who made the games they played, Tim and Chris Stamper were the exception. Most of this fanfare could be attributed to consistently releasing hits (like Pssst, Tranz Am, and Cookie), but a small part stemmed from the brothers' seemingly reclusive behavior. The Stampers rarely gave interviews, never attended developer conferences, and generally showed no interest in coming out from behind the curtain and taking a bow. Were they really as shy as this reputation made them out to be? Probably not, but it also didn't really matter, because they were much more interested in making games than talking about making them. The Stampers believed that the quality of a game directly correlated to how much time was spent making it, which is why they famously worked eighteen-hour days, seven-day weeks, and three-hundred-and-sixty-four-day years (they took off Christmas). Given this incredible work ethic, fans anticipated great games for many years to come, but the Stampers shocked the industry in 1985 when they sold their publishing label and suddenly stopped making games.

They must have burned out, their fans rationalized. After three years (and only three days off), the Stampers must have gone just a little bit insane. That was the only logical explanation, and that theory certainly made a lot of sense, except that it was the complete opposite of what had actually happened. The Stampers weren't burned out; no, they were actually more fired up than ever. It's just that they were no longer lit by the Sinclair ZX Spectrum, but rather by a Japanese console called the Family Computer.

One year earlier, Tim and Chris Stamper had gotten hold of Nintendo's 8-bit system and were immediately convinced that this represented the future of videogames. The only problem with this new interest was that, at the time, Nintendo didn't grant licenses to developers outside of Japan. Since the Stampers were incapable of suddenly becoming Japanese, they decided to do the next best thing: reverse-engineer the console, teach themselves how to make games for this system, and then travel to Kyoto and convince Nintendo that they were worthy. They did all of this under a subdivision of their company code-named "Rare." Not long afterward, they visited Nintendo and were not only granted a license to make games but given an unrestricted one that enabled them to release as many as they wanted. In 1987, they made two games for Nintendo, and in 1988 they made four. In 1989 that total shot up to sixteen, and by 1990, with eighteen games, they were making more than any other developer.

This much productivity and this many quality games had been enormously profitable for both companies. Naturally, Nintendo assumed this relationship would continue for many years to come, but when they launched the SNES, the Stampers appeared to have little interest in programming for Nintendo's new 16-bit system. And similar to what had happened in 1985, the Stampers appeared to have mysteriously lost interest in the videogame market. This time they must have burned out for real, right? What other explanation could there be? But like before, there was a good explanation, and when Tony Harman learned it, he was absolutely stunned.

All he saw was about ten frames of a three-dimensional boxer, but it was enough to do the trick. The smoothness of the gaming environment, how quickly it rendered in real time—it was incredible. This was it, this was the one. But the boxer had to go. To make a great game they would need an iconic character. So they either had to start creating one from scratch or see if Miyamoto might be in a giving mood.

51.

THE LAST AND BEST OF THE PETER PANS

It was a bright, breezy day in June, one of those warm bronze afternoons where the sun makes you squint just enough to take notice from time to time. It was that kind of day, but even if it hadn't been, that's how Al Nilsen would have remembered it, because that's just the kind of weather that goes perfectly with bittersweet memories.

"I can't believe you're really leaving us!" Fornasier exclaimed, shaking her head.

Neither could Nilsen. It was all happening so fast, but it felt like events were moving in slow motion. How was that even possible? When would it stop feeling this way? And why hadn't someone descended from the heavens to put a stop to all this? "Me neither," Nilsen said. "But I guess it's time to move on."

They were standing on the patio of the Sofitel, joined by dozens of Sega veterans waiting for their chance to say goodbye. About twenty feet in front of them was the black-blue lagoon of Redwood Shores, which added a sense of tranquillity to the moment that turned most of the day's sadness into a soft, reflective optimism. Look at what we did together. Look at how we took ideas, or even pieces of ideas, and just went ahead and banged them into all sorts of crazy things. That was us, we did that, and there will be even greater things ahead. Best of luck, I mean it, and you better stay in touch.

"You did a damn fine job," Paul Rioux said, sternly shaking Nilsen's hand.

"It just will not be the same," Shinobu Toyoda said, patting his back.

"Viacom's lucky to have you," Ed Volkwein said with a grin.

Nilsen had accepted a job as Viacom's vice president of strategic marketing. In this role he would still have one foot in the videogame industry (supervising the marketing plans for games based on properties from Nickelodeon, MTV, Paramount, and Showtime), and he'd have the other foot in the entertainment industry (expanding Nickelodeon's licensing business). Most important, however, he'd have both feet on the ground and not ten thousand feet up in the air.

"Leaving is one thing," Ellen Beth Van Buskirk said, looking as vibrant as ever, "but did you really have to go all the way to New York? You know that's not really walking distance from here, don't you?"

"But," Nilsen began and then stopped. It took him a second to summon the right tone. "But," he said, jovially now, "it looks so close on the map."

In truth, that was the worst part about the new job—well, that and having to sport a fake smile or get into a mini-argument with anyone who said he was so lucky to be moving to New York. He didn't like New York. So many people, so much impatience in the streets!

Van Buskirk shook her head. "Is this maybe just another one of your elaborate and unexpected master plans? I'm standing here knowing that I'm supposed to say goodbye, but I keep looking up, half expecting to see a plane skywriting something like 'Just joking!'"

"Well, EB," Nilsen replied, "I think that's a very valid concern. So if I were you, I just would never bother with a goodbye."

"I can live with that."

"How's Bob?"

"Great," Van Buskirk said with a faint blush, primarily referring to her budding relationship with the handsome and argumentative business consultant she'd met at the airport four months earlier. It was always nice to see people who deserved to be happy actually feeling that way, Nilsen thought. And, strangely enough, this was a rare sight.

After Van Buskirk there were several others. Richard Burns. Michael

Latham. Deb Hart. And all the others who had helped the company grow from a question to the answer, from Sega? to Sega! Seeing the parade of faces would occasionally take him to the brink of sadness, but every time he started to feel that way, there would be some small reminder that things were no longer the same: shrugs of complacency, salary complaints, grumbles that certain people didn't know the difference between marketing and simply spending money. A few people even hugged him goodbye and then whispered in his ear that they'd be leaving soon too, that they were just waiting to get their bonus before moving on.

"Al, Al, Al," Kalinske said, walking over with that dapper smile. "I finally figured it out."

"Figured what out?"

Kalinske looked over his shoulder as if about to reveal a precious secret. "Figured out why you're really leaving," he said in a half whisper.

"Oh, yeah?" Nilsen asked. "Can you tell me? Because I'm not so sure anymore."

"I have to admit," Kalinske said, "that it was tough to piece together. You're not the type of person who would bolt for more money. And you're not the type of person who is impressed by climbing the corporate ladder."

"True," Nilsen admitted. "But I do sometimes love visualizing an actual corporate ladder and imagining men going up and down it in fancy suits and ties."

"That's exactly the type of comment that helped me arrive at the answer."

"Which is?"

"I don't know how many years ago it happened—though it couldn't have been all that long ago—you were just a little boy. A regular child, about yea high," Kalinske said, holding a hand right above his waist. "And for whatever reason—a girl, a job, a *hedgehog*—you made a wish, just like Tom Hanks in that movie *Big*. The next morning, voilà—you woke up and you were an adult. So all these years people have been calling you things like a 'big kid' or 'boy trapped in a man's body,' and you must have just been laughing so hard, knowing it was the truth." Kalinske nodded, evidently proud of his theory. "But eventually, just like in the movie, there comes a time when you need to return to just being a kid. And I guess that time is now. So I just want to let you know that your secret is safe with me, okay?"

Nilsen wagged his head up and down. He would miss Tom most of all. Well, after Sonic, of course. But it would be close.

"Just remember," Kalinske said, "it's never too late to change your mind and come back."

Nilsen nodded, even though it wasn't really true. The place had changed, and that's why he had to go. After his recent conversation with Kalinske about the difficulties of his new job, there had been a few changes, but they just hadn't been enough. And at first this had made him upset (at himself, at Kalinske, and at anyone standing in front of him at that moment), but then eventually he realized that it was nobody's fault. There was no change that could be made to his job that would put the puzzle back together.

Things were just different. And as he stood by the lagoon and listened to Kalinske proudly talk about his wife, his kids, and everything that Sega had planned, Nilsen was overcome by a thrust of sadness. The feeling was sharp and it was deep, but he couldn't figure out if he was feeling this way because the magic that had made this place so damn special was now all but gone, or if it was because hardly anybody else had realized it yet.

PART FIVE

THE TORTOISE AND THE HARE

52.

NEXT GEN

Three men stood at a podium, each one bursting with a palpable sense of pride.

To the right was Ed McCracken, the president of Silicon Graphics. Dressed in a brown suit and tan tie, he appeared somewhat uncomfortable in the spotlight. But beneath his uneven smile was a pristine confidence, confirming that this announcement was really as revolutionary as those in attendance had suspected. "It's great to be able to work with the industry leader," McCracken said before going on to describe his new partner as "one of the best marketing companies in any industry."

To the left was Jim Clark, the founder of Silicon Graphics. Although he looked like an adult version of Charlie Brown, this man was no peanut. He knew how to control a room, use smiles as effectively as pauses, and describe big, complex things as if they made all the sense in the world. "We're going to integrate graphics, computing technology, software, compression, and encryption—all of the technologies needed to make all of this happen—and we're going to put it all on one chip."

And between these two men, representing one of the world's largest videogame companies, was a calm, confident, steely-eyed executive, Howard Lincoln. "I'm here this morning to announce the next generation in Nintendo home entertainment products," Lincoln said. "A product whose improved gameplay will, simply put, be stunning. Nintendo, together with Silicon Graphics and MIPS Technologies (a subsidiary of SGI), have entered into a worldwide joint development and licensing agreement under which our companies will develop this new and unique product."

To demonstrate the gaming power of this proposed new product, a splashy video showed off 3-D graphics of a shuttle hurtling through space, a jet jittering through Paris, and a race car revving its way through a speedway.

"It will utilize the real-time, 3-D graphics technology for which Silicon Graphics is world-famous," Lincoln boasted, "and will feature a new, true 64-bit MIPS multimedia media chipset. All of this will be combined with Nintendo's unequaled expertise in videogame creation, technology, and marketing."

"A full quantum step of game performance," McCracken added.

"And it's going to be far more powerful than anything out there in the market today," Clark added as well. "Far more powerful—I guarantee it!"

Although it's easy for executives to boast about having the Best Thing Ever, the 3-D demo was proof that this was about more than just words. The graphics on display weren't quite lifelike, as there was an obvious animated quality to them, but the shapes, colors, and movements were closer to reality than anything that had ever come before.

Perhaps, then, it should come as no surprise that this new console, scheduled for release in 1995, would soon be named Project Reality. But for now, this revolutionary system had no name, nor did it even need one. The only names that really mattered were Silicon Graphics, Nintendo, and not-Sega.

A couple months earlier:

Tom Kalinske stared at the phone on his desk, hoping that something might miraculously change between now and when he dialed the number, but after realizing that hope was no substitute for acceptance, he finally picked up the phone.

"Is Jim in?" Kalinske asked when Jim Clark's secretary picked up. When she went to look for her boss, the founder of Silicon Graphics, Kalinske fought off the urge to go over exactly what he would say. Normally, preparation was a good thing, but here it would only make him sound rehearsed, and that was the last thing Kalinske wanted. He wanted Clark to hear the raw nerves in his voice, for whatever that might be worth.

"Hey, Tom," Clark said, sounding aloof but upbeat, as he often did. "What's going on? Have you come with good news?"

Raw. Honest. Aggravated. "Sadly, the opposite."

"What do you mean?"

"I've just heard back from Japan," Kalinske said with a muted growl in his voice, "and the response . . . well, it's very disappointing."

After the meeting with Silicon Graphics, Kalinske had been over the moon about the idea of forging forward together. To convince Sega of Japan that this was the right move, he worked with Toyoda, Miller, and Glen to present the case for SGI. It had many of the same benefits of working with Sony but would still give SOJ the autonomy they desired. They probably wouldn't love the idea of working with an American company, but there was no denying that what Silicon Graphics could build would be more powerful than anything being developed in Japan (including the Hitachi chipset that was currently in place to power the Saturn) and that SGI's chipset would undoubtedly be cheaper. It was a no-brainer, a guarantee that Sega would be a major player in the next generation, but SOJ didn't see it that way. They had called Kalinske earlier today to notify him that they were not interested in working with SGI. What? Why? The chip is too big. Huh? It's too big. Too big for what? Sorry, no thank you. Goodbye.

"Tom," Clark said, sounding shocked, "I know very little about what your R&D team has planned, only what Joe Miller shared with me, but I can all but guarantee you that our chipset is going to be faster, more powerful, and cheaper than what they have."

"I know," Kalinske said. "I'd be surprised if you weren't correct about that."

"All right," Clark said after a pause. "All right."

"No, it's not all right," Kalinske replied. "And I apologize for wasting your time."

"All right," Clark repeated. "What am I supposed to do now?"

Good question—and one that Kalinske had recently began to ask himself about his own future. With Sega of Japan's rejection of deals with Sony and now Silicon Graphics, Kalinske had seriously begun to doubt that this fairy tale would have a happy ending. He thought seriously about resigning, but he just couldn't deal with the thought of abandoning his team. If SOJ planned to slowly dismantle SOA, then the least he could do was stay here to fight alongside Rioux, Toyoda, Fornasier, Glen, Miller, and everyone else he'd sold on his vision.

"Jim, here's what I'm going to do," Kalinske said, now skimming through his Rolodex. "I think I have the name of someone who might be very interested in hearing about what SGI has to offer."

Kalinske scrolled past the beginning of the alphabet, slowing down as he approached the letter L. "Do you have a pen ready?" He kept skimming until he got to the contact information he was searching for: Lincoln, Howard.

53.

MAN'S BEST FRIEND

Like a lifelong bachelor looking at photographs of his ex-girlfriend's wedding, Tom Kalinske couldn't fully believe that the marriage between Nintendo and Silicon Graphics had taken place . . . except for the fact that it made all the sense in the world.

"How could this happen?" Kalinske asked, but he didn't expect a reply, nor did he receive one. Although there wasn't really a suitable answer to the question, in this case the lack of response was due to the fact that Kalinske was speaking to his dog—a frisky Airedale named Chutney, who over the past few months had come to redefine the meaning and importance of the phrase "man's best friend." Long nights at the office and even more travel than usual had slowly distanced Kalinske from his family. They were still priority number one, and he continued to make the time for soccer matches and school plays, but as the constant hustle at work weighed on him more and more, he could feel the emotional connection eroding ever so slightly. Or maybe that was all in his head. Maybe the girls were just getting older and didn't see their old man as the knight anymore, and maybe Karen was just exhausted from raising another child. Whatever the case, Chutney continued to be a source of stability in a house full of change. And these late-night walks were a great source of maintaining his sanity, giving him a chance to bask in unconditional affection and soliloquize without feeling self-conscious.

"The thing is," Kalinske mused to his beautiful Airedale, "everyone is so happy, their spirits are flying so high; you should see the smiles around the office. But I can see what's bubbling beneath the surface, and it's no good, Chutney. No good at all."

Although Sega was experiencing its best year ever, the company's long-term prospects were looking rather bleak. Not just because of the failed partnerships with Sony and SGI, but more so because Kalinske felt that his ability to lead had been neutered, and for no good reason at all. At Mattel, like most companies, corporate politics was a big part of the job; but at least there, he felt like he knew what his political rivals wanted. He may not have agreed, but he knew where they stood and could even respect their point of view. But here? What was it that Sega of Japan was looking to accomplish? Did they simply want to cut off their nose to spite their face? And where did Nakayama stand in all of this? From everything Fischer had reported, it appeared that Nakayama supported Kalinske, so was this invisible coup so strong that not even the force of Nakayama-san could stop it?

Chutney suddenly started scratching at a patch of grass and then playfully rolling around on it. It was nice to see that uninhibited joy and feel like at least someone was capable of seizing the moment. Kalinske knew that was what he needed to do—seize the moment, that is, not roll around in the grass—and mentally prepared himself to accept everything that had already happened and focus on that which he could control.

Sega was having another great year, and there would be a lot of good things coming in the months ahead. Mortal Kombat in September. The Thanksgiving Day parade in November, at which the gigantic Sonic balloon would make its debut. And likely another Christmas of besting Nintendo. This time, however, they needed to do more than just outsell their competitor. Sega needed to build a big enough lead to create some kind of a cushion for the inevitable fall. And as Chutney rolled around the yard, occasionally silhouetted by the silvery moonlight, those famous words from Mortal Kombat echoed in Kalinske's head: Finish him.

Like most nine-year-old boys alive in the fall of 1993, Chris Andresen couldn't help but slowly lean forward on the couch whenever an ad for Mortal Kombat came on the television. It was a pretty normal commercial, no special effects or anything like that, but there was just something about it. Probably it was the shouting; there was a whole lot of that. The whole commercial, really, was just all these kids running really fast around New York City and shouting "Mortal Kombat." On second thought, "shouting" wasn't really the right word. "Chanting" was more like it—they were chanting it in their toughest tough-guy voice (like the one everyone uses when they're sitting in the back of the bus), and they just kept saying it over and over as the kids took over the city.

And the coolest part about Mortal Kombat (besides the special moves, the crazy fatalities, and all of that stuff) was that the game would be coming out on Sega and Nintendo. Usually the games only came out for one system, or came out first on one of them, but Mortal Kombat was going to be on both Genesis and SNES. So depending on which version was better, it seemed like there would finally be an end to the shouting matches about Sega and Nintendo. Mortal Kombat was going to be the best game ever, so the best version of the best game ever would settle this once and for all. Finally.

That Mortal Kombat ad was usually enough to rev up someone's afternoon, but today Chris Andresen was in for a double feature. Shortly after the one for MK, they also played one of Sega's newest commercials. This one wasn't for any particular game, but instead it was a comparison between Game Boy and Game Gear. But it wasn't just like this-one-does-this and that-one-does-that; it really hammered home the point. It started off in black and white, with the camera going back and forth between both of the handhelds. And then, just when you started to notice the noise of a dog breathing loud, the announcer came in and said, "If you were color-blind and had an IQ less than 12, then you wouldn't care which portable you had." Right after that, the camera switched to color and they showed that the whole time it was actually a dog who was looking at the Game Boy and the Game Gear. And then when the dog was trying to tell the difference between the portable systems, the announcer came back to say, "Of course, you wouldn't care if you drank from the toilet either." Try hearing that for the first time and not falling over laughing.

Either way, between the two commercials, it was enough to make any boy wonder: would Santa Claus be unemployed if it weren't for Sega and Nintendo?

• • •

"Dogs slurping water out of the toilet? It's a terrible stereotype, don't you think?" Peter Main asked, but like Tom Kalinske earlier, he didn't receive a reply, nor did he expect one. It was a chilly autumn morning, somewhere between six and seven o'clock, and Main was on his morning walk with Kasi, a striking, 100-plus-pound black Labrador that he had purchased back at a Boys and Girls Club auction in 1991.

This was how he began each day, and it was also what helped get him through the hard ones. Without these walks, without these forty-five minutes to pal around with Kasi through the neighborhood's topsy-turvy hills, he'd have been a different type of man entirely. Angrier, crankier, and unable to let go of the things that made him want to punch the wall. He needed these forty-five minutes to regroup, rejuvenate, and remember what this game of life was all about. It was about winning, winning some more, and making sure to have a ball, but it took these morning walks to remind him that accomplishing that often required taking the long view. That meant no shortcuts, no impulsive responses, and no trading away tomorrows for todays.

"How about another lap?" Main asked, leading his beloved canine up the verdant incline. He wasn't a patient man by nature, but he had become that way through discipline, and these walks served as his daily reminder about the importance of this virtue. And they had become particularly invaluable as Sega continued to test Peter Main's patience. Whether it was pricing, advertising, or those curious numbers that Tom Kalinske pulled out of his ass, it was always something with those guys. For the most part, this was a good thing. It upped the ante and kept Main and his colleagues on their toes. When Nintendo had 90 percent of the market, that had been a damn fine time and very much deserved, but there's a reason that kids fantasize about coming up to bat with two strikes in the bottom of the ninth. A hero needs his moment, and Sega had set the stage for Nintendo. But keeping with the baseball analogy (Canadians may prefer hockey metaphors, but by now Main had taught himself to look, walk, and talk like an American), Sega had started throwing spitballs. The Mario vs. Sonic stuff? Fine, that was kind of clever. So were those first few Next Level ads; they didn't make much sense, but they were cool and that was the point. But this latest nonsense about drinking out of the toilet . . . that just crossed the line. Sega wanted to come after Nintendo? Fine. Compare products, compare prices, even compare the two companies' images. But don't just lob grenades and run away giggling.

Perhaps if Main had known that the commercial in question was actually titled "Tom's Dog," and that his rival at Sega adored an Airedale named Chutney as much as he loved his Labrador Kasi, he might have realized that the two men had more in common than not. Under different circumstances, perhaps they could have been friends, but as it turned out, that was not in the cards. They were destined to stand on opposite sides of the aisle, each gradually defining himself by what the other was not, and each likely gaining more from the other in opposition then they ever would have in friendship.

"You all right, girl?" Main asked after Kasi flinched away from the patch of flowers she had been investigating. "Kasi?" She briefly looked back at him and then returned to snooping through the flowers. As she did, Main saw what he imagined was the reason for her reaction: a bee loitering over her shoulder. It must have buzzed past her ears, stirred her a bit. Main thought about shooing the little noisemaker away, but Kasi seemed fine now, and that was good enough for him.

It was almost time to head back home, anyhow. The morning was about patience and playing the long-term game. It was about dropping yesterday's baggage and heading into the office with open arms. Yesterday, when he'd seen that stupid ad with the dog, he'd been tempted to respond with fire and fury. But as much as he'd wanted to do so, he'd known that the time wasn't right. Getting involved in a pissing match would end up weakening Nintendo as much as it did Sega. It would cost a lot of money and, more important, cause Nintendo to lose focus on the consumer. The better play was to stand back and let Sega tire themselves out; the ads would surely only get more outrageous, and their tech-obsessed mind-set would inevitably lead them to introduce more products than they could possibly support. And then, when Kalinske was busy juggling fifty mediocre things at once and Nintendo had its next top-grade product ready to launch, the world would see that there's no amount of B's, C's, and D's that can ever add up to an A+. In other words, the bee can buzz all it wants, but if it ain't carrying any honey, then that sucker is best left ignored.

"Time to head home," Main said to Kasi. "But don't you worry, girl. We've got a standing date set for tomorrow."

54.

NIGHT TRAPPED

"Mortal Kombat?" Bill Andresen asked his son. "What's that?"

"It's the best game ever," Chris Andresen replied very matter-of-factly.

They were in the kitchen, having a family dinner, and Andresen looked to his wife to see if she might have any idea what in the world their son was talking about.

"It's a videogame," she said. "He talks about it all the time."

"Because it's the best game ever," Chris Andresen repeated. "And it shows why Sega crushes Nintendo." Mortal Monday had come and gone and, as Kalinske had hoped, Sega annihilated Nintendo. Mortal Kombat on the Genesis outsold the SNES version and, in the process, Sega had finally, impossibly, and amazingly surpassed Nintendo. They now had 55 percent of the market, but more important than the numbers were conversations like these, where nine-year-olds were convinced that Sega was now king.

Andresen nodded. He liked the idea of buying his son a videogame. Decades ago, he used to really enjoy playing in the arcades, and he fondly remembered when his wife had first introduced him to Atari's home version of Pong. "Okay. I'll look into it."

In the coming days, Andresen looked into this Mortal Kombat and was appalled by what he found. It's not as if he expected it to be anything like

Pong, but this game was just incredibly inappropriate—the violence, the blood, the glorification of cruelty . . . And Mortal Kombat was just the first clue about what was really going on with videogames. There was one called Street Fighter and another called Streets of Rage, and those weren't nearly as bad as the newest ones coming out on CD, like Night Trap and Sewer Shark.

After this brief personal journey into the world that his son apparently adored, Andresen decided to share these findings with his boss: Senator Joseph Lieberman, a Democrat from Connecticut.

As it turned out, Lieberman was just as appalled as Andresen, his chief of staff, who went on to explain that in addition to this horrific Mortal Kombat, there were plenty of other, similar games out there. "Look, Senator," Andresen reasoned, "I realize that some people might call our reaction here prudish, but that's not really the point. These games are becoming more and more like movies, and I can't help but wonder if parents have any idea what they're buying for their kids."

This conversation and their mutual concern led Lieberman and Andresen to delve further into the matter. Although they weren't the only ones in Washington sniffing around this issue, they quickly became the most serious about getting the government involved. But before taking any action, they needed to better understand what exactly was going on. And to do so, they looked to the top of the industry and found the two companies most responsible for what was going on: Sega and Nintendo.

Perrin Kaplan, who had worked on Capitol Hill prior to joining Nintendo, still had contacts in D.C., and was informed about what was going on. This was much more than Arthur Pober talking about whispers; these were distinct conversations going on in the shadows. And this made Kaplan intent on educating Senator Lieberman about what Nintendo did. So she invited him to come visit Redmond. Kaplan expertly worked to assuage any doubts he had about making the trip west, and any lingering concerns were likely erased by Senator Slade Gorton of Washington, who had worked with Nintendo on the Mariners deal and was also a close friend of Senator Lieberman. Once the plans were made, Lieberman and members of his staff visited NOA headquarters and got a crash course on Nintendo and the kinds of games that the company produced. Kaplan talked about how Nintendo had resurrected the industry through meticulous quality control, how NOA vigorously vetted (and sometimes censored) its games to ensure a wholesome quality, and how Mortal Kombat was a great example of the difference between Nintendo and its primary competitor. This was not a slight on Sega, but merely the facts, and Senator Lieberman appreciated the information enough to roll up his sleeves, pick up a controller, and spend some time trying out the Super Nintendo.

After returning to D.C., Senator Lieberman gave the issue further consideration. On November 17, 1993, he distributed the following letter to members of Congress:

> Dear Colleague:
>
> A woman is stalked in her home, mutilated and murdered. A man is instructed to "finish" his opponent in a martial arts contest and chooses to rip his opponents still beating heart out of her body. These examples may sound like shocking cases stemming from the current epidemic of violence plaguing the United States. In fact, they are witnessed by children every day. Worse, children participate, since these examples are drawn from some of the most popular and most disturbing of a new generation of video games.
>
> Gone are the days when video games were just Pac-Man and other quaint characters. Advances in technology permit the latest video games to use real actors and actresses to depict murder, mutilation and disfigurement in an extremely graphic manner. And the technology is rapidly becoming ever more realistic. Today's graphic games may seem mild compared to the CD-ROM and virtual reality systems, which will change the market very soon. These games will come right into our living rooms if a pilot cable channel for video games slated to begin next January proves successful. At a time when real violence is threatening to tear the fabric of our country, these games glorify the most depraved acts of cruelty. While parents across the country are trying to teach their children to abhor violence, these games encourage children to enjoy violence. The *Washington Post* recently quoted a 14 year old boy on the appeal of one particularly disturbing video game: "It's violent. It's real. You can

freeze the guy, cut him up, shoot him with a hook that has a rope so you can pull him. I like the moves and stuff."

The video game industry has not addressed the danger presented by their latest creations. In fact, some have made violence a selling point. One game was introduced to great fanfare on "Mortal Monday." Currently, parents have a very difficult time knowing which video games are slightly violent and which are truly repugnant. No uniform system exists for warning a concerned parent about the violent content of a video game, which she is buying for her child.

I plan to introduce legislation and hold a hearing on this topic. The legislation will create the National Independent Council for Entertainment in Video Devices ("NICEVID"), which would oversee a two stage response to the threat of joystick violence. The first stage is a one year window in which the Council will encourage and work with the video game industry to address the problem through voluntary means. If the industry fails to take credible steps within that one year period, the Council will be charged with mandating what type of information concerning the contents of a video game should be available when a person buys a video game. At that time, the Council will have the authority to consider a variety of responses, including establishing a mandatory rating system, warning labels explaining the content of the product, or point of sale warnings.

Today, more than a third of all American households have video game systems. Almost two-thirds of children between the ages of six and 14 play video games. Parents deserve to know what they are buying for their children. If you would like to cosponsor this legislation, please contact me or have your staff contact Sloan Walker of the Subcommittee on Regulation and Government Regulation at 224-3993.

Sincerely,

Joseph I. Lieberman

As Kalinske read these words, a brittle numbness set in. All the darkest thoughts he'd had over the past two years, those moments of doubt about himself, Sega, or the videogame business at large, were being held up to a mirror the size of the Washington Monument. His toy industry days really were behind him; the man who had once empowered children to imaginatively become Masters of the Universe would now be seen as a peddler of smut, the drug pusher who gets children hooked. The first one's free, kiddies; that's why we put Sonic in the box.

Wait a minute. No. Those were just the doubts creeping in again, a variant on the universal doubts that all humans are prone to have when the sky blocks out the stars. Am I worthy of success? Am I the beneficiary of luck, misconceptions, and the work of others? Have I been a good husband, father, and friend? The letter on Kalinske's desk preyed upon his worst fears, but Senator Lieberman's words didn't magically turn doubts into facts. Well, here were the facts:

- Eighty percent of Sega's players were older than twelve years old.
- Sega was the first (and only) company to institute a ratings system, which had successfully been in place since July.
- Over 90 percent of the titles released for Sega's system were rated GA, for general audiences.
- Sega worked with Scholastic, the teacher-trusted publisher, to create a widely distributed brochure aimed at educating parents about videogames.
- On Sega's own initiative, it established a charitable foundation targeted at giving our youth a better society. In only two years, the foundation had already raised $4.3 million.
- If Sega was guilty, then so was Nintendo.

After considering this last fact, Kalinske decided to write a letter to Nintendo of America's man behind the curtain, and on November 22, 1993, he sent the following:

Dear Mr. Arakawa:

Though we have never had the occasion to meet, I believe recent developments merit this attempt to open a dialogue with you.

As you know, our companies embrace different approaches to handling fighting games and adult-appropriate interactive entertainment. I think the time has come for a comprehensive industry-wide approach to the issue of informing consumers about our products so they can make intelligent purchase decisions. In short, I think it is necessary for our nascent industry to grapple with, and for you and I to proactively lead, that industry to a uniform, responsible solution to this issue, one of which we all can be proud.

As the leading companies in the video game category of the larger interactive media and entertainment industry, you and I can forge a bond of personal commitment to doing what is right for all involved—insuring free choice and enabling people to control what comes into their own homes.

I am urging you in the strongest possible way, and with the greatest respect, to join us at Sega—and the scores of independent software companies who make software for both Nintendo's and Sega's hardware platforms—in adopting the Sega rating system administered by the independent Videogame Rating Council (VRC) which is composed of highly respected PhD's from a variety of disciplines, or to adopt some other similar rating systems and to communicate it on your new products.

You know, as do I, that your company creates and/or permits marketing of software titles for your game systems that portray a level of violence as great as anything in the industry. In fact, the PhD's of education, childhood development, psychology, and sociology on our own VRC tell us any game where the objective is to demolish enemies through martial arts or weapons should be designated for an audience over thirteen years of age (no matter if animated blood is included or not). Nintendo's game development guidelines are an inadequate way: 1) to insure consumer's freedom of choice, 2) to enable consumers to gain enough information that would allow them to exert control over the types of software titles brought into their homes, and 3) to guarantee they are always appropriate for the age of the user in that home.

I know you must appreciate that your guidelines were probably appropriate for this business when it was less sophisticated, and when over three-quarters of the user base was under 17 years old—in the late 1970s and early 1980s. Now, when our technology is so much more sophisticated, and increasingly attractive to adult audiences, it seems to me an industry-wide rating system is the type of responsible self-regulation you and your company should join us in adopting.

I am sure that we could find sponsors in our government's legislative branch who would assist us in getting an anti-trust exemption to collaborate on this matter if that is a concern that is inhibiting your positive response to this entreaty.

Mr. Arakawa, I ask you to please, with an open mind, consider this suggestion and to consider the benefit we all gain by doing the right thing for the consumers—adults and children—who have been so generous in rewarding both our companies with such great success during the last decade.

Attached is additional information on the Sega rating system.

Sincerely,

Thomas J. Kalinske

One thought that raced through Howard Lincoln's mind as he read through Tom Kalinske's letter: boo-hoo. If it weren't for Sega, none of this would be happening! You don't get to stand on the shoulders of Nintendo while trying to climb up the mountain and then ask Nintendo for help when it's time to climb back down. Boo-hoo—that said it all, but it didn't make for a particularly tactful response. Nor did shredding Kalinske's letter into confetti and sending it back to Sega.

Following Senator Lieberman's call to action, strong support in Washington had led to a series of Senate subcommittee hearings slated to begin on December 9. Lincoln would be attending the hearings on behalf of Nintendo, and Kalinske would likely be there to represent Sega. Lincoln and Arakawa believed that the hearings would be the appropriate time to reply to the issues voiced in Kalinske's letter, and so they opted to focus on preparing for that and not waste time becoming Sega's new pen pal.

Although they didn't write back to Kalinske, there was a Nintendo mailing that indirectly amounted to something of a reply—one that accomplished the take-that sensation of a confetti party but remained elegant and tactful while also serving a purpose. When the drama had first started to unfold in Washington, D.C., Perrin Kaplan had made a VHS tape containing the most graphic scenes from Sega's Night Trap, a graphic, grisly game that was unlike anything Nintendo would ever offer to its customers. And to demonstrate the difference between the two companies, Kaplan arranged for several hundred copies of these tapes to be sent to representatives in Washington prior to the hearings. If the videogame industry was about to be sent in front of a firing squad, then she wanted to at least make sure the riflemen knew where they should be aiming.

55.

IT'S JUST WINDY . . . NOT A METAPHOR

At forty-eight feet tall and twenty-six feet wide, Sonic The Hedgehog soared through the sky. Swiftly and serenely, Sega's beloved mascot made his way through the chaos of Manhattan, the newest larger-than-life balloon in Macy's Thanksgiving Day Parade through a deal struck by Tom Abramson.

"Where is he?" Ashley Kalinske asked from the window of a building downtown.

"Yeah," Nicole Kalinske added, eager to see her dad's friend Sonic.

"He's coming," Tom Kalinske told his daughters, "and he can't wait to see you. But in the meantime, don't forget to stuff your faces with food." Behind the window, in the room full of retailers, merchandisers, and other friends of Sega, was a wonderful buffet that added an exclamation mark to the occasion. It had been a hectic past few weeks, what with preparing for the Senate hearings and the media storm that would inevitably follow, but Thanksgiving offered a nice chance to take step back and enjoy Sega's success with his family. Kalinske had flown out to New York with his wife and kids a day earlier, enjoyed a quick romp around the Big Apple, and ended the day by taking his girls over to Central Park West, where they got to see the balloons being blown up outside the Museum of Natural History.

"Is that him?" Nicole asked, ignoring the table full of cookies.

"No, it's just Clifford the Big Red Dog," Ashley said, disappointed.

Kalinske put his arms around the girls and was about to remind them not to worry, but at that moment he was informed that there was an urgent phone call from Brenda Lynch that required his attention. He excused himself and walked over to the reception area to take the call. "Is everything okay?" he asked right off the bat, but of course it was not.

Apparently Sonic's incredible aerodynamics had backfired, and a vicious gust of wind had rammed him right into a lamppost at West 58th Street and Broadway. Because of the balloon's size and velocity, a light fixture had fallen and hit an off-duty Suffolk County police captain named Joseph D. Kistinger. The medics on hand believed that he had broken his shoulder and were about to rush him to the hospital.

"Is this a joke?" Kalinske asked incredulously, glancing at his daughters stationed by the window.

"No, I'm sorry," Lynch said. "It's just windy, that's all. Bad luck for us."

Maybe it really was just bad luck, something to laugh about a few years down the line, but in the moment it felt like something more than that. Fate? Karma? Metaphorical proof of a company in decline? To the outside world, decline would have been the last word to describe Sega. After the release of Mortal Kombat, Sega had pulled past Nintendo and was hotter than ever. Kalinske, of course, was thrilled by this, but ever since SOJ had thwarted the potential deals with Sony and SGI, he'd started to notice cracks in the foundation. Little things, usually, but things that he hadn't paid much attention to before. For example, Rioux had been pushing harder than ever for permission to manufacture hardware in North America. It would be cheaper this way, and save time on shipping, but SOJ would not allow it. They manufactured all of the hardware, that's just how it was. At this point, however, the cultural differences with SOJ were to be expected, but it was an unexpected incident that had him most concerned.

In early 1993, Sega had made the decision to split up Sonic 3 into two separate games. This was a calculated risk, one that they likely would have taken again, but between that decision and now, one of the game's best assets had been eliminated: Michael Jackson. While the company was working on the sound track for Sonic 3, allegations surfaced that Jackson had molested a thirteen-year-old boy. True or not, the stigma was too crippling to ignore, so a decision was made to unwind the deal with Jackson and remove any association he had with the game. At the time, that too seemed to be just a case of bad luck. But between that, the Senate calling subcommittee hearings, and the Sonic balloon fiasco, bad luck seemed to be an unfortunate trend.

"What would you like me to do?" Lynch asked.

The only good part about receiving this lousy news from Brenda Lynch was that, well, he was receiving it from Brenda Lynch, and when it came to spinning stories, that woman was a black widow. "Get down to the hospital," Kalinske instructed, "and try to make this significantly less bad."

"I'll do you one better," Lynch replied. "How about making this into something good?" Before Kalinske could answer, she was already gone. Lynch ran through the snow-paced streets of New York City, leaving behind her husband and two children so that she could get to the hospital and twist bad luck into something good.

In the following days, the press broadcast all sorts of stories about Sega's balloon going rogue; but instead of focusing on an officer down, they were almost all about how Sonic was simply too damn fast to ever slow down.

56.

COMBAT PAY

Bill White hadn't necessarily planned to pull out the gun, but when he noticed the way everyone in the room was looking at him, he felt like he had no choice.

The chips had been stacked against him from the beginning, ever since Senator Herbert Kohl (D-Wisc.) called the session to order. "This meeting," he had begun stoically, "is a joint meeting between Senator Lieberman's Governmental Affairs Subcommittee of Government Regulation and Information and my Judiciary Subcommittee on Juvenile Justice." That first sentence had sounded harmless, like run-of-the-mill politico-speak, but after that it became open season. "Before I turn the meeting over to Senator Lieberman, I want to make one point: today is the first day of Hanukkah and we have already begun the Christmas season. It is a time when we think about peace on earth and goodwill towards all people and also about giving gifts to our friends and our loved ones. But it is also a time when we need to take a close hard look at just what it is that we are actually buying for our kids. That is why we are holding this hearing on violent video games at this time. That is why we intend to introduce legislation on violent videogames as soon as Congress returns. Senator Lieberman?"

"Thank you very much, Senator Kohl," Lieberman had said, and then got right down to it. "Every day the news brings more and more images of violence, torture, and sexual aggression. Violence and violent images permeate more and more aspects of our lives and I think it's time to draw the line. I know that one place parents want us to draw the line is with violence and videogames." And then he said it, the sound bite that would forever define these hearings: "Like the Grinch who stole Christmas these violent videogames threaten to rob this particular holiday season of a spirit of goodwill."

After the good senators had set the tone for the hearings, a parade of assorted experts delivered testimony before the representatives from Sega and Nintendo who were brought in to defend their livelihood. Parker Page, president of the Children's Television Resource and Education Center, spoke first. He suggested that the federal government fund an independent research into the psychological effects of videogames, as not enough was known at the time, and he recommended that until more was known, the industry ought to put a cap on the amount of violence allowed. After that, the floor went to Eugene Provenzo, a sociology professor at the University of Miami, who had recently published a book on the subject (*Video Kids: Making Sense of Nintendo*). He indicated that his extensive research had revealed that beneath the fun of a seemingly harmless diversion lurked a number of insidious themes: "During the past decade, the videogame industry has developed games whose social content has been overwhelmingly violent, sexist, and racist." Following Provenzo came Robert Chase, vice president of the National Education Association. He began by warning against the dangers of censorship, but then seemed to upend his own thesis by explaining the incalculable negative effects of this new form of entertainment: "Electronic games, because they are active rather than passive, can do more than desensitize impressionable children to violence. They actually encourage violence as the resolution of first resort by rewarding participants for killing one's opponents in the most grisly ways imaginable." The last of the experts was Marilyn Droz, vice president of the National Coalition on Television Violence, who simultaneously complained about too much videogame violence and not enough videogames for girls.

After these experts had delivered their diverse pleas for caution, a group of industry representatives were invited to take a seat in front of the senators. This panel of five included Dawn Wiener (Video Software Dealers Association), Craig Johnson (Amusement and Music Operations Association), Ilene Rosenthal (Software Publishers Association), and the main attractions: Howard Lincoln (Nintendo) and Bill White (Sega), who were seated beside each other, only adding to the tension.

Howard Lincoln spoke first, and from the moment he opened his mouth it was clear that there was no environment on earth that could ever make him appear unpoised. The man had nerves of steel through and through, which appeared even more impressive in contrast to White's natural reticence. "In the past year, some very violent and offensive games have reached the market," Lincoln described, "and, of course, I am speaking about Mortal Kombat and Night Trap." He spoke briskly but calmly about the conscious business decision that Nintendo made back in 1985 not to allow violence or pornography, eschewing profits in favor of ensuring quality content. This was the Nintendo way, a relentless commitment to values, and as proof that it hadn't wavered since then, he cited the company's decision to censor Mortal Kombat. And not only had the company left money on the table by making this decision, they'd also received thousands of angry calls and letters from parents around the country who criticized Nintendo for censorship. If Senator Lieberman hadn't already seemed to favor Nintendo, then Lincoln's comment about the blowback clinched it. Which made it even tougher when the man from Sega spoke next.

White did everything possible to defend Sega's actions ("The average Sega user is almost nineteen years old") and appealed to those in the room as both a father ("of two boys, age five and eight") and a former Nintendo employee ("having worked there for five years"), but no matter what he said, it appeared the die had already been cast. He was the villain, and his attempts to appear anything less than evil made his villainy all the more obvious.

The longer this went on, the worse things got for White. Some of this was due to Lincoln's well-crafted expressions of exasperation ("I can't sit here and allow you to be told that somehow the videogame business has transformed today from children to adults. It hasn't been, and Mr. White, who is a former Nintendo employee, knows the demographics as well as I do"), some of it was due to Senator Lieberman's continual praise of Nintendo (which he felt had been a "damn sight better than the competition"), and some of it was due to the way that clips from some of Sega's latest commercials seemed to offend just about everyone in the room. And as the hearings transformed into a well-dressed, well-worded pile-on session, White saw no other option to defend his and Sega's honor, so he took it out. The gun.

He held it up proudly, like a championship belt, Nintendo's big gray Super Scope bazooka. White had brought it with him to Washington without advising anyone from Sega, and kept it under the desk in case things got out of hand. As the eyes in the room squinted at him with disdain, he saw no other option, and so he pulled it out. "A rapid-fire machine gun," White said decisively. "And they have no rating on that product."

It should have been Bill White's moment of triumph, delivering the kind of personal screw-you that creates a lifelong pocket of confidence, but he had flubbed his words. Upon lifting up the bazooka, White had said accusingly, "I may also point out that Sega produces this product," when obviously he had meant to fire a kill shot at Nintendo by naming it as the producer. Was this actually an illuminating Freudian slip, indicating that the only executive who had worked at both Sega and Nintendo saw, deep down, that the two companies were interchangeable? Perhaps, but it didn't matter. Bill White had fired his silver bullet, and as noble as the effort had been, nobody in the room appeared to be particularly moved by it, causing the shot to come right right back at him.

"No!" Kalinske exclaimed when he heard White jumble his words. He was back in Redwood Shores, watching it live on C-SPAN from the television in his office, and despite the flub, he was very pleased with how White had handled himself: perpetually poised, meticulously defensive, and just the right amount of angry.

And so the following day, when White returned and walked through the office doors at Sega of America, he was greeted by a standing ovation. Peter Main's former protégé had delivered an admirable performance for Sega, and as a result, Senator Lieberman and the other members of the committee had decided to give the industry a chance to regulate themselves before the government chose to intervene. There would be additional hearings in March 1994, to check in and see if enough progress had been made, but for now the threat had been deterred.

For Kalinske, this turn of events meant even more than just Sega's safety, and it relieved a lump on his conscience that had been growing for some time. Whether he was the true villain, for having allowed such games to be released, or the actual hero, for creating the first rating system, didn't matter anymore. It was time to move forward, and he would be doing so unscathed. He had Bill White to thank for that.

To show his gratitude, Kalinske gave White the largest Christmas bonus that he had ever received. When he asked his boss what it was for, Kalinske smiled and explained that it hadn't been his idea. "As you know, a week or two after your triumphant performance, I went down to Capitol Hill to speak with many of the same men who had grilled you. One of them, of course, was Senator Lieberman. I asked him what he thought about how you had handled yourself, and do you know what he said?"

"B-plus?"

Kalinske laughed. "He said, 'For what he put up with, that kid deserves combat pay.'" He motioned with his chin to the bonus check. "Well, as an upstanding American, I've been taught that it's wise to listen to your government representatives. So there you go!"

57.

LIFE ON MARS

Tom Kalinske, Paul Rioux, Shinobu Toyoda, and Joe Miller were all in Vegas for the 1994 winter CES, and they could all be found slowly walking down one of the exquisite hallways in the Alexis Park Hotel as they approached Nakayama's suite.

"Do you think he watched the hearings?" Miller asked.

"Watched them?" Kalinske asked. "Or will admit to watching them?"

"I know that he saw at least a part," Toyoda put in, anticlimactically ending the group's speculation about Sega of Japan's volatile ruler.

When they arrived at his suite, they were ushered in to sit down with Nakayama-san and his chief lieutenants at SOJ to discuss the future. In the hours that followed, a lot of strategies for the year ahead were tossed around, but amongst SOJ's diversity of strategies one thing was clear: they were ready to kill the Genesis.

"Look," Kalinske said, "I think that staying on the cutting edge is as important as everyone else in the room, but not at the expense of something that's still working. I mean, we're barely even in the middle of the console's life cycle!"

This statement was generally accurate, but it was also false, which is what accounted for part of the schism in the room. Kalinske considered the summer of 1991 to be the de facto American launch of the Genesis. That was when SOA had dropped the price, introduced Sonic, and started doing the kind of marketing that would redefine Sega. If that was considered the starting point of the life cycle, then this was really only year three. Sega of Japan, on the other hand, had released the Mega Drive in August 1988, never enjoyed the second wind that SOA had, and consequently viewed this as something like year six. Ultimately, though, this all boiled down to the fact that despite SOA's earth-shattering success with the Genesis, SOJ had continued to strike out with their Mega Drive. As a result, Nakayama proposed moving forward with a pair of systems to take Sega into the next generation:

Project Saturn: A CD-based system with 32-bit technology that would effectively serve as Sega's next-generation console. No surprise here. This was the SOJ-only system that had been selected in favor of developing something with Sony and Silicon Graphics.

Project Mars: A cartridge-based system with 32-bit-like capabilities. This was basically a souped-up version of the Genesis.

Kalinske understood the merits of the Saturn, which would eventually replace the Genesis as Sega's next-generation console, but what exactly was Mars supposed to be? SOJ had been pitching this "32-bit-like" device for almost a year now (he had been shown something very similar to this back when he'd first seen the Pico), but as much as he enjoyed pushing the envelope with new products, he wasn't convinced that Sega would have the resources to fully support this. Nor were Rioux, Toyoda, and Miller (as well as Nilsen, who was adamantly against this until the day he left). Sega was already asking developers to create games for the Genesis, the Game Gear, the Sega CD, and the upcoming Pico. That already pushed the limit, and that's why the guys from SOA couldn't help but look at one another, spooked by the idea of adding more hardware to the mix.

To clarify the supposed necessity, Nakayama explained that Mars would bridge the gap between the Genesis and the Saturn. Kalinske didn't understand what gap needed to be bridged: the Genesis was still selling well and the Saturn would be out in just over a year. But Nakayama explained that this wasn't an issue up for debate. It was now starting to become clear that

Kalinske was no longer the chief decision maker. Sega of Japan was sick of being compared to the much more successful SOA (and sick of watching the company's name get dragged through the mud in Washington), so they had decided that it was time to take back the company, at whatever cost. So Kalinske had better buck up and get used to life on Mars.

Frustrating as the meeting in Nakayama's suite may have been, it wasn't the only get-together between those with opposing ideas for the future. In an effort to prevent Washington from putting the videogame industry at the whims of the FCC, representatives from Sega, Nintendo, and the prominent players in the software industry agreed to a secret, off-site meeting in Vegas. The cloak-and-dagger aspects may have felt a bit excessive, but they were necessary to avoid allegations of market collusion.

The first few hours consisted of little more than shouting and swearing. This is all your fault! No, it's all yours! Fuck you. No, fuck you! But beneath the infighting, the meeting provided an interesting opportunity for the folks from Sega and Nintendo to finally sit down in the same room. Tom Kalinske, Shinobu Toyoda, and Bill White; Howard Lincoln, Peter Main, and George Harrison. All the alpha dogs had willingly come to the trough.

"We've all come here for a common goal," Kalinske explained, "and the answer seems rather simple. Why doesn't everyone simply adopt Sega's rating system?"

"How is that the simple solution?" Peter Main asked.

"Why adopt a system at all?" Lincoln asked. "Nintendo already closely reviews all of its content."

"How many times do you plan to repeat that?" White wondered aloud, and the shouting and swearing escalated once again. In the middle of the bickering, Arthur Pober arrived to discuss the possibility of using his expertise to create an industry-wide rating system—something very similar to what he had created for Sega, but which would address the concerns of the other companies at the table. Shortly after surveying what was going on in the meeting, however, Pober was ready to call it a day. "I have no interest in this," he declared as he walked toward the exit. "You are all men: act like it."

Bill White chased after him, urging him to reconsider, but Pober was adamant and said to call him at the hotel if they ever managed to get their houses in order.

Later in the day, while watching some terrible movie that must only ever get watched from hotel rooms, Arthur Pober received a call: everyone was now on their best behavior and they would like for him to return. He accepted, put on his jacket, and looked at the clock. It had only taken a few hours. Not awful. Perhaps there was hope for them yet.

Pober's optimism was confirmed when he got back to the meeting site. The guys from Sega and Nintendo no doubt despised one another, but beneath the anger and irritation they were all incredibly smart men. And they were smart enough to know that the he-said/she-said crap gets old real fast.

"Look," Kalinske said, "it's obvious our companies have different approaches to doing business, but while we're stuck together in this room, we need to suspend the past."

"Tom's right," Lincoln said in support. "We're all in this thing together, and we need to come out of this with a united front. I don't like it any more than all of you folks do, but I like it a hell of a lot more than the alternative."

It became clear to Pober rather quickly that in a room full of leaders, Kalinske and Lincoln were the ones whose voices carried weight. And that was a good thing, because he'd worked with them before and knew them both to be capable of one of the rarest of feats: not letting personal feelings or selfish desires cloud their judgment. And so by the end of the day, while nobody would be fool enough to say that Sega and Nintendo (or Kalinske and Lincoln) had become friends, they had certainly become friendlier than they'd been before, and demonstrated that they were willing to put down their swords in order to fight for the greater good.

"What do you think?" Kalinske asked Joe Miller, who had come into his office not long after CES to check out the latest prototype of the Mars project.

"I think what you think," Miller said, "just with some more technical language."

"Can it be salvaged?"

"Anything can be salvaged, but the question is at what cost? Regardless,

I've been speaking with a couple of the guys at SOJ, and if this is a road they are intent on going down, I have to seriously suggest that we consider doing this as an add-on."

Kalinske shrugged. "Pitch it as the next Sega CD?"

Miller nodded. "Yes, except I would note that the movie-like quality of those titles and the unique look of a compact disc make it an easier pitch."

"Yeah, well," Kalinske said, "if we've got to cut off either our hands or just our thumbs, the thumbs seem like the better choice."

Kalinske relayed this feedback to Nakayama, who was now preoccupied with Project Saturn, which had run into a lot of developmental issues. SOJ needed all hands on deck to fix the Saturn and decided that Miller and his SOA crew should finish up the Mars themselves. As a "compromise," they let him do it as an add-on. It was now Sega of America's problem to salvage a product they didn't even want to exist in the first place.

Some merely saw this as a hiccup in Sega's plans for world domination; after all, on the surface Sega still had a bigger market share than Nintendo. But there were others internally who could see that the tsunami was headed their way and it was time to get out. Two of those men were Richard Burns and Doug Glen, who had decided that it was time to ride off into the sunset and into another opportunity. Both losses hurt Kalinske, as did remembering that Glen had a reputation for leaving companies that had reached their tipping point in order to join the next big thing.

Whats that? Whatcha working on? Mind if I take a look?

Harman had recently grown accustomed to the constant interruption of questions like these from his colleagues, but it was starting to become a distraction. He knew there was nothing but good intentions behind these inquiries, and he wouldn't have expected any less from Nintendo's congenial open-door atmosphere, but as Rare got closer and closer to finishing the game that he hoped would finally crush Sega, staying focused was more important than ever. That's why he decided he had to speak with Arakawa.

"I see," Arakawa said, weighing the situation. "What shall be done?"

The most obvious solution would have been for Harman to work out of Rare's office in Leicestershire until the game was complete, but that also would have defeated the purpose of building up NOA's product development capabilities. So in the spirit of this objective, he offered an alternative solution. "With your permission," Harman explained, "I would like to build a sealed-off area dedicated solely to development."

"Okay," Arakawa replied, surprising Harman.

"You don't need me to provide additional explanation?"

Arakawa shook his head. The decision was made.

"I just want to make sure you fully understand what I'm asking for," Harman said. "Basically, it would be the equivalent of a top-secret, game-centric fort somewhere in the middle of our office."

"Yes," Arakawa said. "This is a good idea."

Harman smiled serenely, fantasizing about what exactly his treehouse for adults should look like.

When Glen, Burns, and other members of the team left, Kalinske did the same thing as always: smile, nod, and wish them the best of luck, never revealing the icebergs of discontent below that winning smile. That's what he did during the good days (like the "Hedgehog Day" event that Fornasier arranged to take in Punxsutawney on Groundhog Day 1994), and that's what he did during the bad days (like when he saw specs for Sony's upcoming PS-X console).

It was rare that anyone could get far enough under Kalinske's skin to unleash the full spectrum of emotions, but Peter Main had made a career out of accomplishing rare feats. And in February 1994, there was something about the way that Main painted Kalinske as a charlatan that got so far under the president of Sega's skin that it sawed deep down into his bones.

They were at Piper Jaffray's annual industry conference in New York, where a couple hundred bankers and retailers gathered each year to hear about the profits and potential of the videogame industry. Speeches would be made by a variety of analysts, representatives from premier software companies (like Acclaim and Electronic Arts), and of course Peter Main and Tom Kalinske. Just like at the CES shows, it was always a war of words between the two, but this conference each February (along with the Gerard Klauer Mattison one each October) offered an unusual chance for both men to engage in that war from the same room. It was especially exciting for those in attendance to watch as they slung arrows, argued over numbers, and couldn't stop glaring at each other.

Normally, Kalinske enjoyed playing up the rivalry and reacting to Main as if they were mortal enemies. But this year, because of everything going on inside Sega, playing a game of heroes and villains just didn't seem as appealing. He was sick of the charade, year after year, and wanted Nintendo to just go away. Hadn't he done enough to vanquish them by now? Or at least relieve them of some of their arrogance before Sega went into decline?

Despite the fact that since the release of Mortal Kombat Sega had become the leader in 16-bit, Peter Main still talked about Kalinske as if he were some rube from the toy industry. "Now, after I'm done speaking," Main said, "I'm sure my competitor will try to fatten you up with all sorts of numbers. But I'd like you to ask yourself what these numbers really mean, and where he came up with some of these magical statistics. Oh, and while you're at it, find out if he has any idea what the difference is between sell-in and sell-through."

After the snickers of laughter and a few never-short anecdotes from Main, Kalinske finally got his chance to speak. His words carried none of that for-the-greater-good attitude that they'd held at the offsite meeting in Vegas. "Is it just me, or does anyone else feel like calling up Dr. Kevorkian whenever Peter Main is done speaking?" And from there, Kalinske didn't relent. "Did you ever notice how Peter Main or any of those guys at Nintendo never have the guts to say my name, or even the word 'Sega'? It's always 'our competitor' or 'another company in the industry,' but I suppose we should take this as a compliment. After all, it's the same way my daughters refer to the boogeyman and other things that scare them."

Although Main, like Kalinske, had a talent for removing personal feelings from judgment, he did not share Kalinske's skill at removing them from his face. He looked angry, as well he should be, because just as with that dog-drinking-from-the-toilet commercial, Kalinske wanted to turn everything into a popularity contest.

"But we shouldn't be surprised," Kalinske continued. "I'd be scared too if I were them. Peter wants to talk about numbers, so let me pass along some figures that I notice he forgot to mention. In November, the Nintendo Company announced a drop in earnings for the first time in ten years."

As murmurs rippled throughout the audience, Kalinske's eyes glanced at the window, and he noticed that the snow was still coming down hard. There was a storm passing through the city and they were expecting a few inches or more, but as the relentless whiteness kept falling, he couldn't help reflecting that there was a certain beauty to it. Of course, it wouldn't make finding a cab any easier, but to a guest from California, the snow came as a nice surprise.

"Anyway," Kalinske continued, "as I was saying, the numbers look pretty grim. Peter also forgot to mention that Nintendo's first-half pretax profits dropped 24 percent from the same time last year. Meanwhile, for those in the audience who like to compare and contrast, Sega's half-year pretax profits are up by 4.3 percent."

Kalinske went on to describe the many reasons he believed the two companies were going in opposite directions, and did so with that poke-the-bear flair throughout. "Now, I realize that there are some of you listening," he continued, looking to convert any lingering doubters in the crowd, "who might not believe a word that I've just said. Forget the marketing, you might say; forget my management style, the work of my team, and Sega's commitment to remaining on the cutting edge. That's not the reason that we have surpassed Nintendo, you might think; it's all just because we were first to market way back when. Obviously I don't agree with that analysis, but even if that's how you feel, I've got news for you: it's going to happen like that once again."

It was true. Nintendo had made a grave mistake by letting Sega beat them to the 16-bit market, and they had conceded over the past few months that Sega was going to speed past them once again. Nintendo assured the press that everything was going extremely well with Project Reality and that the new hardware would apparently retail for under $250, but that as part of the company's commitment to excellence there would be a slight delay. Project Reality wouldn't hit stores until late 1995, which would be one year after the 3DO and likely several months after the Sega Saturn (as well as Sony's hardware system, if they decided to take the leap and enter the market with a system they were now calling the PS-X). Kalinske still wasn't particularly excited about SOJ's mandate for Saturn and Mars, but at least they'd get to market quickly and provide Sega with additional opportunities to try to finish off Nintendo. And in keeping with the spirit of knocking them off the ledge, he didn't stop needling them throughout his speech.

When Kalinske finished and the after-conference mingling had come to a close, it was snowing harder than before—there must have been at least eight inches on the ground already, maybe even ten. Just then Kalinske caught the tail end of a conversation between Main and Michael Goldstein, the president of Toys "R" Us, as they made they their way toward the elevator. Main was still trying to cast aspersions on Kalinske's sales figures, and so Kalinske decided to intervene.

"Do you really think I make this stuff up?" Kalinske asked, following the two of them into the elevator.

"Why, hello, Tom," Goldstein said, happy to have both men together to settle this.

"It wouldn't be the first time," Main replied, staring past Goldstein at his competitor. "And I hardly suspect it will be the last."

"How about this?" Kalinske suggested as the elevator descended. "When we leave New York, why don't you come back to Redwood Shores with me? I'll show you the data we used to come up with these numbers."

"What the hell is that going to accomplish?" Main asked. "I doubt your numbers, not your ability to create a paper trail."

"Gentlemen, please," Goldstein interrupted, but the sparring continued even as they reached the lobby, and it had nearly gotten physical by the time they stepped out onto the snow-filled streets. "Without Nintendo there wouldn't even be a goddamned videogame industry!" Peter Main declared, now inches away from his competitor's face.

"What the hell do I care?" Kalinske asked, moving even closer. "Do you expect me to be grateful or something? This is business, not charity."

"It was a business until you came in here with all your bullshit!"

"Then it must hurt even more, to be losing to nothing more than bullshit."

Before any punches could be thrown, Goldstein wedged himself between Main and Kalinske. "Come on, it's time to go back to your hotels."

Both men wanted nothing more than for Goldstein to suddenly vanish, leaving them to try to transfer years of unrelated frustrations into each other's face, but the president of Toys "R" Us refused to give them this gift. He tried to flag down a cab so that he could send one of them away, but it proved very difficult because of the snow.

Eventually, after several failed attempts to hail a taxi, Kalinske decided to walk uptown, and he slogged away through the thick snow—maybe because he felt this made him the bigger man, or maybe because he was just too brittle to maintain all the anger. But more likely it was because by walking uptown on this one-way street, he'd be able to intercept a cab that otherwise might have gone to Peter Main.

58.

ROSES ARE RED

Throughout most of February 1994, New York City was puddled in snow and slush. It made getting around town much more difficult than usual, but the blizzard-like weather wasn't enough to keep Al Nilsen and Ellen Beth Van Buskirk apart. Since Nilsen had moved to New York for Viacom, and Van Buskirk had moved there as well to work at Sega Channel, they tried to stay in touch and get together for coffee every so often. Not only did these connections confirm to both that this friendship was of the lifelong variety, but whenever they got together it had the strange effect of rekindling that old Sega magic. Alone each of them was a smart, confident, and clever individual, but together they became something else. Over the years it would always be that way, not just with the two of them, but also with Kalinske, Rioux, Toyoda, Fornasier, or any of the others who had unknowingly forged this unbreakable bond. Perhaps that was what made team success such a beautiful thing; the wins and loses were inherently impermanent, but that feeling among those on the roller-coaster ride would somehow sustain forever.

"Remember that time we stayed up at CES all night and created the Core System?" Van Buskirk mused. "And remember how annoyed those Nintendo guys looked the next morning when we came out before them with $99?"

"I'm still not convinced that night followed the normal rules of time," Nilsen said. "I've gone over it in my head several times since, and I've come to the conclusion that between two and four a.m., at least eight hours actually passed."

"Oh," Van Buskirk added, happily tapping the table, "and remember how strangely talented Diane was when it came to breaking and entering?"

"How could I ever forget? How is she? Have you spoken to her?"

"She's great. Enjoying life, enjoying work, and enjoying whatever comes in between. Actually, and please don't repeat this, she and Don are trying to get pregnant."

"Of course not," Nilsen said.

"Oh, remember when we all had to pull that Aladdin EPROM heist?"

"Um, I'm sorry to interrupt this trip down memory lane," Nilsen said, "but I'm going to have to pull you over and check your license."

"My license to go down memory lane?" Van Buskirk asked.

"Yes," Nilsen said. "I believe it's called a nostalgia permit."

"Oh, sure. And why is it that my license has been revoked?"

"Because you still work for Sega!"

"True. But not really. Sega Channel is different."

Van Buskirk had expected the transition to be something like being traded in baseball, going from, say, the San Francisco Giants to the New York Mets (definitely not the Yankees; she was forever a woman who despised the designated hitter). But moving from Sega of America to Sega Channel turned out to be more like being traded from the San Francisco baseball Giants to the New York football Giants. It was a completely different sport and not nearly as fulfilling. She still completely believed in the vision of the company—it was a brilliant idea whose time would certainly come—but for the moment the cable operators made it impossible in the present. By June 1994, Sega Channel would have a total of twenty-one companies signed up to carry the service, but the cable companies were so boringly bureaucratic and so disinterested in making the system work well or easy to use that it was going to be a struggle. Van Buskirk was up for the fight, but she missed all of her friends back in California. Well, except for Nilsen.

"How are Beavis and Butt-head?" Van Buskirk asked.

"Oh, you know," Nilsen replied. "They're always trying to get me into trouble, but so far I've managed to resist their bad influence."

"Good for you!"

"Thanks, EB! Wait, am I still allowed to call you that?"

"Why not?"

"I don't have to call you EBVBK?"

One month earlier, on January 15, Ellen Beth Van Buskirk had married Bob Knapp, the man she had met almost a year earlier at Gate 81. Nilsen was there, and so were Rioux, Fornasier, Race, Glen, and Schroeder (the whole gang, except for Kalinske, who had suffered a tennis injury earlier that day). It was a nice little reunion and made for an even more memorable wedding day.

"I told you I wasn't going to take his name," Van Buskirk reminded Nilsen. "Otherwise my name would just turn into a tongue twister."

They had a laugh as they mentioned favorite tongue-twisters and challenged each other to see who could repeat them faster. After racing each other through Sally's efforts to get into the seashell business, Van Buskirk remembered that she had brought something she wanted to show him. "Have you seen this?" she asked, pulling out a copy of the February 21 issue of *Business Week*.

"Oh, wow," Nilsen said, smiling at the issue. On the cover was a drawing full of Sega characters: Sonic, Tails, ToeJam, the race car from Daytona, and a few others. Above them, the name Sega was in bright yellow letters, with a laudatory subtitle: "The $4 Billion Company That Stung Nintendo Is Making a Risky Push into the Exploding World of High-Tech Entertainment."

"I need to run," Van Buskirk said, gathering her mittens and preparing to brave the snow. "But you can borrow that if you'd like."

"Fantastic," Nilsen said, and bade her farewell. But before heading back to the land of Beavis and Butt-head, he stayed in the coffee shop with the magazine to spend a few more minutes with Sonic and friends.

The story was as wonderful as the cover indicated, but it was something at the very end of the story that caught his attention. After all the pages of gushing about Sega, Sonic, Kalinske, and Nakayama, there was an article written by Neil Gross and Robert D. Hof after a recent interview they had conducted with Nintendo's Hiroshi Yamauchi. The article itself wasn't all that interesting, but what was interesting was that although it appeared in *Business Week* and would be read by millions, it seemed as if it were written exclusively for Yamauchi's son-in-law:

> After months of being bloodied by Sega Enterprises Ltd. in North America, Nintendo Co. President Hiroshi Yamauchi is taking off the gloves. In a talk with *Business Week* at his spartan headquarters in Kyoto, Yamauchi revealed a plan to recoup market share. Many details remain vague. But the stern, 66-year-old patriarch of Japan's biggest game company made one thing clear: He'll make big change in the way Nintendo manages its U.S. Operations, promotes its products, and develops its game.
>
> His first priority is fixing the disaster in the U.S. market, where Nintendo's share of the 16-bit market plummeted from 60 percent at the end of 1992 to 37 percent a year later, according to Goldman Sachs & Co. With surprising candor, Yamauchi lays part of the blame on his son-in-law, Nintendo of America Inc. President Minoru Arakawa. When Sega started running comparative ads in 1990, Nintendo failed to respond. In effect, says Yamauchi, Arakawa "allowed Sega to brand our games as children's toys. It was a serious mistake."
>
> Yamauchi expects Arakawa to change his style—and hand more responsibility to senior American staffers. "I'm giving him another chance," says the president. "But even in Japan, if results don't improve, you can't stay in a job." Informed of those comments, Arakawa issued a statement promising that "1994 will be the most aggressive marketing year Nintendo of America has ever seen."
>
> What stopped Yamauchi from acting sooner is that the U.S. unit did so well under Arakawa's early stewardship. From 1987 to 1991, Nintendo's exports in the U.S. grew eightfold. Americans have snapped up over 65 million Nintendo game machines since they went on sale in 1985. But growth has slowed in the past year and profits are falling.

Big words, Nilsen thought as he finished, but he had trouble imaging that Nintendo would actually do anything different. Yamauchi might be upset, but it wasn't like he was going to fire his son-in-law.

Nilsen was correct, Minoru Arakawa would not be fired, but Yamauchi did end up firing a warning shot at his son-in-law. Days after the interview,

Howard Lincoln received a call from Yamauchi and was informed that he was Nintendo of America's new chairman (making him comanager of NOA with Arakawa). It was just a title, and Lincoln had always been steering the ship right there with Arakawa, but it meant that for the first time in years, Nintendo was willing to change. And that was significant.

A couple of months later, Perrin Kaplan was in San Francisco to visit Silicon Graphics and discuss ways to position Project Reality. She was joined there by Don Varyu, a former news director (he had won the 1988 Edward R. Murrow Award for the best local news operation in the country) who had transitioned into public relations and been working with NOA since 1991. He was a lover of telling stories, a man who knew how to combine big ideas with small anecdotes, and as a result of this passion and talent he had earned Nintendo's trust and was often involved in the company's big-picture messaging. He was also usually the one who worked with Peter Main and Howard Lincoln to craft their speeches for press conferences, analysts' meetings, and the CES shows.

"I'm sure that Perrin already has at least a million golden ideas," Varyu explained to the guys at SGI, "but I think something we should keep in the back of our minds throughout is the cautionary tale of 3DO."

The 32-bit 3DO console had only been out for a few months, but it had failed so badly that it already felt like a thing of the past. A large part of the failure was the system's ridiculous price tag, $699, but another part was that it had been marketed so heavily as the technological answer to everything that nobody thought of it when it came to a specific anything. It was a videogame system, a movie player, and a CD-ROM device, but nobody thought of it as the videogame system, the movie player, or the CD-ROM device. Even more important than taking it as a lesson in the value of specificity, Nintendo interpreted the failure of the 3DO as proof that consumers need more than just technology. For years, Sega had been goading Nintendo into entering a technological arms race, and although they had thus far resisted, there was always that temptation to consider the what-ifs. But after 3DO, Nintendo felt even more confident in their longer-term, higher-quality strategy, particularly when it came to demonstrating patience with Project Reality.

As Kaplan and Varyu started discussing the broad strokes of the messaging behind Nintendo's new console, a secretary from SGI interrupted them with an urgent message. "Howard needs to speak with you immediately," she advised Kaplan, who went outside to take the call. Minutes later she returned, looking half astonished, half petrified. "You're not going to believe what Howard wants to do."

"Probably not," Varyu replied. "But now I'm very curious to find out."

"How do you feel about poetry?" Kaplan asked, still with that peculiar look on her face. "Particularly poems that rhyme."

"Uh-oh," Varyu replied, now sporting the same expression as Kaplan.

Earlier that morning, Howard Lincoln had come across a quote from Tom Kalinske in which he claimed to be amazed that Nintendo "would so irresponsibly drag retailers and the entire video game . . . industry through the mud in their efforts to slow down our momentum." Lincoln was stunned by the quote; it represented exactly how he felt, except about Sega. They were the ones who had caused this whole mess and threatened to ruin the industry with their blatant disregard for integrity. Not only was Kalinske's quote patently ridiculous, but the timing of it was almost as offensive.

About six weeks earlier, Sega, Nintendo, and the industry's other major players had come to an agreement to set up a national ratings board, which would later become the Entertainment Software Ratings Board (ESRB). Dr. Arthur Pober would head the effort and work to get the operation up and running by the end of year. To supervise the ratings board and also usher the industry into maturity, Kalinske had played a vital role in creating the Interactive Digital Software Association (IDSA). Accomplishing both of these things in such a short period of time was no small task, but it had the desired effect of impressing Senator Lieberman and the other representatives, so following the second round of hearings in March, the politicians decided to give the videogame industry the autonomy to regulate itself. There was no doubt that Kalinske deserved a significant amount of credit for rallying the industry, but so did Lincoln. Pulling this off required unity, compromise, and string-pulling on the part of both leaders, but they had gotten it done and managed to save their respective companies from the clutches of Congress. Lincoln harbored no illusions that they were now friends, nor did he believe that Sega and Nintendo would ever blissfully hold hands and waltz into the next generation of videogames, but come on . . . for Kalinske to say what he'd said, and to say it so soon? Give me a break.

Lincoln was mad and wanted a way to fire back at Kalinske. In the past that outlet had often been litigation, but here that made no sense. The most efficient and effective response would be to give an interview or issue some kind of press release. But both of these methods felt too tame, too expected, and too unmemorable. Lincoln would say something, and then Kalinske would say something back, and the word count between the two would pile up faster than dead Montagues and Capulets. What Lincoln needed was a way to end the conversation and make sure Nintendo had the final word—and, ideally, have some fun in the process. Over the past couple of years, and particularly the past couple of months, Nintendo had stopped remembering to have fun. That likely had a lot to do with Sega taking a bite out of Nintendo's market share, but it was time to remember what Nintendo was all about.

When asked what she did for a living, Perrin Kaplan used to say, "I sell joy," and she'd been 100 percent correct. Nintendo was about wandering through the Mushroom Kingdom and saving the princess. That was what they did for a living, and it was time to bring back the fun. Yamauchi had appointed Lincoln as chairman to start making changes, and there was no better way to demonstrate this new era at Nintendo than by publicly doing something fun.

Kaplan loved this attitude and was with Lincoln every step of the way, except when it came time to demonstrate the fun.

"A poem?" Varyu confirmed.

"Actually," she revised, "a roses-are-red poem."

"That's terrible."

"Isn't it?"

"Absolutely."

"Good, because that's what I told him."

"And what did he say?"

"He said I was right."

Varyu sighed with relief. What an embarrassment that would have been! After a relieved chuckle, he and Kaplan got back to discussing Project Reality. Things were going well until an hour later, when the secretary returned with news that Howard Lincoln was on the line again. This time Varyu went into the hallway with her and stated his case for why this was a terrible idea. Although Lincoln was angry with Kalinske and upset by their reaction, he could see the reason in their logic and thanked both of them for their advice.

Once again, they felt glad to get back to their discussions about how to position Nintendo's new 64-bit system: sophisticated but accessible, a high-powered machine but one that didn't look intimidating. Everyone was on the same page and they appeared to be getting somewhere until, for the third time, Howard Lincoln called to discuss poetry.

"It's still not a good idea," Kaplan responded in faux exasperation. "And since we last spoke, I think I've come up with at least ten more reasons why!"

"Howard, buddy," Varyu added, "I admire the effort, but you just can't do it."

"Well, actually," Lincoln replied, sounding as calm and calculating as he always did, "I can. I'm the chairman, and I'm doing it."

The following day Nintendo issued a press release with the poem that the new chairman had written:

Dear Tom,
Roses are red,
violets are blue,
so you had a bad day,
boo hoo hoo hoo.
All my best,
Howard

59.

BLAST FROM THE PAST

"Now tell me," Olafsson said from his office in New York, as he interviewed candidates for the president's job at Sony Computer Entertainment of America (SCEA), "what have you been up to since departing Sega?"

The candidate chuckled. "How much time do you have?"

Olafsson liked this guy already, and his answers further convinced Olafsson that he was the right man to run SCEA, which had been formed in May 1994 as the formal division within Sony to launch its new videogame console, PS-X. Much like he did with Sony Imagesoft and Psygnosis, Olafsson would oversee this new division, but whoever became president would be given the space to operate as he or she saw fit. Although the Sony name carried a lot of weight, there was skepticism about this venture from retailers, from distributors, and most of all from consumers. Whoever became responsible for launching the PS-X would need to be an intuitive thinker, an inspiring leader, and a smooth talker—and, as crass as it may sound, would need to have balls of steel. "But on a serious note," Olafsson continued, "where has the world taken you?"

The candidate, Steve Race, smirked. He loved to tell a good story. "Let's see . . . After Sega I did a few turnaround assignments for some venture capital firms that I've kept in touch with over the years. Nothing wild, nothing lasting more than a couple of months, until I got a call from Philips. They wanted me to come in and evaluate this product called DBI. Have you ever seen this thing? I'd sprinkle in a few details about the piece of shit, but it's a waste of goddamn brain space if you ask me. Anyway, I came to the conclusion very quickly that this was a noncompetitive product and they ought to shut it down. They became we, ultimately me, and I was given the exciting task of dismantling this needlessly elaborate house of cards."

"Is that something you take pleasure in?"

Race shrugged. "I'm good at it, I guess. I seem to do well with putting things together, or with taking things apart."

"Beginnings and endings," Olafsson said. "Well, that pretty much runs the spectrum, then. Do you not possess any areas of weakness?"

"Oh, I've got plenty of those," Race said with a grin.

"Would you like to share?"

"Nah," Race replied. "If you bring me in, you'll find out soon enough."

The interview progressed very nicely from there. Olafsson appreciated Race's run-through-a-wall mentality and his ability to convince others to follow him through. This would be particularly valuable given that Sony would be selling a product nobody had ever seen, heard of, or thought about before. Race too was taken with Olafsson, whom he quickly nicknamed the "Nordic Knucklehead."

As Race heard more about the budgets that would be involved, and Sony's commitment to aggressive marketing, he began to covet the chance to once again shake up the videogame industry with his heat-seeking competitive strategies and snarky brand of shenanigans. But, as at Sega, there was a lingering concern.

"To be perfectly honest," Race explained, "I'm very reluctant to go and work for a Japanese company. That kind of conservative micromanaging is not really for me. So can you tell me a little about what that dynamic would look like?"

"This is an excellent question," Olafsson replied, "and I believe you will be pleased with the answer. Essentially, Sony corporate does not fully believe in what we are doing. They have agreed to a pay a few bills and open a couple of connections, but it seems to me like they might be very happy if we just went ahead and failed spectacularly. This sounds like embellishment, perhaps, but there is a hard edge of truth to it. For example, they would prefer we don't use the Sony name on the product, and they have already prohibited its usage on the packing."

"Sounds like a serious bummer."

"It is," Olafsson agreed. "But with this imposed distance comes a certain level of freedom. One which I believe would suit you quite well."

This was mostly truth, but it was also what Olafsson knew Race needed to hear. He also knew that given Race's reputation for either flaming out or getting bored easily, he likely wouldn't be at the company long enough for this to become a major issue. At this specific time, Race was the right man for the job, and so Olafsson offered to protect him from the Japanese if this man, who had had a finger in Atari, Nintendo, and Sega, was willing to slap on an eye patch and come sail Sony's pirate ship into the unknown.

"One more question," Olafsson said, "which I had failed to ask earlier. Sega: do you have any ill will toward them?"

"No," Race replied. "The opposite, actually. I think rather highly of Kalinske and what he's got going on over there."

"And it will be no problem for you to compete with former colleagues?"

"Not at all," Race said, shaking his head in a way that made it appear weightless. "As the saying goes, the devil you know is more fun to do battle with than the devil you don't."

Olafsson wasn't quite sure the saying went this way, but he couldn't argue with the logic of SCEA's soon-to-be president, and he wouldn't dare do anything to dampen that wonderful fighting spirit.

While Olafsson was at work reeling in a blast from the past, Kalinske took the opportunity to welcome one of his own: Mike Fischer, the American at SOJ who had been such a great subtle peacemaker over the years.

Fischer had been on the fence about leaving Sega of Japan, but after his father's recent heart attack and the quintuple bypass that followed, it felt like the right time to move back to the United States. And so, in the spring of 1994, Fischer had left SOJ to join SOA and help Kalinske oversee SOE. With Sega's rapid ascension in Europe, Fischer's move appeared to have come at the perfect time, but as excited as Kalinske was to have a trusted colleague building European relations, he worried about what this meant for his Japanese relations. Fischer had always alerted him to things on SOJ's radar, but now he would be flying blind.

"Let me ask you a question," Kalinske began, giving Fischer a tour of Sega's headquarters at Redwood Shores, "are you at all opposed to my cloning you? That way you can work both here and in Japan?"

"Not at all," Fischer said with a smile, "as long as I or, rather, my clone and I, are getting paid double."

"Sure," Kalinske replied, "I'll just have to check with HR about that."

Although Fischer was being treated to the same wit as usual, he sensed something askew in Kalinske's tone. "Tom, is everything okay?"

Ninety-nine times out of a hundred Kalinske answered that question with an optimistic nonresponse, but there was something about Fischer that always relaxed his demeanor. "I don't know," Kalinske said, pausing to let a group of smiling employees from product developed pass by. "These are just weird times, I suppose."

"What kind of weird?"

Kalinske didn't quite know how to put it into words, but it was a very specific feeling. "It's hard to say," he described, scouring his mind for the best example of this wordless thing. "But it's happening more and more lately. Okay, I realize that sounds ominous, so let me just give you an example. Earlier today I was looking at the most recent software numbers."

"Were they good?"

"No," Kalinske replied. "They were fantastic."

Fantastic was an understatement. Of the top ten videogames sold in March 1994, eight of them were for the Genesis (and four of them were published directly by Sega):

1. NBA Jam (Genesis)
2. NBA Jam (SNES)
3. Sonic 3 (Genesis)
4. Mortal Kombat (Genesis)
5. Aladdin (Genesis)
6. Ken Griffey Jr. Baseball (SNES)
7. NHL '94 (Genesis)
8. NBA Showdown '94 (Genesis)

9. World Series Baseball (Genesis)
10. Sonic Spinball (Genesis)

"That's terrific, Tom," Fischer declared. "You should be proud!"

"We should be proud," Kalinske corrected, "but that's the point I'm trying to make. When I look at those figures I should be overjoyed. And part of me is, believe me, but another part can't help but fast-forward to a couple of years from now and wonder if Sega will even have one game in the top ten. Have you seen the Saturn?"

"I'm sure the developers are still getting used to the environment," Fischer explained. "Besides, concerns like that are normal. We all occasionally suffer from fear of success. It's human."

"That's what I thought at first," Kalinske replied. "But what if it's not actually fear of success, and actually fear of something lurking beneath that success? Almost like it's a perfect day at the beach, the sun is shining and the water is warm, but right there beneath the surface there's a very hungry shark."

"Interesting," Fischer said, bobbing his head side to side. "Well, if that were the case, and I don't think it is, then you have two choices: enjoy it while it lasts, or learn how to outswim the shark."

What if it's already too late, Kalinske wanted to ask. He looked around the office, taking great pride in having built such a powerful team. Eight out of ten titles in the top ten. Outstanding, but what if it really was already too late? Instead of asking this, however, Kalinske showed Fischer to his new office and officially welcomed him to Sega of America.

"Huh?" Minoru Arakawa quietly murmured when his key card did not work. He tried it once more, but again was denied, and it slowly dawned on Arakawa that he had not been granted access to the third-floor office space that Tony Harman had taken to calling the treehouse. "Tony!" Arakawa playfully shouted, while gently knocking on the door and laughing to himself. "Tony!"

Finally, Harman poked his head out the door. "Mr. A.? What are you doing here?"

"I heard you have something special to show me," Arakawa replied. "But first I must ask why my keycard does not work."

"Oh, that," Harman said with a sheepish grin. "I never thought you'd want to come in. But since you're here, let me show you around."

Physically, Nintendo's "treehouse" did not quite live up to its name. It was a nearly 2,000-square-foot area with a few four-foot-high cubicles, a small conference room, and an L-shaped bench in the back of the office. But a closer look revealed that, philosophically, it was everything an adult treehouse should be. Beside those cubicles there were some old arcade cabinets, hanging in the conference room were storyboards of games-to-be, and piled on that bench in the back was a pyramid of hardware systems. It was a fun place to work no doubt, but there was lots of work to be done. This new space was where NOA now localized NCL's Japanese games, reviewed third-party titles, and, for Tony Harman (as well as his external producers Ken Lobb and Brian Ullrich), it was a place to develop games without constantly being asked "Whatcha working on?"

After a quick tour of the treehouse, Harman showed Arakawa what he had come to see: the game that Rare had been working on. It was still about ten to twelve months from completion and didn't even have a name (its codename was "Country," because Rare's studio was located on the English countryside) but the Stamper brothers had been doing such an incredible job with the game that Harman wanted his boss to have a look. When Arakawa finally sat down and watched a demo of the game, he looked like he had seen a ghost. And in many ways he had. Because what Arakawa saw was the hulking body of an old friend, one who had been defeated by Mario so many years earlier. Thirteen years ago, this character had rescued Nintendo of America from obscurity, and now he was back again to haunt the competition. After all these years the wonderful beast had returned: Donkey Kong.

"This is incredible!" Arakawa declared. "The 64-bit games will look this good?"

"What do you mean?" Harman asked.

"Do you expect many of the games will appear in this quality on Ultra 64?"

"Oh!" Harman said, pleasantly taken aback. "No, Mr. A. This game is for the Super Nintendo."

The look on Arakawa's face was equal to incredible multiplied by impossible. This game, which would later be called Donkey Kong Country, was only 16-bits? Summer CES 1994 was only six weeks away, but Arakawa was convinced that it was worth undoing everything to feature this and only this. Don James, who designed Nintendo's fantastical display, would later be asked to make this change and somehow, in only a matter of weeks, create a 30,000-square-foot booth with a volcanic island flaming in the middle. Thousands would come to marvel at this amazing game, the one that would extend the lifespan of the SNES and prove that Nintendo could rise back to the top without needing to rely on violence, name-calling, or flashy marketing. Make no mistake, there would be tons of marketing for the game: Peter Main, George Harrison, and Perrin Kaplan would be allowed to go to town. The difference was just that unlike their competitor, there would be substance behind the style.

Normally, Arakawa's initial order for a game that Nintendo deemed to be an A+ title was one million units, but for Donkey Kong Country he ordered four million right off the bat. This is what Nintendo had been waiting for after all, and it was finally time for the tortoise to open the war chest and strap on a jetpack. Did it take longer than Arakawa expected for Nintendo to make their big move? Maybe, but he was not at all surprised that the time had finally arrived. 'The only thing that surprised him in the end was that inside of the tortoiseshell there was a gorilla hidden inside.

"Tony?"

"Yes?"

"You did good,"

"Thank you, Mr. A. Any time."

60

KINGS OF THE JUNGLE

In a continued effort to reclaim the fun, and in honor of what Nintendo hoped would be a game-changing surprise, the guys in Redmond decided to do something a little different for the 1994 summer CES. Instead of just writing a straightforward speech for Peter Main or Howard Lincoln, Don Varyu scripted an elaborate skit to perform in front of developers, distributors, and retailers on June 23, 1994. And to finally usher Nintendo out of its dark ages, the performance fittingly began in complete and utter darkness.

BRUCE DONALDSON

Randy, how much further do we have to go?

RANDY PERETZMAN

Relax, Bruce, the map says we're almost there.

A tiny halo of light appears, and then another, so small it's hard to determine the source until Nintendo's Bruce Donaldson and Randy Peretzman move through the audience on their way toward the stage.

BRUCE DONALDSON

You hear them say these marketing guys are pretty far out and out of touch with reality, but this is really far out. We're in the middle of no place.

RANDY PERETZMAN

Nope, no place at all.

Some lights turn on, still dim but with a bright auburn hue, and the intrepid explorers Donaldson and Peretzman are revealed to be dressed in full safari gear.

BRUCE DONALDSON

You know, I bet there's not a dry martini within a thousand miles of this place.

RANDY PERETZMAN

Shhh, there's the tent right over there.

The stage becomes fully lit, revealing an elaborate campsite in the middle of nowhere. And then from the tent emerges Peter Main, a cigar dangling between his lips with just the right amount of nonchalance.

PETER MAIN

Hey hey, aren't you a sight for sore eyes.

George Harrison steps out of the tent next, and then Nintendo's powers all shake hands.

BRUCE DONALDSON

Dr. Livingwell, I presume.

PETER MAIN

Life has its ups and downs. Whoever said you couldn't make the best of a bad deal? Come here and sit down for a minute.

GEORGE HARRISON

You know, Randy, we have almost everything we need here. And anything that we want to hear from the States, we have all this equipment.

Harrison tilts his head toward electrical thingamajigs and a large television screen.

PETER MAIN

George has been working on this for five or six months, but the only problem is that we don't get any sports news. So Bruce, tell me: the Sonics really did go all the way, didn't they?

The audience breaks out in laughter.

BRUCE DONALDSON

Man, you're completely out of touch with reality. It must be the heat out here.

GEORGE HARRISON

Heat? Try running the marketing department. You get a second-degree burn these days just answering the phone.

The audience breaks out in laughter again, genuine this time.

PETER MAIN

That's why we left. The guys in Japan were screaming, the guys in Redmond were screaming, the reps were raising hell; man, it was time to clear out and go to ground.

BRUCE DONALDSON

So this is the solution? To run away from it all?

PETER MAIN

Brucey, running away? You gotta understand, we're not running away from anything. We just decided to come out here on a little hunting trip. As a matter of fact, there's some of those little suckers again.

Main grabs a shotgun and fires twice, and suddenly a bloodied image of Sonic The Hedgehog appears on the large TV. The crowd loves it—this is so precisely their humor—and Peter Main picks up a nearby martini as he basks in the adoration.

BRUCE DONALDSON

And that's what you're after? Flying hedgehogs?

PETER MAIN

Only in a manner of speaking, my good man. That's just the small stuff. Let me put it this way. You remember it was about ten years ago when Nintendo took its first trip out here to the videogame jungle, looking for that beast that would put us back on the map. What we came up with ten years ago was that big ape called Donkey Kong, who took us from a single coin-op machine to a company doing over a hundred million dollars in just a couple of months. Well, we're on the same kind of hunt again. And we're going to find that next group of eight-hundred-pound gorillas, and I mean big, hairy, muscular games. And I gotta tell you, this expedition is right on track. It's leading us back to where we belong as undisputed king of the jungle.

Peter Main smiles, nods, and slides the cigar back into his mouth.

BRUCE DONALDSON

Yeah, I heard that before, but I gotta tell you, Peter: a lot of people back in the States think that Nintendo and this whole business is going right down the river. I mean, we all knew it was going to be a transition year, but nobody thought we were going to see numbers like this.

PETER MAIN

Well, no question it's become a real jungle. The competition's out there like a bunch of drunken headshrinkers. I mean, there's a whole group out there saying every day that the current technology is tired and over. But let's not lose sight of the big picture, guys. First of all, this videogame industry ain't small potatoes . . .

A graphic appears on the television, indicating the industry has grown to $15 billion worldwide ($6 billion in the United States).

PETER MAIN

And secondly, yeah, Nintendo had a bad year. I mean, what the heck? Sales dropped, profits dropped, but we still made a half a billion dollars in profit, which was about five times what that other guy made, because as he ran around the world trying to buy market share, he gave it all away and ended up barely breaking even.

It dawns on the audience, almost all at once, that this is more than just a silly skit: it's a satire about the war between Sega and Nintendo.

PETER MAIN

And finally, I gotta tell you the future's going to be bright. I mean, look at all those big strategic thinkers who are standing in line: AT&T, Sony, Panasonic, JVC, even that guy *Schpiel*berg over there in Hollywood, all waiting to get into this business. It can't be so bad.

BRUCE DONALDSON

You guys look awfully calm for the predicament you're in.

Peter Main suavely removes the cigar from his mouth. And at that point, it must also dawn on the audience that Main should never be without a cigar. It just fits him so well.

PETER MAIN

Well, Bruce, old buddy, you're right. Twenty points ain't exactly chopped liver. But you gotta be patient. Be patient a little bit. First of all, this is June, and we still got over two-thirds of the business year ahead of us. And I gotta tell you, we're going to make some really big things happen over these next six months. So why don't you just cool your jets a little bit. I mean sit back, relax, have a drink. We've got some real reason for optimism. And to start out with, I want to have my good man George of the Jungle here fill you in on a couple of discoveries he's made since we're out here. And believe me, it's going to help you relax.

With a spear in hand, Harrison proudly talks about an amazing discovery he made one day while wandering through the outback.

GEORGE HARRISON

It was the "Lost Temple of Pac-Man." Apparently, it was a sacred cave that had been accidentally covered over when the natives started building condos. But even so, I found a way inside, and there before me was the secret to the lost civilizations of videogames. There before me were the carcasses of Atari and Coleco. And there stood a magic cauldron, still at a full boil. Once I looked at it, I knew exactly what it was. A native crystal ball with all the secrets of the videogame industry.

RANDY PERETZMAN

What were you smoking in there?

GEORGE HARRISON

What the cauldron said was exactly what we've been hearing for months in our own market research. I asked the cauldron what the players want, and every few seconds, not surprisingly, the answer was "good games." But then came a group of answers that probably in hindsight shouldn't have surprised us. They not only wanted great games, but also wanted to be associated with a system that makes them feel cool. They want to be associated with hardware that's advanced and up to date. In short, they wanted fun and they wanted image. So I asked, what does all this mean for the Super NES? And the cauldron boiled for a while and it got cloudy. It was clear there was work to be done.

PETER MAIN

That's right, and that was all the information we needed to get started in a whole new direction.

GEORGE HARRISON

The first thing we did was to set out to try and find a stronger Nintendo message and even stronger delivery. Our TV campaigns have to be outrageous, and they have to be seen in more places than ever before. Places that even your reps might see them. And at the same time, you're going to see an explosion of Nintendo ads in the gaming magazines. We decided once and for all to break our competitor's lock on that key media.

Harrison then goes on to talk about everything that the new, postdenial Nintendo has in store for the months ahead: bigger events, better communication, and a recommitment to the fun that made Nintendo a household name.

GEORGE HARRISON

I know this is a real change of attitude, but we can't afford to miss out anymore.

And just like that, Nintendo awakens and finds a balance between the tortoise and the hare; slow and deliberate, friendly and flashy . . .

GEORGE HARRISON

Finally, we decided to get some street smarts on the whole violence issue. Is there a market out there for more sophisticated games? Of course. So we'll accept the ability of a ratings system to allow the consumers to make informed choices. From now on, we'll see no repeats of the Mortal Kombat incident, where we saw customers buy what they thought was a more attractive product from our competitor. Those days are behind us. We've modified the internal guidelines, and I can tell you now that Mortal Kombat 2 will be the same whether it's on our platform or anyone else's.

. . . and cunning, always cunning. With the newly created ratings board, Nintendo now had a built-in bunker in which it could duck for cover.

PETER MAIN

Hey, isn't that all right? I mean, thanks, George. All of that good stuff, now that's quite a start.

Thunderous applause takes hold of the room.

To his left and to his right, everyone applauded except for him. Al Nilsen was impressed by the performance, which exuded an almost Sonic-worthy level of cleverness, but he knew all too well that words were only that. Had Nintendo really learned from its mistakes? Had they finally awoken? Would they really be able to merge that long-term, stay-the-course attitude with the short-term shifts of this post-Sega world?

Nilsen was skeptical, but less so about ten minutes later when Don Coyner, Nintendo's director of advertising, appeared onstage to discuss the company's new marketing campaign. "Look who I found thrashing around the swamp," George Harrison said, introducing the cameo. "It's Don Coyner, and he said he had a revelation out there."

"Don, I didn't know you were out here," Donaldson added. "What are you doing in the middle of the jungle?"

"Well," Coyner began, with a long overdue twinkle in his eye, "I got so tired of all those retailers and sales reps complaining about Nintendo's advertising that I decided to come out here and fix it for myself."

Finally, Coyner thought, as he finally had the opportunity to speak these words. Finally, finally, finally. After spending a large, stressful portion of the past few years trying to convince Arakawa, Lincoln, and Main to let Nintendo to take off the gloves—tastefully, of course—they were finally willing to do something different.

There were a variety of reasons that Nintendo was now willing to spruce things up, but the straw that broke the camel's back turned out literally to be a camel. In late 1993 Coyner initiated a series of research studies in which gamers were given a stack of twenty pictures depicting animals and then asked to use these images to answer a series of questions about videogames. Questions like: which animals best represent Nintendo (and which ones best represent Sega)? The results were nearly unanimous: Cheetahs, gazelles, and other speedy creatures were associated with Sega, whereas elephants, camels, and other slow-movers supposedly embodied Nintendo. Arakawa was normally distrustful of focus groups and marketing data, but something about the images made a dent this time, and that enabled Coyner to move forward with what he was now about to unveil. "Kids want more excitement from our advertising. But the answer was not to imitate the competition. We have to do them one better," he said, and with that he introduced Nintendo's "Play It Loud" campaign.

On the surface, "Play It Loud" appeared to be exactly what Coyner had claimed it would not: a Sega ripoff. With quicker cuts, louder music, and an aggressive feeling, it had much in common with its competitor's spots. But upon closer examination, that wasn't quite the case. The colors were still bright, the game footage was still plentiful, and although there was mischief, there was also a sense of optimism. "We're going to give the kids an anthem," Coyner proudly explained, "one that says: 'You can't be young forever, so live it large, live it free, and play it loud.'"

Good but not great, Nilsen thought. But even with his high standards, he couldn't help but echo Coyner's sentiment: finally. Nintendo should have done this, or something like it, several years ago. But better late than never, right?

"Anyway," Peter Main said, stepping in, "when Arakawa and Lincoln sent us out into the jungle, they said they weren't going to take no for an answer. They wanted to know if we could put behind us this lily-white image on game content. Well, I think we've just shown that the answer to that is yes."

Tattered applause swept through the room, but still nothing from Nilsen. That guy was a tough one to please. "Then, of course, there are the games," Main continued. "It's no secret we were weak on this in the second half of last year. And yes, Griffey and Metroid have started to turn the tide, but they're not enough. We thought we'd found a great combination of old and new, but when we unloaded the cages back in Redmond, the response was: 'Gee, that's terrific, but it ain't enough, and don't you come back home until you get one more drop-dead-in-your-tracks killer attraction.' They reminded me that ten years ago they shook the entire country with the big feet of Donkey Kong. 'Can't you guys go out there and find us that one huge hit that will once more make Nintendo undisputed king of the jungle?' And so that's exactly what we did."

Main took a couple of steps closer to the audience, finding his rhythm and then appearing to near a crescendo to this whole performance. "Ladies and gentlemen, allow me to introduce you to the eighth wonder of the videogame world. The best game ever created by Nintendo or anyone else. The new stunning industry breakthrough . . . Donkey Kong Country."

With these words, Peter Main introduced a clip of the atomic bomb Nintendo needed to turn the tide in its war with Sega. A game that never would have existed had Tony Harman not willed it into existence on the basis of the scenario he'd presented to Yamauchi of "one less bad commercial." This was it, the true game-changer that both returned Nintendo to their roots and also took them one step ahead.

Nintendo was back. And as everyone in the room went wild with applause, this time Nilsen was the loudest of them all.

"Bravo," Kmart's Senior Buyer said two days later, as he approached Peter Main at Nintendo's booth. Although he was a famously difficult man to please, on this day his face was filled with nothing but pleasantries.

"Well hello there, old friend," Main said with an easygoing grin. "I'll take a bravo whenever I can get one, but to what do I owe the honor?"

"Where do I even begin?" the buyer from Kmart replied, forcing a chuckle. "But really, that game you all showed off at the presentation, it's out of this world."

"It's something else, isn't it?"

"It's gonna put you guys back on top!"

"Whoa now," Main cautioned. "One thing at a time."

"No, I mean it," the buyer declared. "And in the spirit of putting my money where my mouth is, Kmart is ready to make a very serious commitment."

Before Main even realized what was happening, several Kmart employees emerged from the trade show's hubbub and stepped forward with a large check mounted on a plywood frame. The check was made out to Nintendo of America for $32 million.

"We'd like to officially place an order for a million units!" Kmart's senior buyer announced. "What do you say?"

For a moment, Main was speechless. Not only was the gesture wonderfully surprising, but selling a quantity like this, right off the bat, would make Donkey Kong Country one of the bestselling games of 1994. It would also give them enormous momentum going into the Christmas season (the game would be released in November) and almost singlehandedly propel Nintendo back past Sega.

"I think you know exactly what I'm going to say," Main said, that grin still sliding across his face. "This is a terrific gesture, and I'm truly honored, but there ain't no way we can allocate that much product to you guys. Let me crunch the numbers with my guys back in Redmond, and I'll get back to you with what Kmart's allocation will be. How does that sound?"

How did it sound? Well, it sounded like although the videogame industry had dramatically changed these past few years, Nintendo was still the same: dedicated to great quality, and even greater quality control.

61.

AND THEN THERE WERE THREE

"Come one, come all," Stretch Anderson proclaimed, poking his head out from the first-class partition, "and see the amazing, outstanding, Mr. Steve Race!"

Stretch Anderson, Sony's newest director of operations, was a lanky man by nature, and he seemed even lankier when he was on an airplane. He made his way down the aisle in economy class to see who wanted to speak with the man behind the curtain. He waved for one of the eager faces to come forward. "You'll have ten minutes with the man," Anderson said, ushering this stranger toward the first-class seating, "and not a minute more." As soon as he scooted the gentleman past the partition, he turned around and looked for the next person to poach. The flight was packed, mostly full of guys who were heading back to California after a week spent at the 1994 Summer CES in Chicago. That made Anderson's job even easier, all these employees from Sega and West Coast developers—it was like shooting fish in a barrel. "All right, who wants to go next?"

Several hands shot into the air, all of them hoping for ten minutes with the Great Oz at Sony and, if those minutes went well, maybe a job offer. It wasn't so much that these people on the flight were unhappy with their current jobs, but Sony was offering big money, and based on the recently released hardware specs, people had begun to take the PS-X seriously. Ken Kutaragi's masterpiece had dazzled at the Consumer Electronics Show. It was the world's first true 3-D console, and able to process 360,000 polygons per second. In addition to the technical specifications, there was another surprising reason to get behind Sony: the software.

Like Tom Kalinske and Peter Main before him, Olafsson realized that the key to selling hardware lay in the software. But unlike Sega and Nintendo, Sony lacked an arcade pedigree and would struggle greatly to churn out first-party titles. Therefore, the secret to Sony's success would hinge on successfully recruiting third-party developers. This is why Olafsson's "overpaying" for Psygnosis had absolutely been worth it. Not only did the acquisition help stabilize Sony's first-party roster, but the tech-savvy studio created an easy-to-use development kit with the intention of enticing developers to favor Sony over Sega (whose kit was said to be extremely complicated) and Nintendo (who was notoriously stingy with providing development tools). That was one way that Olafsson hoped to inspire a lemming-like following toward Sony. The other was Electronic Arts, the same company that way back had hoisted skeptics onto Sega's bandwagon. Olafsson and Jim Whims, SCEA's newly appointed VP of sales, made a deal with EA's CEO, Larry Probst: if EA could provide five titles in time for the launch, then for the lifetime of the console, they would only pay a $2 royalty (as opposed to the standard toll of $10). Another sweetheart deal for Electronic Arts, and this time they didn't even have to reverse-engineer a console. Sony had slowly been building an army over the past few years, and now that people had started to pay attention, they were stunned by how formidable the force appeared to be.

"Who else wants in?" Anderson asked, making his rounds. Several hands rose, the faces below them revealing whimsical excitement and no trace of remorse.

"Maybe you should think about leaving too," Michael Milken suggested to Kalinske when they met for lunch in the summer of 1994. "Come join Larry and me—you'd be an absolutely perfect fit."

"That's flattering," Kalinske said, allowing himself for a moment to consider the possibility. Michael Milken, the renowned junk bond kingpin who had celebrated the eighties by shaking up Wall Street—which led to his starting off the nineties by serving twenty-two months in prison—was now a free man, and he wanted to do something good with his life. No, something great. And, like Kalinske, he'd always had tremendous passion for combining education with technology, which had led him to start a new company, with his brother and Oracle CEO Larry Ellison, which did just that.

"I've talked about it with Larry," Milken explained, "and we'd really like for you to run it. What do you think?" Milken had admired his lunch date ever since the two met in the early eighties, but it was something Kalinske had done at the end of the decade that really got his attention. On April 23, 1989, when the country was busy blaming its financial problems on a recently-indicted Michael Milken, Kalinske wrote an op-ed to the *Los Angeles Times* declaring that "The U.S. government should have given Michael Milken a commendation and gold medal for the help he's provided our country's economy instead of indicting him." That was the kind of thing someone never forgets, and it further convinced Milken that Kalinske was the right guy to run his new venture.

"What do I think?" Kalinske asked. "It sounds great, absolutely fantastic, but I can't leave Sega. Especially not now."

"Why not? You said it's a sinking ship."

"I said it feels like a sinking ship," Kalinske corrected. "That doesn't mean things can't be fixed. Plus, even if the ship sinks, I should be there to go down with it."

"A martyr?"

"Hardly. I owe it to my company and the employees."

"How is that not being a martyr?"

Kalinske gave this a moment of consideration. "Because in the videogame industry nobody cares about history. I'll just be the guy who was there for the fall. No one will remember anything before that."

"Hmmm," Milken said, trying to decide whether to continue pushing. "Sounds to me like all the more reason to get out of there."

"You're probably right, but you're not changing my mind," Kalinske said, before falling into a smile. "At least not yet."

• • •

"So," Peter Main wondered aloud, with a grin befitting his befuddled excitement, "does anyone have half a clue about how the hell this is supposed to work?"

"Hardly," Howard Lincoln said, also sporting a dangerously excited grin. "But I'm feeling rather confident that we'll all be able to figure this out."

"Yes," Minoru Arakawa added with a soft blink of his eyes, which roughly translated to mean "we always do, don't we?"

It was October 19, 1994, and Nintendo of America's three amigos were bracing themselves to boldly go where no videogame executives had gone before: cyberspace. Six days earlier, NOA had announced that in an effort to make Donkey Kong Country the biggest game ever, Nintendo would become the first videogame to utilize online technology for a new product launch. Nintendo's three-month online campaign would be available exclusively on CompuServe, the leading worldwide Internet service with 2.3 million members. To attract attention and bridge the gap between corporation and consumer, the campaign would kick off with a live, one-hour chat hosted by Arakawa, Lincoln, and Main.

"Now, what I'd like to know," Main mused, "is after we post a reply, how will it know that we're done with that question and onto the next one?"

"Don't worry," assured Ron Luks, CompuServe's systems operator (SYSOP), who had come out to Redmond to moderate the conversation. "There's some lingo that we'll use to keep everything running smoothly. So just sit back, relax, and do what you all do best." Luks looked at his watch, the seconds ticking down to 6:00 p.m. "Ready?"

Nearly ten years earlier, Arakawa, Lincoln, and Main were plotting how to get a product nobody had ever heard of into stores everywhere. Slowly at first, and then quickly as credibility grew, these men succeeded in spectacular fashion. Inch by inch, they willed their way into more than 20,000 stores, and Nintendo products were available in just about every retail space imaginable. But now there was a whole new space to think about, one that was much harder to imagine because it didn't physically exist. It was there but it was not; it was everywhere and nowhere at the same time.

The technocrats spoke about the Internet as if it were a foregone conclusion, and maybe it would be, but it was hard to accept this certainty and not wonder what happened to that foregone conclusion of multimedia. What about virtual reality? Or any of the other Next Big Things that so quickly shrunk into extinction over the years? Maybe the Internet would catch on, or maybe it would be the latest casualty in the graveyard of foregone conclusions, but either way Nintendo would be there. Those days of wait-and-see were over, allowing scrappy go-getters take what should have been their own. Arakawa, Lincoln, and Main were ready to look forward. Not out of panic or desperation, but with the same thing that always propelled them forward: patience, persistence, and the numerical beauty of a well-calculated risk. Today a new quest began, a journey into the beyond. And just as they had done before, they set out to slowly take over this intangible new space, inch by theoretical inch.

"Ready?" Lincoln repeated to himself. "Always and never." Arakawa and Main nodded, signaling to Luks that they were ready to get started with whatever came next.

The Nintendo of America conference is beginning

RON LUKS/SYSOP: Welcome everyone. On behalf of CompuServe and the Video Game Publishers Forums, I would like to welcome our members to the first Nintendo online conference. My name is Ron Luks, and I'll be the moderator for tonight's conference. Now, I'd like to introduce our three guests of honor.

RON LUKS/SYSOP: First is Mr. Minoru Arakawa, who serves as President of Nintendo of America (His screen name will be "MR.A").

RON LUKS/SYSOP: Next is Howard Lincoln, who joined Nintendo as Sr. Vice President in 1983, and was appointed Chairman of NOA in Feb 1994 ("HOWARD").

RON LUKS/SYSOP: And last but not least we have Peter Main, who has directed all sales and marketing activities for Nintendo products since 1987 ("PETER").

RON LUKS/SYSOP: Before we open up for questions, I'd like to introduce Howard Lincoln and have him provide a quick introduction to Donkey Kong country and its new technology. Howard, GA (GO AHEAD).

HOWARD: First Ron, on behalf of Minoru Arakawa and Peter Main, I'd like to let everyone know how excited we are to be here tonight. Conversing online is a new concept and we are looking forward to communicating using this new medium.

HOWARD: Needless to say, we're very excited about this new game, which in our estimation will set the gold standard for 16-bit game play. We've been able to take the computer graphics technology that brought the dinosaurs to life in "Jurassic Park" and put it into a 16-bit video game. We call this process Advanced Computer Modeling, and the results are outstanding.

RON LUKS/SYSOP: Thanks Howard. So without any further delay. Let me open the floor to the first question. And remember, to get in line to ask a question type "/QUE"

Moderator recognizes question #1 from Christian Mueller

CHRISTIAN MUELLER: DKC has superb graphics, but we've seen such gameplay before. Do you think the consumer will accept "old games" even when they have superior graphics? GA

MR. A: Thank you for asking us about this. This is a brand new Donkey Kong game. Not only are the graphics superior but the gameplay is outstanding in each of 100 levels. The gameplay is outstanding. GA

Moderator recognizes question #2 from Jer Horwitz

JER HORWITZ: This conference is a wonderful idea. Thanks for doing it! I remember when I first saw DKC at the Summer CES show. My writers and I were all astonished that it was a SNES game . . . I'm wondering, what will the next game be to use the ACM technology? GA

PETER: That's a good question. UniRacers, coming on December 12th also utilizes ACM technology to generate over 8,000 frames of animation. We also have other games currently in development which you will hear more about shortly. GA

The online chat continued for nearly an hour, and ended with the question that everyone under eighteen had been wondering since first picking up a Nintendo controller.

Moderator recognizes question #23 from Chocobo

JEN KUIPER: Ever since Nintendo started the revolution that changed video games from a hobby to an industry, I have seriously wanted to work for you people.

Who do I contact for more info? I'm there dude! PLAY IT LOUD!! GA

PETER: (All three gentlemen laughing) We are always looking for talented people so just drop us a line attention: Bev Mitchell, and we will be right back to you.

RON LUKS/SYSOP: A nice note to end the conference. Thanks for your questions. I would once again like to thank Minoru Arakawa, Howard Lincoln, and Peter Main for joining us tonight. We enjoyed having you online.

HOWARD: Thank you Ron. On behalf of Mr. Arakawa and Peter, I'd also like to thank the conference participants. We've enjoyed being online with all of you and look forward to doing this again. GA

RON LUKS/SYSOP: Before we sign off, I'd like to remind everyone of some additional Nintendo activities going on during the next few weeks. Starting on Monday, October 24, video clips (in both Mac and Windows formats) will be available for downloading from our online Nintendo promotional area. ("GO NINTENDO") This will include never before seen footage.

RON LUKS/SYSOP: Also starting next week, we'll be holding a Donkey Kong Trivia Contest. The grand prize winner will get an all-expense paid trip to Nintendo. Donkey Kong country Game Paks and other prizes will be awarded. Consult the contest details in the promotional area.

RON LUKS/SYSOP: Then on November 8th, Nintendo will host another online conference with the video game designers and technicians behind Donkey Kong Country. Here's your chance to get the inside track on the development of a video game.

RON LUKS/SYSOP: Goodnight everybody. Thanks for attending.

The conference has ended

"What about now?" Milken asked again at the end of 1994. "Are you ready to allow yourself to enjoy working again?"

"Nope, not yet," Kalinske said, noticeably less optimistic than he had been before. With no autonomy when it came to hardware, he focused even more on trumpeting Sega's software. This had always been the primary focus, but Kalinske could sense the inklings of desperation. Decent games were marketed like they were good ones, the good games were launched like they were the greatest thing ever, and because Sega was now supporting so many platforms, the software was more and more being rushed to market. Kalinske was beginning to feel like a prisoner, locked in a cell by Sega of Japan, and he couldn't help but laugh at the irony when he, Rioux, and Toyoda were dressed in prison stripes to promote the release of Sonic & Knuckles at a videogame competition held in Alcatraz. He had tried to remain upbeat, and most of the time he continued to find a way, but on November 24, 1994, those bubbles of hope seemed to burst all at once. That was the day when Sega released the 32X and Nintendo released Donkey Kong Country, which perfectly epitomized the different approaches between the two companies. Sega offered another new gizmo, which was expensive and would require entirely new software (which had been hard to get developers to make because Sega just had too many different systems). Nintendo offered a breakthrough game in the middle of the console's life cycle, which not only reinvigorated the SNES but hinted that there would be more like this going forward.

Although the 32X would lose momentum rather quickly, Sega's reputation and market prowess had managed to create large demand for the launch. But even this caused a problem for Sega internally and externally. SOA had rushed the 32X to market in order to meet Nakayama-san's Christmas deadline, but SOJ had failed to produce the requested number of units. SOA had wanted a million, but SOJ, who refused to relinquish the hardware manufacturing responsibilities, only provided four hundred thousand units. Not only did this upset the underserviced retailers, but it made the launch look weak, which became something of a self-fulfilling prophecy. For the first time since Kalinske took over the company, a Sega product was met with mediocre reviews. He took no comfort in watching the failure of something he had said would fail; instead, he chose to keep fighting for his team. That meant putting all his eggs into Sega Saturn's basket. The hardware was lousy and developers were shying away, but Kalinske felt like SOA's marketing muscle could turn this into something special, something that would even outdo Sonic 2sday. So his team decided that on September 2, 1995, they would launch the console on what they now called "Sega Saturnday," shooting for the biggest console launch ever, and gathering the kind of momentum that 32X had sorely missed.

Meanwhile, Nintendo was living not in the future but fully in the moment. The launch of Donkey Kong Country was the most successful in videogame history, selling more than seven million copies in only six months and helping Nintendo reclaim Christmas. Additionally, Nintendo had taken a page out of Sega's playbook and launched their "Play It Loud" marketing campaign, which beautifully merged Nintendo's straight-arrow mentality with Sega's edgy attitude.

This was a blow to Sega, but what Nintendo was doing on the marketing front hardly compared to what Sony was up to. With Race and Anderson in charge, the plan was to continue to push the envelope further than Sega ever had: edgier, wackier, cooler. Under Kalinske, Sega had indeed slain Nintendo and was the new king, so Race wanted to position Sega the same way that Sega had once positioned Nintendo: as the lame establishment that had forgotten how to have fun. Sony started running competitive ads against

Sega: If you buy a Saturn, then your head must be up Uranus. This was the first time in Race's career that he'd worked for a company so flush with cash, and he was having the time of his life doing anything and everything he wanted. It's like having money in high school, he had been heard to say, and the confidence he exuded made him a shoo-in for homecoming king.

"But I can't stand to see you like this," Michael Milken said to his friend. "When did you become a masochist?"

"A masochist?" Kalinske asked. "So I'm not a martyr anymore?"

"A man can be both," Milken explained.

Kalinske shook his head. "It's not that, Michael. But I just can't go ahead and give up here. There's too much at stake. And as long as we do something big with the new system, then it will be like all this stress never happened."

"But I thought you said it's no good."

"I'm not a technical guy—what the hell do I know?" Kalinske asked. "So it doesn't have the right number of polygons or whatnot. As long as we have the right games and the right messaging, then it could work."

"The right messaging?" Milken laughed. "Just when I think you're an okay guy, you remind me that you're still a marketing maven through and through."

Kalinske shrugged. "It's all just telling stories."

And the story that Sega needed to tell now was that its players were even younger and hipper than Sony's crowd. This led Kalinske to make a tough decision and find a new athlete to become the face of Sega Sports and the center of the subbrand's football franchise. Although this effectively ended his personal and professional relationship with Joe Montana, Kalinske felt like it was the right thing to do. But it was hard to feel good about the transition as Fornasier reluctantly passed along negative reports about Montana's replacement, Deion Sanders. Montana was unanimously described by those he worked with as a "class act," while everyone who came into contact with Sanders described working with him as a "nightmare." He didn't show up at meetings, wouldn't speak directly to those he was working with (his preferred form of communication was through his entourage), and when he did speak, it was always in the third person. But Sega had no choice but to put up with him: he was the face of Sega Sports for the beginning of the new console war.

"All right," Milken said. "You're not there yet. But let's keep talking, okay?"

"Sure," Kalinske said. "Even a marketing maven is wise enough to realize it would be foolish to turn down free sessions of flattery and therapy."

Kalinske had not yet reached his breaking point, but Steve Race did everything in his power to push him there. "Relax, Tom, it's just a joke," Race said, putting an arm around Kalinske's shoulder.

They were in Las Vegas for the 1995 winter Consumer Electronics Show, which would be the last one that Sega ever attended. Thank God, just about everyone thought, after all the years that the CES treated videogames like the fad they didn't turn out to be. Going forward, the videogame industry would have its own trade show: the Electronic Entertainment Expo (E3), the first of which would take place on May 15 of that year in Los Angeles. This was the direct result of the videogame trade association that Kalinske had spearheaded, and it was there that Sega, Sony, and Nintendo would all unveil their next-generation consoles. But that still lay in the future. Today Kalinske had awakened in Vegas to find himself a victim of the latest prank from the guys at Sony. Apparently, in the middle of the night, all of the Sonic memorabilia displayed throughout the hotel where he was staying had been defaced: displays torn down, posters given mustaches, deflated Sonic balloons strewn throughout the lobby. Obviously Race (or Race's guys) were behind this, but if Kalinske had had any doubts, they vanished when he saw the hotel's pool, which was overflowing with large black balloons inscribed with the letters "PS-X."

Kalinske looked at Race and shook his head. "I know it's a joke," he said, "but don't you think this one went a little too far? I don't really care, but between you and me, I don't need anything else lowering morale."

"They're just balloons!"

"Steve," Kalinske said, mustering all the patience he could find, "please stop."

"Tom," Race said, matching the pattern of his speech, "no."

Race gave Kalinske another pat on the shoulder, then walked away. Later that evening, Race would prove to be a man of his word when he arranged for Sega's evening party to be overrun with thousands of napkins that said "PS-X Welcomes Sega to CES!" Kalinske shook his head and readied himself to lead his guys into their final Consumer Electronics Show. But before turning around, he made sure to take a moment to put on a brave face.

Kalinske wasn't the only one learning to perfect a brave face. Diane Fornasier had become quite skilled herself, although for entirely different reasons. She was twelve weeks pregnant and didn't want to let anyone know because she was afraid that if people found out, she might lose her job. Not lose her job as in getting fired—she knew that wouldn't happen, especially not with Kalinske in charge—but slowly get looped out of plans, conversations, and strategies to the point that her value to Sega would drop to zero. Therefore, until the moment was right, she planned to hide the morning sickness, act slightly embarrassed to be gaining weight, and continue to accept drinks of alcohol and pour out the contents when nobody was looking.

By the second week in February, though, she could no longer keep the secret. At her age, the pregnancy was already considered high risk, and those odds got even worse when she learned that she was having triplets. Not long after revealing the news, she went to see the doctor and found out that at least two of the triplets had passed away. She worried that this was her fault because she had kept working at Sega, but the doctor assured her that was not at all the case. In light of the circumstances, however, he recommended she get a week of bed rest.

Although the doctor had confirmed that the miscarriages had nothing to do with work, Fornasier was filled with guilt. Yet still, even feeling that way, she wanted to work. If it had been any other job, she'd likely have exited a while back, but this was Sega. This was the company that had taught her to take, that had given her the life and career she'd had always hoped was possible. And with the current hardware transition and so much at stake, she felt particularly obliged to repay Sega. This was significant because it demonstrated the unbending dedication that many felt toward Kalinske and the company; it's also significant because, in a nutshell, Fornasier's commitment touched upon the reason Kalinske hadn't left to go with Milken. If she hadn't given up, and neither had Rioux, Toyoda, and so many others, then how the hell could he give up?

After a week of bed rest, Fornasier went in to speak with Kalinske. She wanted to go back to work, but not quite yet and not full-time. She didn't know if something like that might be possible, or how that kind of arrangement might work, but if it interfered with any of her responsibilities, she promised to find a way to—

"Diane," Kalinske said, "don't worry about any of that. Take all the time you need. I mean that. Whatever arrangement works for you, I'll make it work for us."

"But what about—"

"Nope," Kalinske said, cutting her off. "We're not even going to continue this conversation until you put to rest all that doubt. Whatever you need. I mean it."

Fornasier was touched, and relieved in a way that she hadn't been in months. "Thank you, Tom."

"I mean this so incredibly sincerely: it's the least we could do," he said, staring at her until he was sure it had sunk in. "Besides, you have to admit that the timing works out not so bad, with Mike joining us."

Mike was Mike Ribeiro, a marketing wunderkind from Hilton Hotels who had recently joined SOA to fill Ed Volkwein's position and run the marketing department. For many who had been with Sega since the beginning of the decade and watched the company grow, Ribeiro's arrival symbolized the final piece of Sega's transformation from scrappy go-getter to just another company. This likely had much more to do with the timing of his arrival and the unwinnable situation that he had been thrust into than with his personality or talents; after all, Sega had plenty of candidates to choose from (at least among those Sony hadn't already gotten). But Ribeiro differed from his predecessors in one significant way: for him, it really could have been widgets. Through the years, some members of Sega's marketing team had loved playing videogames, while others didn't really embrace the pastime, but regardless where anyone fell on the gamer spectrum, they all appreciated the art of product development. That's why even those who didn't like playing games loved selling them; they were selling stories, adventures, friends from an alternate reality. But for Ribeiro, all that really mattered was the marketing. In essence, he represented to those at Sega the same thing that Sega itself had represented to Nintendo: style not only superseded substance, but it did so to the point that substance was almost moot.

Normally, Kalinske considered a mentality like that dangerous, but when it came to Saturn, an inferior system in just about every way possible, this was exactly what Sega needed to make a final stand on Saturnday.

"You must cancel it," Nakayama advised Kalinske while the two were at a steak restaurant in the Bay Area.

"I don't think you understand what you're saying," Kalinske replied.

"We must go sooner," he said, meaning to move up the launch of the Saturn.

"Nakayama-san," Kalinske said, still trying to wrap his mind around this, "with all due respect, we've spent an uncountable amount of hours coordinating the launch of this new system, and it all hinges upon that release date."

"No," Nakayama said with something that resembled a pout. He turned his attention to carving out a large piece of steak, seemingly losing interest in the conversation.

"We can't just change our plans like that," Kalinske said. How could Nakayama be blind to what was going on at SOA? Sony was coming, and not only did they have a better system, but they were aggressively poaching Kalinske's employees. Just days earlier Sega had honored a Third Party Analyst named Kirby Fong as employee of the year; twenty-four hours later, he left for a higher-paying job at Sony. "Especially not with the new trade show only weeks away. We need to put our best foot forward."

"That's what I am thinking," Nakayama said, in a sweeter voice that seemed to indicate he believed they were now on the same page. "It makes for a perfect opportunity to launch our new Saturn."

"Wait, you mean to actually launch it there?"

Nakayama nodded slowly. "See? It's perfect."

Kalinske vehemently opposed this idea; it made absolutely no sense. But in the course of his explaining why launching at E3 would effectively kill Saturn in the United States, he started to see what had caused Nakayama to feel this way. SOJ had launched the Saturn on November 22, 1994, and thus far the results had been rather strong. With 170,000 systems sold on just the first day, the Saturn appeared well on its way to becoming Sega's most successful console in Japan. Why had the Saturn triumphed where the Mega Drive and Master System had not? One reason was certainly the software. Although only five titles were available at launch, one of them was Virtua Fighter, which was currently the most popular arcade game in Japan. Another reason for the success could be SOJ's rowdier and more aggressive advertising campaign. And another reason, the one Kalinske assumed Nakayama suddenly valued most, was that Saturn had entered the market two weeks before Sony's PlayStation (which, notably, still outsold the Saturn).

"With Genesis, we went first," Nakayama explained, "and you have seen how we reaped this reward. We must embrace the opportunity and also do the same."

"Nakayama-san," Kalinske said, "I don't quite understand what's been going on over the past year, but up until now I've gone along with the plan. This, however, just can't happen."

"It is not for you to decide!"

"Yes, it is!"

"You must better understand the situation!"

The conversation escalated, the same tone at different volumes, until Nakayama got up and walked out on Kalinske, who was left alone at the table. How could Nakayama be acting like this? Even in the past, when they'd disagreed, Nakayama had always made efforts to compromise, or at least make it seem that way. But this time it was different. Or maybe it wasn't. Maybe this was just the first time since the tide had turned that Kalinske refused to budge. Either way, if this was what Nakayama wanted, then this was what would happen. If Kalinske didn't launch the Saturn at E3, then he'd be replaced by someone who would. For a second that didn't sound like such a bad fate, but he just wasn't ready to quit. The situation looked dire, but maybe, just maybe, Nakayama's way would work out. The Saturn had indeed been successful in Japan, so perhaps it wasn't such a bad system after all. And as he sat there alone in the restaurant, Kalinske suffered from the hidden problem that faces men of great inspiration: they can talk anyone into anything, even themselves.

62.

FORK IN THE ROAD

All we are is a collection of moments, Kalinske thought as he walked onto the stage at the Los Angeles Convention Center. It was May 11, 1995, the world's very first Electronic Entertainment Expo, and given how far he and those around him had come, it was hard not to feel nostalgic. Some of those moments are big, others small, and a few curiously change size over time. For this one, however, this milestone, brimming with 41,000 fans and 420 exhibitors, it wasn't its size that really mattered, but merely the fact that it would be shared by all the right people: the forefathers of the videogame industry. Their careers now spanned various points along the industry's spectrum, but they were all there, and that's what made this moment eternally special. People like Nolan Bushnell, the founder of Atari, who was now senior consultant for a small game developer called PlayNet; Howard Phillips, Nintendo's former Game Master, who was currently in between jobs; and Michael Katz, who could have probably helped Phillips, as he had become one of the videogame industry's first headhunters (with a particular penchant for moving employees from Sega to Sony).

"I'd like to take a moment," Kalinske began vibrantly, officially kicking off this groundbreaking event, "right at the beginning of my remarks, to convey one of the more important pieces of news you're likely to hear while at E3. Tomorrow, May 12, 1995, will be Yogi Berra's birthday. His seventieth. So we all need to prepare ourselves for a heavier-than-usual onslaught of Yogiisms from radio and TV commentators. If I may, I'd like to kick off the birthday observance by repeating my own personal favorite of his: 'When I come to a fork in the road, I take it.'

"Now that's not only vintage Berra," Kalinske continued, "but it's also good for kicking off speeches. It can even be construed without an awful lot of effort to be representative of what I'm about to say today. I am going to discuss choices. I am going to discuss change."

There were about thirty thousand people on hand for the inaugural show. Some were the forefathers, some were members of the industry's new generation, but most were just people from around the country who loved videogames. The revolution had indeed been pixelated; videogames were not just a fad, not just for kids, but a source of art, story, and entertainment for anyone of any age at any time.

"E3 works as a pretty good symbol of some of the changes our industry is experiencing," Kalinske continued. "Here is this great big show designed solely for interactive entertainment. CES, great success that it's been, was never really designed for us. It related better to an older culture. It forced some of the most creative companies on earth to at least figuratively put on gray suits and, well, fit into a sort of TV-buying, furniture-selling mode."

Twenty years ago this industry had not existed; ten years ago Kalinske had never thought he'd be a part of it; five years ago Sega was a punch line. That's just how it goes when the track's being built while the train is already in motion. "But even that worked okay until recently," Kalinske explained. "Now, interactive entertainment has become far more than just an annex to a bigger, broader electronics business. And frankly, I don't miss the endless rows of car stereos, speakers, and cellular phones. We've broken out to become a whole new category, a whole new culture for that matter. This business resists hard-and-fast rules. It defies conventional wisdom."

Kalinske then described what made the videogame industry unique, what made it superbly unpredictable, and what tomorrow might or might not bring. But along this wild roller-coaster ride, there was one thing that would not change. "Suspension of disbelief. It's always been the fundamental component of diversion, whether that diversion is books, movies, or the theater. Advances in gaming mean we will come to supply that component more effectively than any other medium. The interactive entertainment business is going to allow the Walter Mitty in all of us to finally realize our dreams. We are going to *become* great football players, race car drivers, or aviators. We are going to move into and occupy new worlds that were formerly only available to us in dreams."

There was an elusive poetry to everything Kalinske said that made those in the audience momentarily see things just like he did, but as much as they enjoyed the lyricism of life, what they enjoyed even more were videogames. And Kalinske did not fail to deliver, providing new details about the Sega Channel, upcoming Genesis games, and of course the Sega Saturn.

"The kind of technology under the hood of every Sega Saturn will deliver the leap forward our new consumers demand. Sega Saturn employs an orchestra of engines. No less than three Hitachi 32-bit RISC processors are used for main power, two video display processors generate character and gameplay images, and a digital signal processor yields up to thirty-two voices at CD-quality audio."

Five years later and Kalinske still had no idea what any of that meant, but apparently the thousands before him did, and loved what he had to say. Maybe the Saturn would be a hit after all. Maybe he hadn't just talked himself into the possibility, but rather convinced himself of what others would see. "You will be hearing a lot from us, and a lot from everybody else, as our business approaches this new level. In advertising, for example, as our industry gears up for its 32-bit launches, we're going to be spending over $100 million on marketing, and believe me, we don't spend that kind of money unless we're confident of success."

A hundred million dollars! When Kalinske first took over, the marketing budget had been under $5 million. That number, and how it evolved to this point (1991: $20 million; 1993: $76 million; 1994: $100 million), was enough to make both Sega and Nintendo think that *their* philosophy had been proven right. Kalinske would say that it took money to make money, that the budget had risen in proportion with success, and what a small price it was to pay for global name recognition. And Peter Main would counter that style couldn't compensate for lack of substance, that the budget had risen as product quality had diminished, and that while Sega might now have a global name, it should think long and hard about what its reputation was.

"Lastly, I wanted to come back to Yogi Berra," Kalinske said, relishing his last moments onstage. Would he be back next year? Five years from now?

Ten? So much of that would depend on what he had to say next. "I think when Yogi said 'fork in the road,' he was talking about opportunity. When he sees opportunity, he takes it. And so we do. We're taking all the opportunities we can to make this business soar. And since I began my remarks with an announcement, I may as well finish with another. Yesterday we started our rollout of Sega Saturn. We're at retail today in eighteen hundred Toys 'R' Us, Babbage's, Software Etc., and Electronics Boutique stores around the U.S. and Canada."

Upon hearing this, the crowd went wild. The next generation was already here? What a wonderful surprise! Kalinske knew this crowd was biased, but he hoped their sentiments would be shared by most, because they certainly weren't shared by the retailers.

Following his last meeting with Nakayama, the one where the temperamental tyrant had walked out in the middle of dinner, Kalinske had the unenviable task of informing his team about the new launch plan. And, as always, he delivered the news with a buoyant smile and an excess of can-do spirit. After the fury that followed, when it became clear that this was a command and not a conversation, the only thing left to really decide was the matter of distribution. Because the release had now been moved up by four months, there would not be nearly enough product to satisfy retailer demand. Only about half a million units would be ready in time, less than a quarter of what had been planned for the launch. This left two options: give each retailer only a quarter of their order, or give some retailers the full order at the expense of others who would get nothing.

Most of the employees were too upset by what had just happened to make rational points, but Bill White stepped up to the plate and made a case for choosing the latter option. The smaller retailers would be furious and make all kinds of threats, but as long as the product was a success, then they would reluctantly forgive. That was just the way they were; he'd seen it a million times at Nintendo. And so, between two terrible choices, this was the route Sega went. This was the strategy deemed best for building momentum, but it would also likely reverse the delicately crafted retailer relationships that Kalinske had worked so hard to build over the past five years. Once upon a time, Wal-Mart had refused to carry Sega; now Sega would be refusing to stock Wal-Mart. Would this be the ironic ending to all Sega had accomplished, or something that would be soon be forgotten on the long road ahead?

"Saturn will be the steal of the summer at $399," Kalinske said as he neared the end of his speech. "We'll have ten software titles at retail in the next few days, twenty by August. Our total rollout will take the summer to complete, but we're starting today in-store and on prime-time television nationwide. Sega is not only here now, it's out there."

One last look at the crowd, an image to remember the moment with forever, and that was it. "Thank you," Kalinske finished. "It's been an honor to speak with you today."

"Thoughts?" Kalinske asked about fifteen minutes later as he, Paul Rioux, and Shinobu Toyoda took seats in the front row of a large conference room moments before Sony's presentation was set to begin.

"I feel better than before," Toyoda said. "But it is still too early to feel certain."

Kalinske nodded and then turned to Paul, who appeared to be unmoved by the crowd's reaction to Saturn's early release. "Yeah, they clapped a lot, but what else were they supposed to do?" Rioux said with a shrug.

Paul was always the pessimist, and Shinobu was a realist every step of the way. That made Kalinske the optimistic, and following his speech, he felt pretty good about that role. These last two years had been such a struggle, fighting wars on so many fronts (versus Nintendo, Sega of Japan, and Capitol Hill), but all of those stressful moments would be water under the bridge if Saturn could carry the torch that Genesis had been running with for years. Kalinske had been skeptical, and in many ways he still felt that way, but when he thought about the horses in this race, who was really going to beat the boys from Redwood Shores? Nintendo? Project Reality, which had been renamed Ultra 64, wouldn't be out until next year. In the past year they'd reinvented themselves and had a nice run, but would that really excuse them for making the same mistake, coming to the party late once again? Or would it be Sony, then, who would beat Sega (with a console that they should have been launching together)? Supposedly Sony had the better hardware between the two, but nobody buys a console to display a copy of the technical specifications on the refrigerator. Olafsson had done an incredible job of building a strong game library, and Race had excelled at all the pyrotechnics that went into choreographing a launch (building a team, creating an image, getting into stores, and so on), but the truth was that in the videogame world Sony was an unknown entity. They just didn't have the same reputation as Sega or Nintendo, and that would be a big hurdle to overcome. Could they do it? Absolutely, but Kalinske felt slightly better about his chances after delivering that speech. Until Nintendo was ready to enter the race, it would be Sega vs. Sony, and with both systems likely to be similarly priced (it was assumed that Sony's PlayStation would cost $399, the same as the Saturn), then it would be up to the consumer to decide who lived and who died.

After a short video at the front of the room sought to explain how Sony was now "into the game," Jim Whims, SCEA's VP of sales, stood at the podium to introduce the day's speaker. "In 1991, Sony established a new division to focus on the burgeoning multimedia marketplace," Whims explained. "It was called Sony Electronic Publishing Company, and Michael Schulhof appointed Olaf Olafsson the head of this operation. He is in charge of Sony Computer Entertainment of America, as well as Europe, Sony Imagesoft, and Psygnosis. He also oversees CD operations for Sony's manufacturing business. At the same time, Olaf finds time to write. He's written three novels, a collection of short stories, and a play. His latest novel, *Absolution*, has been published by Random House. And the story doesn't end there—Olaf is also a physicist from Brandeis University. Without any further delay, I'm pleased to introduce to you Olaf Olafsson."

Smiling, smiling, but never too much, Olafsson addressed the room. "Thank you very much," he humbly began. "You can tell from the video that we don't represent a conventional videogame company. Today I want to tell you why that might be the best news our industry could possibly receive."

Lots of applause, and then some more. Maybe Rioux was right after all. "First I want to welcome all of you to our first trade show," Olafsson began. "Five billion in sales has moved us out of the CES parking lot. It's no coincidence that the introduction of Sony's PlayStation coincides with this development. Now, more than ever, is the time for a definitive technology."

But before discussing the future, Olafsson wanted to discuss the 16-bit generation, those videogames of yesterday. He talked about what worked (sports and riddle-based puzzles), what did not (attempts at realism), and how this compared to what was available on the personal computer (more, but computer games were less accessible). All of this, combined with so many recent technological and cultural changes, had created a new kind of consumer. "The Digital Kid," Olafsson explained. "And the Digital Kid expects technology to be different every single day. He can't remember anything before MTV and the PC. He eats shock rays for breakfast, the Internet is his lunch, and if we continue to serve him two-dimensional games, he'll leave the table. So the question is, how can we become dinner?"

"Maybe we ought to fry them?" Kalinske whispered to Rioux and Toyoda.

Olafsson described the many ways that Sony could reach this consumer and carve out a piece of the next frontier. There were a lot of reasons to believe it might all work out. Sony clearly understood the market, they had over four hundred developers set up with development kits and working on games, and with nearly two million systems sold, the PlayStation had surpassed Saturn in Japan. "All of this is great," Olafsson affirmed, "but let me get you the best news of all: aggressive pricing. Now, some of you might actually want to know what that price is. And since it's a beautiful day here in Los Angeles, I'm going to ask SCEA president Steve Race to join me for a brief presentation."

Through the flood of applause (after all, the price was what everyone really cared about most), Race trotted up to the podium. In his hand was a stack of papers, which appeared to be his speech, and on his face was a devilish grin, which had absolutely nothing to do with this presentation. He was looking forward to his speech, and actually knew it well enough to recite it from memory, but whatever blow this might strike on Sega, it wouldn't compare to what he'd just done to Nintendo.

In the days leading up to the big event, with anyone and everyone associated with the industry in Los Angeles, Sony, Sega, and Nintendo all hosted grand parties to show the world they had arrived. Race, playing the role of diplomatic class act, sent an invitation to Nintendo's Howard Lincoln, graciously inviting him to attend Sony's event, which would be held on the studio lot. Lincoln, surprised but impressed by Race's attempt to bury the hatchet, responded by sending him an invitation to Nintendo's event, which would be held the night before E3 was set to begin. Upon receiving Nintendo's invitation, however, Race's eyes gleamed with a naughty idea, which led to him and Stretch Anderson making five thousand copies of the invitation. And so at this very moment, as Race was set to begin his speech, members of the Sony team were in downtown LA, handing out the invitations to the weirdest and wackiest people they could find.

Race looked at the audience, nodded, and then delivered one of the best speeches of his long, storied, and unpredictable career. "Two hundred ninety-nine," he said, and then nodded again and walked away from the podium.

That was it; that was all. Two hundred ninety-nine, he had said, and then he'd walked away. Why bother with all the other stuff? That's all they wanted to know anyway. And whatever unsatisfied expectation for words they might be feeling would surely be replaced with shock at this low, low price.

As Steve Race sat back down, Tom Kalinske turned to Paul Rioux and delivered a speech of his own. And like Race's, the words were few but just about summed up the situation completely. "Oh, shit," he said.

63.

KILLER INSTINCT

In a sharp gray suit, boyishly combed graying hair, and an uncharacteristically cheeky smile, Howard Lincoln stepped in front of a blue podium. "Thank you, and good afternoon everyone," he said, greeting an audience of industry players at the historic Los Angeles Theater.

The smile grew larger. "Welcome to Tinseltown. The city of hopes and dreams, the place where telling the difference between the real world and make-believe is always a challenge. And consequently there couldn't be a better venue for an industry facing the same kind of challenge: separating dreams from reality, the contenders from the pretenders, not only for this coming Christmas but for an entire new generation of videogame play."

As Lincoln switched gears, Tom Kalinske, still stinging from the shortest speech he'd ever heard, took a seat in one of the back rows. Kalinske knew that coming to see Nintendo's presentation was unlikely to improve his mood, but for some reason it felt important for him to be there. For the next ninety minutes he would put down the sword, suppress any sarcastic thoughts, and just observe the men who had once resurrected the industry that everyone was celebrating today. A moment of admiration, a tip of the cap, and then back to wholeheartedly despising Nintendo before the end of the afternoon.

"Now, everyone knows that the videogame world is changing," Lincoln said. "But it takes a little investigation to figure out exactly how. So here are the headlines . . ."

There was something about Howard Lincoln, his stately but affable demeanor perhaps, that made almost anything he said sound like a State of the Union address. For this reason, he was the perfect guy to give a rundown of where the industry stood ten years after it had been resurrected by Nintendo.

He began by acknowledging the downturn of the industry. Since Christmas 1993, the industry had slowed, but it had been slowing faster for some than for others. "The sector is off by 26 percent through the first three months of the year," he said, "but by far the biggest part of that hit is being taken by Nintendo."

At this, the audience rippled with laughter—and, surprisingly, so did Kalinske. It was true, and of the hundreds of so-called insiders packing the room, he felt like he was the only one who actually knew why. And the fact that the biggest threat to Sega was actually Sega itself, and not just in some metaphorical sense but in the most pathetically literal way possible, was enough to make him laugh as well. Over the past year, Nintendo's 16-bit sales were actually up 2 percent, while Sega's had plummeted 43 percent, falling for nine straight months in a row.

"Yet despite that downturn, Donkey Kong Country remains one of the biggest-selling games of all time, with worldwide sales now in excess of 7.5 million cartridges. So in one sense, I suppose nothing really has changed at all. Great software still sells great."

What was so amazing about Nintendo's newest blockbuster title went beyond the game itself. Despite all the technobabble about next-generation hardware, polygons, and CPUs, Donkey Kong Country was a reminder that there were always breakthroughs to be made with what existed already. It was a concept that perfectly mirrored Nintendo's recent renaissance: old dog, new tricks. Tony Harman was already scouring the globe for new old things, Gail Tilden was gradually bringing *Nintendo Power* into the Internet era, and Shigeru Miyamoto was patiently making new worlds in 64 bits. All of these benefits pointed to the primary advantage that Donkey Kong Country gave Nintendo: more time.

"And there's one last headline that's definitely making news," Lincoln explained. "The revised release date of Nintendo Ultra 64. From the start, we targeted a fall 1995 launch, but the world's only true 64-bit platform will now hit North American and European shelves in April 1996, for reasons that we'll elaborate on in just a minute."

Nintendo was never a company to take its foot off the pedal, but this runaway hit gave them the ability to continue moving at their own pace. With the Super NES making a comeback like no other and Sega seemingly more interested in the next than the now, Nintendo felt no need to rush the Ultra 64 system. "This concept," Lincoln explained, "that the name of the game is the game, that it's about hits and not bits, is certainly nothing new to these Nintendo meetings, but we don't just say it. We base our whole business on it. And that is exactly why we're delaying our launch of Ultra 64 in North America and Europe until next April."

More time. Time was the most valuable commodity possible in the fast-moving videogame world. More time to have software available at launch, more time to implement marketing plans, and more time to let Sony come into the market and start feasting on Sega. In the long run, Sony's entry wouldn't be a good thing for Nintendo, but with similar demographic targets, Sony would inadvertently help Nintendo win the last years of the 16-bit battles.

After a video presentation that featured software publishers gushing about the power and possibilities of Ultra 64, Howard Lincoln prepared to hand the afternoon over to Peter Main. But before stepping aside, he left the audience with one final message, which in a way felt like it had been plucked from Kalinske's ill-fated conversation with Nakayama. "First to market doesn't mean much," Lincoln started. "It's what you do, not when you do it. And most of you know the long-term viability of Genesis in America was not related to the time when that platform was launched; it was related to far more important factors. . . ."

If Lincoln had known Kalinske was there, locking eyes with his competitor likely would have made for the perfect final punctuation mark to the history the two had been writing over these past several years. Unfortunately, the moment was not meant to be, and the two would need to find another way to attach an image to whatever this had been all about. And on a positive note of past, present, and future, Nintendo of America's chairman left the stage to Peter Main.

"Howard, you're absolutely right," Main began, moving around the stage with an ease that contrasted with Lincoln's placid stiffness. "I can tell you as I stand here that even without Ultra 64 in this year's holiday picture, there are going to be a lot of happy gamers buying Nintendo products for the balance of this year—and I'm sure that's going to make a lot of folks in this room extremely happy."

Even when Main and Lincoln said the same thing, it always sounded different. They were members of the same family but had different ways of relating. One spoke like a lead-by-example father, the other like everyone's favorite uncle. Main ran through the usual gamut of topics (sales, merchandising, promotions, etc.) but also had the honor of unveiling a game that seemed to complete Nintendo's transformation. "Fasten your seat belts, folks, because our commitment to millions of Nintendo players is about to broadside the competition. Strap yourselves in for an assault that's going to nail the 16-bit title for Nintendo this year. Are you ready? Let's rumble."

Sirens blared, obnoxiously loud, but that discomfort was soothed by what appeared on the screen next: images and gameplay from Rare's new game Killer Instinct, the first fighting game that Nintendo had ever developed. It was like Mortal Kombat, but slightly less grisly, and it made good on Nintendo's pledge from one year earlier to stop giving advantages to the competition. The days of being squeaky clean and keeping to themselves were over, but what had replaced them was not necessarily what anyone would have expected. Nintendo was still Nintendo; they hadn't sold out or traded integrity for a quick buck. Rather, they had found a way to do what most iconic companies and characters do so poorly: evolve.

But out from the trenches Nintendo had come, stronger now than they had ever been, even when they'd had 90 percent of the market. Survivors, all of them, refocused and recommitted and reinvigorated. Arakawa's long-term philosophy had been vindicated, Main's alchemy of old and new had proved to be the magic elixir Nintendo needed, and Lincoln's gamble twenty years ago to leave his career at a law firm for a roller-coaster ride unlike any other had paid off. These three, and all the others who were responsible for slowly but surely propelling Nintendo forward, had warped down a pipe together and come out the other end stronger than before. They were still the same ragtag squad that had beaten impossible odds with a launch in New York; they just happened to look a little different now. Armed with a killer instinct, they stood tall beside Mario and Luigi, ready to fight whoever and whatever might confront them next.

64.

GAME OVER

Ten months later, Kalinske sat at his desk and stared out the window. He had been doing a lot of this lately, not by choice, but it was a side effect of failure and, in truth, seemed about as productive as anything else going on at Sega.

As he had predicted, the Saturn was a flop. Since the surprise launch in May, Sega had sold eight hundred thousand units of hardware, less than half of what Sony's PlayStation had sold (and the PlayStation had been available for only half as long as the Saturn). This dismal experience was accompanied by an untold number of frustrations, but none more grating than the fact that he had seen it coming. Kalinske had tried to work with Sony, and then Silicon Graphics, and even with Sega of Japan for what would have been Sega Saturn. He and his team had redefined the videogame industry with the Genesis, then been forced to sit back and watch as it was methodically destroyed by people who should have been on his side.

To put it like that made him sound naive, and maybe he had been at times, but if so, then shouldn't he have regrets? Things he would have done differently. People and products he had trusted too much. Maybe the problem had been that he had been fighting the wrong war, all that time believing it was Sega vs. Nintendo, when really the more treacherous battles were between Sega of America and Sega of Japan. But without the triumphs of Sega vs. Nintendo, there wouldn't even have been an SOA vs. SOJ; Sega of America would still be the same fifty-person operation that Kalinske had taken over in 1990. He thought back over the years, rewriting history in his head and playing out scenarios, but even then he couldn't find an alternative ending to this story. He knew all the whos and whats and whys, but never could figure out how things could have been different. And that haunting mystery was his ultimate frustration.

He could feel history rewriting itself, and it made him sick. Sega was no longer the alternative upstart, but just the new giant ready to fall, the way Nintendo and Atari had before them. Sega CD, 32X, and the Sega Channel would be viewed not as dynamic innovations but as flops that took Sega's attention away from the oncoming freight train. Kalinske felt he would no longer be perceived as the mastermind of a company that had gone from a 5 percent market share to 55 percent in three years, but instead thought of as just some guy who'd happened to be lucky enough to ride the wave of Sonic and reap the benefits of hawking games full of blood, sex, and violence. He would be remembered as the guy in the right place at the right time who'd benefited from everything Michael Katz had done. Or maybe, and probably even more likely, he wouldn't be remembered at all. History was indeed rewriting itself, and history was cruel.

Perhaps Kalinske would have seen some small silver lining if Sega's defeat had meant big things for those he admired most at Sony. That, however, was hardly the case. Although Sony would go on to sell ten million PlayStation systems by the end of 1996 (with more than half of those sales occurring in the United States), most of SCEA's key executives would be fired or let go within a year of the launch. The curious timing of this purge (which included Steve Race, Olaf Olafsson, and Michael Schulhof) would later be questioned in a September 23, 1996, *Forbes* article titled "Great Job: You're Fired." Why these men were let go will forever be something that only a select few truly know, but to this day Steve Race has no problem sharing his own theory, explaining that "the way to attain success at Sony was to either get into, or come out of, a Japanese vagina." Despite the salty language, there would be no hint of bitterness (neither then nor today), and Steve Race would soon forget all about Sony and simply do what he did best:

shrug it off, move on, and find a new company to turn around. And from his quiet office in Redwood Shores, Tom Kalinske couldn't help but feel it was time for him to do the same.

"Are you busy?" Fornasier asked, interrupting his thoughts.

The sight of a friendly face made Kalinske perk up. "Never for you," he said. "You're one of the good guys. So take a seat, tell me what's going on."

"Thank you," she said, her voice still chipper despite the faded light in her eyes. She'd had that hollow look ever since June 1995, when her son Troy, the only one of the triplets to survive the pregnancy, passed away ten days after his birth. The doctors had told her that these things sometimes happen with high-risk pregnancies; it was an absolute tragedy, but not one that could have been avoided. They said the words, and presented medical data, but she would always partially blame herself for working through the pregnancy. Unlike Kalinske, who could never figure out the how of his problem, Fornasier believed that she had figured it out for hers, even though she was repeatedly told there was no reason for her to feel guilty. Still, nothing could ever change the thoughts in that unseeably small part of her mind; she had made the ultimate sacrifice, and would forever have to live with that.

"What can I help you with?" Kalinske asked, suddenly hopeful that he might be able to do something productive with his day.

"Tom," she said, shaking her head, "I'm just in a real bad spot here."

"In what way?"

Logically, she knew it was time to leave Sega, but emotionally she just couldn't let go. The choice had become easier to make after Paul Rioux left in June, Tom Abramson resigned in February, and Nakayama had fired Goodby, Silverstein & Partners just a few weeks ago. But even with all that, she still couldn't bring herself to do it. She was one of the last remaining dinosaurs, but just couldn't face extinction—that is, unless she was not alone. And she knew Kalinske's contract would be up in June.

"I know that you're legally not allowed to tell me," Fornasier explained, "but if you could give me any sort of indication on whether you're staying at Sega, that would really help a lot. Just to know that you'd be here, and that we could build another team . . ."

"I know what you're asking, but it's hard for me to answer."

She was aware that he wasn't allowed to talk about which way he was leaning, because Sega was a publicly traded company in Japan. Still, she had hoped that asking him in person might reveal some sort of answer. She felt kind of bad about putting him in this position, but she just didn't know what else to do. Her job had become almost unbearable; she just couldn't take any more, particularly any interactions with Paul Rioux's replacement.

When Rioux had stepped down, his position was filled by someone from Sega of Japan named Makota Kaneshiro whose best talent, as far as she could tell, was not caring what anyone else had to say. This had really started to wear her down with the release of a game called NiGHTS into Dreams, which was the first non-Sonic game being made by the Sonic development team. The graphics of the game were beautiful, but conceptually it was a tough sell. Inspired by Jungian philosophy, it starred an androgynous fairy who guides a young boy and girl through a colorful but dreary dream world. She protested that the game would have trouble finding an audience in the United States, but he was adamant that Sega of America put all their eggs into this basket. It was by the prestigious Sonic Team, it had beautiful graphics, and people in Japan would understand it. She didn't disagree with any of this, but she didn't believe the audience here would connect. But he had little interest in her point of view. Although Fornasier had not been at SOA during the creation of Sonic, she couldn't help but think about how perfectly this situation epitomized the change that had taken place.

For the most part, she had made peace with the dynamic, but something Makota had said to her last week wouldn't stop rattling around her head. They had been discussing plans for the second annual E3 show and strategizing about the best ways to highlight various titles. Long discussions about what went where and which titles got how much attention had led to a meticulously crafted plan. But the following morning, when Makota distributed plans for the show, everything had been changed. The new strategy, unsurprisingly, featured all NiGHTS, all day long. She confronted him about this, hoping for at least an explanation, and he told her that she was mistaken.

"But everything is completely different," she said.

"No, no, you approved this, Ms. Diane," he said. "You have just forgotten."

She shook her head and wanted to let the issue go, but her feet simply wouldn't turn and walk away. She was sick of this happening, over and over, and she was sick of the new but unspoken corporate culture of giving in.

What had happened to the Sega that had taught her to take? More important, what had happened to her that she was okay with no longer being that person? "Makota-san," she said, now refusing to budge, "I apologize if I am speaking out of turn, and I have tremendous respect for all the good work you do, but I feel like I would be doing a disservice to my job if I did not stand here and fight for this."

When she finished, Makota-san smiled widely and appeared to finally understand where she was coming from. Maybe this particular situation wouldn't change, nor would the next, but there was now this understanding, and that had to be worth something. And then after a moment, nodding slightly, he replied, "Ms. Diane, you need to learn to be less passionate about your work."

Suddenly her feet started working again, and as she walked through Sega's increasingly unfamiliar hallways, she decided that she very much needed to speak with Tom Kalinske.

"I'm sorry for asking," Fornasier said to Kalinske. "I shouldn't have put you in that position. It's not right."

"Don't even think twice," he told her. "I'd have done the same."

"Thanks, Tom."

"But even though I can't discuss my contract with you, I can give you some advice if you would like."

"Yeah, of course. Always."

"Good," he said. "I would just recommend that you don't make your decision dependent on me. I don't mean to imply anything by that, one way or another, but you need to do what's best for you. You owe that to yourself."

Somehow, with those words, it was as if a curse had been broken, and her logic and emotions were intertwined and functioning once again. "Thank you," Fornasier said, with an assertiveness befitting someone who had accomplished all she had done. Shortly after this, she would go on to work for Paul Rioux, who had recently been named president of New Media at Universal Studios. There, as VP of marketing and business administration, Fornasier would seal a multiple-title contract with Sony to license Spyro and Crash Bandicoot. But before any of that, and coming to terms with life outside of Sega, she thanked Kalinske for everything he had done with a farewell hug.

• • •

"No, stop, don't go," Kalinske said. "I didn't mean it—I need you to stay."

"All right, all right, that's all I needed to hear," his wife replied, and then turned around and swam back in his direction.

Tom, Karen, and the kids had returned to a beach in Hawaii. Once again the sun was shining, and the children were collecting hermit crabs in a bright yellow bucket, but on this trip something there was noticeably different. Tom was different.

Not long after his conversation with Fornasier, Kalinske called Nakayama to resign. By this point, Kalinske had assumed that his leaving Sega was all but a formality, but Nakayama appeared to be genuinely surprised by this and asked if they could speak in person. The following week, Nakayama flew to Redwood Shores and met Kalinske to further discuss the matter. At first their time together felt just like the good old days, but eventually that feeling faded and Kalinske reiterated that it was time for him to look for a new challenge.

"What kind of challenge?" Nakayama asked.

"Something that uses technology to improve education," Kalinske replied.

Nakayama's eyes bubbled with questions until the president of Sega Enterprises blinked and suddenly they all went away. He then gave Kalinske a strange look, perhaps wistful and said, "I understand, thank you."

And that was that, the end of Tom Kalinske's time at Sega. After several years spent turning nothing into something only to watch it slowly be turned back into nothing once again, he would go on to successfully apply video-game technology to education throughout the rest of his storied career (at places like Knowledge Universe, Leapfrog, Blackboard, and many, many more).

"Hold still," Karen said, trying to climb on her husband's shoulders, but accidentally dunking him in the process.

"This is why I sent you away!" Tom exclaimed when he came up for air.

"Okay, I give up," Karen said, floating by his side. "So now what?"

Tom didn't respond right away, but as he looked at his wife a smile slowly grew across his face. This was his beautiful Barbie, the mother of his children, and the one who made his world possible. "Anything we want."

This time, instead of trying to conceal his dislike for the beach, Tom Kalinske was up for anything. He had learned how to adapt, how to enjoy, and how much better it felt to smile and actually mean it. And finally, after all this time, he was getting to finish the family vacation that had been so strangely interrupted six years earlier.

EPILOGUE

When he returned from Hawaii, Tom Kalinske was pleasantly surprised to find a note waiting for him from his longtime rival at Nintendo.

Dear Tom,

I was saddened to learn that you are leaving Sega. You've done a great job over the last six years, both in dramatically increasing Sega's market share (at our expense!) and also in representing the video game industry. You were the driving force behind the formation of the IDSA and the E3 Show. Neither would have happened without your leadership.

Let me wish you the very best of luck in your new venture.

All my best . . .

Sincerely,

Howard C. Lincoln
Chairman
Nintendo of America, Inc.

The letter was marked "Personal and Confidential" and, to this day, remains somewhere safe in Tom Kalinske's home office as a lasting reminder of the epic battle between Sega and Nintendo.

ACKNOWLEDGMENTS

This book would not exist (nor likely would my writing career) without Julian Rosenberg, my mind-bogglingly wonderful lit manager. I'm still not sure how to properly thank you, but I thought that a good start would be to give you the one thing that those in Los Angeles covet above all else: top billing.

And while the globe is spun to Los Angeles, I'd like to extend a giant thanks to the fine gentlemen at Point Grey Pictures—Seth Rogen, Evan Goldberg, and James Weaver—for believing in this book from the very beginning. In the Nintendo of my life, you three are my power glove. I am also forever indebted to Scott Rudin, whose talent for telling stories is second to none, and whose early confidence in my work is one of the greatest compliments I've ever received. Without you and the ever-brilliant Eli Bush, this book never would have reached the next level.

Thank you to my incredible attorney, Lev Ginsburg, and my fantastic agents: Jon Cassir and Dan Rabinow at CAA, and Alex Glass at Trident Media.

Thank you to my amazing editor, Mark Chait, and everyone at HarperCollins for your patience, positivity, and overall perfection.

Thank you to everyone who worked on the documentary based on this book, especially Jonah Tulis (my co-director), Matt Hamachek (our editor), and Seamus Tierney (our cinematographer).

Thank you to my pals who generously read half-finished thoughts and mangled chapters along the way: Josh Benedek, Frank Ceruzzi, Grant DeSimone, Andrew Hirsch, Dan Kim, Josh Kleinman, Dave McGrath, Brian Nathanson, and Jeremy Redleaf.

Thank you to my parents, Robin and Richard, for pretty much being the greatest people in the world (and for supporting my literary desires even after that racially charged novel I wrote in fourth grade).

Thank you to my brother, Dylan, for being the kindest person I know, and forgiving me for how I acted during our childhood (like a dick).

Thank you to Aunt Loren, Uncle Christopher, Jackson and Hunter; Aunt Erica, Uncle Bradley, Tyler and Amelie; and, of course, Grandma, for all those "amazing connections" over the years.

Thank you to Katie for being my daily inspiration. How you put up with me during the writing of this book, I'll never know, but please never stop.

And second-to-last but not least, thank you to my incredible research assistants: Claude Bear, Kiki Bear, Baby Bart, Freggly, Tater Tot, Boots, and the one and only Pipstick. Without you rascals, life would be un*bear*able.

Finally, the biggest thanks of all must go to the people who populate the pages of this book. During the writing of this book, I interviewed more than two hundred former employees of Sega and Nintendo, as well as dozens of other individuals with various vantage points into the videogame industry from this era. Although I am grateful to every single person who shared with me their time and insights, there are a few individuals whom I'd like to mention by name whose above-and-beyond effort really shaped the scope of the narrative.

First and foremost I'd like to thank Tom Kalinske, who gave me the time of day back when I was just a kid with an idea. Actually, he gave me much more than simply the "time of day," because that's just the kind of guy he is. Over the course of our three-year relationship, he has never been anything less than the kind, clever, and cool guy described within the pages of this book.

In addition to Tom, I am particularly indebted to the following individuals: Sam Borofsky, Don Coyner, Cindy Gordon, Mike Fischer, Diane Fornasier, Jeff Goodby, Tony Harman, Karen Kalinske, Howard Lincoln, Brenda Lynch, Peter Main, Sean McGowan, Al Nilsen, Al Nilsen #2 (because he must have been a clone to be as omnipresently helpful as he's been), Olaf Olafsson, Randy Peretzman, Howard Phillips, Arthur Pober, Larry Probst, Steve Race, Paul Rioux, John Sakaley, Gail Tilden, Shinobu Toyoda, Ellen Beth Van Buskirk, and Bill White.

Thank you all for letting me tell your story; it has been the greatest honor of my life.